RECENT DEVELOPMENTS IN POLYMER RESEARCH

Recent Developments in Polymer Research

Anthony V. Hopper
Editor

Nova Science Publishers, Inc.
New York

For permission to use material from this book please contact us:
Telephone 631-231-7269; Fax 631-231-8175
Web Site: http://www.novapublishers.com

LIBRARY OF CONGRESS CATALOGING-IN-PUBLICATION DATA

Recent developments in polymer research / Anthony V. Hopper (editor).
p. cm.
Includes index.
ISBN 1-60021-346-4
1. Polymers--Research. I. Hopper, Anthony V.
TP1087.R43 2006
668.9--dc22 2006021605

Published by Nova Science Publishers, Inc. ✦ New York

Contents

PREFACE

Polymers are substances containing a large number of structural units joined by the same type of linkage. These substances often form into a chain-like structure. Starch, cellulose, and rubber all possess polymeric properties. Today, the polymer industry has grown to be larger than the aluminum, copper and steel industries combined. Polymers already have a range of applications that far exceeds that of any other class of material available to man. Current applications extend from adhesives, coatings, foams, and packaging materials to textile and industrial fibers, elastomers, and structural plastics. Polymers are also used for most composites, electronic devices, biomedical devices, optical devices, and precursors for many newly developed high-tech ceramics. This new book presents leading-edge research in this rapidly-changing and evolving field.

Chapter 1 - A series of polycarbosilazane-based di-, tri- and mixed-valent homometallic (-CoII, -FeII, -FeIII and -FeIIFeIII) and heterobimetallic (-CoIIFeIII) metallopolymers were prepared by incorporating the desired transition metal chloride into polydimethylsilylethylenediamine (PDMSEN) matrix, [–(CH3)2SiNH(CH2CH2)NH–]n in an aprotic solvent under inert atmosphere and continuous sonication. All the obtained metallopolycarbosilazane macromolecules (M-PCSZ) are colored air-stable solids. They were structurally characterized using different spectroscopic (UV-visible and infrared), thermolysis, powder X-ray diffraction (XRD) and magnetic susceptibility techniques. All the results demonstrate that the metal centers are held at the dimethylsilylethylenediamine units of the polycarbosilazane backbone. Furthermore, the XRD data provide strong evidence that the incorporation of these metal ions have caused d-spacing between the PDMSEN chains to expand to permit attachment of metal ions into this coordinating matrix. Interestingly, evidence gathered from electrical and magnetic studies indicate that the mixed-valence PDMSEN-CoIIFeIII and PDMSEN-FeIIFeIII regimes exhibit unique properties not found in the analogous homometallic components, which can be attributed to the electronic/magnetic interaction between the adjacent divalent-trivalent metal centers (-CoIIFeIII and -FeIIFeIII). Electrical conductivity measurements on all these M-PCSZ regimes reveal that electriconic conduction occurs via electron-hopping process in the semiconductor range of 10-8-10-10 Sm-1 with wide range of energy of activation (Eg = 0.64-1.40 eV). The dielectric behavior of the mixed-valence PDMSEN-CoIIFeIII and -FeIIFeIII two relaxation mechanisms at the two metal sites.

Chapter 2 - The synthesis of poly(dibromophenylene oxide)s by thermal decomposition of Bis(2,4,6-tribromophenolato)di(N-methyl imidazole)Cu(II) (C1), Bis(2,4,6-tribromophenolato)di(N-methyl imidazole)Co(II) (C2) and Bis(2,4,6-tribromophenolato)di(N-methyl imidazole) Ni(II) (C3) complexes were achieved by solid state polymerization. The characterizations of the complexes were performed by Fourier Transform Infrared Spectroscopy (FTIR), Differential Scanning Calorimeter (DSC), Mass Spectroscopy (MS), Magnetic Susceptibility and CHN elemental analysis. The effects of temperature and time (at the highest conversion) on percent conversion and intrinsic viscosity of the polymers were examined. Among the synthesized polymers, the highest percent conversion and intrinsic viscosity was achieved by the decomposition of Cu(TBP)2(NMIZ)2 complex, whereas Ni(TBP)2(NMIZ)2 yielded the lowest. Varying the decomposition temperature during 3 h scan showed that the Cu(TBP)2(NMIZ)2 complex revealed lowest and Ni(TBP)2(NMIZ)2 complex revealed the highest decomposition temperature. The Ni(TBP)2(NMIZ)2 (having the highest minimum decomposition temperature) complex was more stable than either Co(TBP)2(NMIZ)2 and Cu(TBP)2(NMIZ)2 complexes. Synthesized polymers were characterized by Fourier Transform Infrared Spectroscopy (FTIR), Differential Scanning Calorimeter (DSC), 1H-Nuclear Magnetic Resonance (1H-NMR), 13C-Nuclear Magnetic Resonance (13C-NMR) , Scanning Electron Microscope (SEM) and viscometric measurements. The poly(dibromophenylene oxides)s, synthesized from Cu, Co and Ni transition metal complexes, having nonchelating ligand N-methyl imidazole, displayed selectivity in the favor of 1-2 and 1-4 additions, taking place at equal rates irrespective of the type of the metal used. All the polymers having Tg values in between 211 0C to 190 0C from Cu to Ni, were rigid and have branched structures.

Chapter 3 - The melting behavior of PBT was investigated after isothermal crystallization from the melt at a fixed temperature. Analyses conducted by differential scanning calorimetry (DSC) revealed a multiple melting behavior. Temperature-modulated differential scanning calorimetry investigations (TMDSC), performed at various modulation periods, showed that the frequency of modulation largely affects the apparent reversing specific heat capacity (cp) measured during melting, but not the total cp. In addition, a detailed examination of the modulated heat-flow-rate curves revealed that the melting process of PBT is the sum of simultaneous events of melting of small and defective crystals and their immediate recrystallization into more stable structures.

Quasi-isothermal TMDSC analyses, conducted for extended times at different temperatures in the whole melting range, revealed different degrees of reversible melting in PBT. The total latent heat involved in the fusion/recrystallization/crystal-perfection processes was found to decrease with time for the quasi-isothermal analyses performed at temperatures below 500 K. At higher temperatures, instead, recrystallization resulted prevailing with respect to fusion, with a increase of the apparent reversing heat capacity as a function of time.

Chapter 4 - In this article, UV-curable polydimethylsiloxane epoxy acrylate (PSEA) was synthesized by hydrosilylation of allyl glycidyl ether with hydrogen-containing polydimethylsiloxane to give polydimethylsiloxane-type epoxy resin which modified with acrylic acid. The influences of the synthetic process were discussed in detail. The properties of the cured film were also investigated by thermogravimetric analyzer, dynamical thermal mechanical analysis, and etc. In order to improve the photo-oxidative stability of photocured PSEA film, photocured PSEA without and with added photostabilizers have been investigated by measuring Fourier transform infrared photoacoustic spectroscopy and X-ray photoelectron spectroscopy. The hindered amine stabilizers based on 2,2,6,6-teteramethyl-piperdine, largely decrease the rate of surface oxidation of photocured PSEA. As an application sample, PSEA, polyethylene glycol urethane acrylate (PGUA) and trimethylolpropane triacrylate (TMPTA) were mixed in different proportions to modify the properties of blends to prepare UV-curable optical fiber coating. Formulation was optimized by the orthogonal table. When the proportion was PGUA: PSEA:TMPTA = 10:4:3, it was used for a single coating (OFC-6). However, the proportion PGUA: PSEA:TMPTA = 8:3:2 was used for a primary buffer coating. The rheological behavior and the selection of photoinitiators for OFC-6 were also discussed. The photopolymerization of OFC-6 was inspected by the FT-IR spectrum. Results show OFC-6 film is completely cured after irradiated 5 s. The thermal behaviors of the cured film were also investigated by TGA, DMA, and etc.

Chapter 5 - Sorption of ordinary fluids on polymer surfaces is of interest in various technologies including packaging and coating technology, membrane separation of gases and vapors, controlled drug release systems, biocompatible materials for biomedical applications and contact lens industry. In this work we use the vapor-sorption equilibrium data to show the degree of solvent upturn in each solvent-polymer system.

For this purpose, seventy-seven isothermal data for copolymer+solvent binaries and for some of their parent homopolymer+solvent binaries have been used in the temperature range of 23.5-80oc.

Solvents studied are acetone, acetonitrile, benzene, 1-butanol, carbon tetrachloride, 1,2-dichloroethane, chloroform, cyclohexane, hexane, methanol, octane, pentane, and toluene.

Copolymers studied are poly(acrylonitrile-co-butadiene), poly(styrene-co-acrylonitrile), poly(styrene-co-butadiene), poly(vinyl acetate-co-ethylene), poly(vinyl acetate-co-vinyl chloride), and poly(n-tetracosane- ethylene oxide -n-tetracosane).All copolymers are random copolymers except one triblock copolymer.

Some homopolymers are also studied: polyacrylonitrile, poly(cis-1,4-butadiene), polyisobutylene, poly(ε-caprolactone), poly(ethylene oxide), n-heptadecane, polystyrene, poly(vinyl acetate), poly(vinyl chloride), poly(vinyl methyl ether), and n-tetracosane.

According to these data sets, solvent weight fraction in the polymer is plotted against solvent vapor activity that is calculated assuming an ideal gas phase of pure solvent vapor neglecting the vapor pressure of the polymer.

For thermodynamic reasons, most polymer pairs are immiscible and their degree of compatibility is of underlying importance to the microphase structure and consequently, to the mechanical properties of the blend. The Flory-Huggins χ interaction parameter for the polymer pair plays a dominant role in explaining critical phase behavior of a compatible pair and in estimating interfacial tension and interfacial thickness for semicompatible or incompatible pairs. Direct measurement of this parameter is not always possible and we use the Flory-Huggins theory to obtain this dimensionless parameter.

Solvent molecules have a tendency to form clusters when absorbed on polymer surfaces. To provide a measure for clustering, the Zimm and Lundberg clustering function Gss/Vs, that is the ratio of the clustering integral to the partial molecular volume, and the quantity φs(Gss/Vs) as the mean number of solvent molecules in the neighborhood of a given solvent molecule is used. Finally the non-ideality factor, Γ, can be calculated from the clustering function.

Chapter 6 - Polyester type shape memory polymers were synthesized to improve their mechanical and shape memory properties and used for the preparation of sandwich type composite materials. Especially, poly(ethylene terephthalate) (PET) and poly(ethylene glycol) (PEG) copolymers with shape memory ability were prepared. After selecting the best composition of PET-PEG copolymer in mechanical properties, cross-linking agent such as glycerine, sorbitol, or maleic anhydride (MAH) was included for cross-linked copolymer, followed by analysis of its effect on mechanical, shape memory, and damping properties. The highest shape recovery was observed for copolymer with 2.5 mol% of glycerine, and the best damping effect indicating vibration control ability was from copolymer with 2.5 mol% of sorbitol. With the optimum copolymers in hand, sandwich-structured epoxy beam composites fabricated from epoxy beam laminate and cross-linked PET-PEG copolymer showed that impact strength increased from 1.9 to 3.7 times depending on the type of copolymer, and damping effect also increased as much as 23 times for the best case as compared to epoxy laminate beam alone. PET-PEG copolymers cross-linked with glycerol and sulfoisophthalate (SP) were also prepared to investigate the feasibility of vibration-control of composite laminate by additional ionic interaction. Composition of glycerol and SP was varied in order to get a copolymer with the best mechanical and shape memory properties. The highest shape recovery was observed for the copolymer with 2.5 mol% of glycerol and 2.5 mol% of SP. The sandwich-type copolymer composite showed improved impact strength (3.5 times) and damping effect (2.6 times) as compared to epoxy laminate beam alone. The resultant sandwich-structured epoxy beam composite can be utilized as structural composite material with vibration control ability and its glass transition temperature can be controlled by adjustment of hard segment content and cross linking agent composition.

Chapter 7 - The oxidation of polymers is accompanied by more or less intense chemiluminescence which intensity is determined by the quality of a polymer, temperature, concentration of oxygen in surrounding medium, presence of additives and stabilizers and by the initial molar mass of the substrate. The potential of the measurements of the light emission which appears during oxidation has not been explored yet fully and the method is mostly used in scientific institutions either as the main technique or it complements other methods used for the study of oxidative degradation of polymers. The review is mostly based on personal experience of the authors with the method and may serve as fast introduction for those who intend to apply the chemiluminescence technique in their own research.

Chapter 8 - Coupling or condensation reagents are species that increase the rate of condensation between two components, such as an amine and a carboxylic acid. The coupling reagent activates one of the components (generally a carboxylic acid) rendering it more susceptible to nucleophilic attack by the other component. The result is formation of a species arising from the coupled components and a by-product – a hydrated coupling reagent derivative. However, use of coupling reagents can also result in the formation of un-wanted products. As a result of the high efficiency of coupling reagents, it has become commonplace to use them in conjunction with coupling additives that limit side reactions of the activated species. For a coupling reagent to be selected for a particular process, either peptide coupling or polymer synthesis, it must fulfil a number of essential criteria. Ideally a coupling reagent should be stable under the reaction conditions used, work efficiently and with high conversion under mild conditions in stoichiometric quantities. It should produce minimum number of side-products and perhaps most importantly, the hydrated by-product should be removed easily by solvent extraction or other simple isolation procedures.

The aim of this review is to provide the reader with a detailed insight into the use of coupling reagents in polymer chemistries. In addition to providing detailed examples of coupling reagent mediated polymerisations, the article covers the mode of action of coupling reagents and assesses the relative advantages and disadvantages associated with conventional polymerisation methods in comparison to coupling reagent mediated approaches.

In: Recent Developments in Polymer Research
Editor: Antony V. Hopper, pp. 1-27
ISBN 1-60021-346-4

Chapter 1

ELECTRICAL AND MAGNETIC BEHAVIOR OF HOMOMETALLIC AND HETEROMETALLIC OF POLYETHYLENEDIAMINECARBOSILAZANE-BASED METALLOPOLYMERS

Isam Arafa[1]*, Hassan El-Ghanem[2,3] and Sufyan Nemrat[2]

[1] Department of Applied Chemistry, Faculty of Science and Arts, Jordan University of Science and Technology

[2] Department of Physics, Faculty of Sciences and Arts, Jordan University of Science and Technology, Jordan

[3] On sabbatical leave at Private College of Dentistry and Pharmacy, Riyadh, Saudi Arabia

ABSTRACT

A series of polycarbosilazane-based di-, tri- and mixed-valent homometallic (-Co^{II}, -Fe^{II}, -Fe^{III} and -$Fe^{II}Fe^{III}$) and heterobimetallic (-$Co^{II}Fe^{III}$) metallopolymers were prepared by incorporating the desired transition metal chloride into polydimethylsilylethylene-diamine (PDMSEN) matrix, $[-(CH_3)_2SiNH(CH_2CH_2)NH-]_n$ in an aprotic solvent under inert atmosphere and continuous sonication. All the obtained metallopolycarbosilazane macromolecules (M-PCSZ) are colored air-stable solids. They were structurally characterized using different spectroscopic (UV-visible and infrared), thermolysis, powder X-ray diffraction (XRD) and magnetic susceptibility techniques. All the results demonstrate that the metal centers are held at the dimethylsilylethylenediamine units of the polycarbosilazane backbone. Furthermore, the XRD data provide strong evidence that the incorporation of these metal ions have caused d-spacing between the PDMSEN chains to expand to permit attachment of metal ions into this coordinating matrix. Interestingly, evidence gathered from electrical and magnetic studies indicate that the mixed-valence PDMSEN-$Co^{II}Fe^{III}$ and PDMSEN-$Fe^{II}Fe^{III}$ regimes exhibit unique

* Corresponding author: Department of Applied Chemistry, Faculty of Science and Arts, Jordan University of Science and Technology, PO.Box 3030, Irbid 22110, Jordan, Facsimile 962-2-7095014, E-mail, isamaraf@just.edu.jo

properties not found in the analogous homometallic components, which can be attributed to the electronic/magnetic interaction between the adjacent divalent-trivalent metal centers (-$Co^{II}Fe^{III}$ and -$Fe^{II}Fe^{III}$). Electrical conductivity measurements on all these M-PCSZ regimes reveal that electriconic conduction occurs via electron-hopping process in the semiconductor range of 10^{-8}-10^{-10} Sm^{-1} with wide range of energy of activation (E_g = 0.64-1.40 eV). The dielectric behavior of the mixed-valence PDMSEN-$Co^{II}Fe^{III}$ and -$Fe^{II}Fe^{III}$ two relaxation mechanisms at the two metal sites.

Keywords: Dielectric constant, metallopolycarbosilzane, metallopolymer, organosilicon polymer, homometallic, heterometallic, macromolecular complexes, carbosilazane

INTRODUCTION

Metallopolymers [1-7], metallodendrimers [8-13], metalloorganosiloxanes [14-16], metallopolyrotaxanes [17], metallomesogenes [18-20], organometallic polymers [21-24], and other transition metal-based macromolecules belong to a rapidly developing class of advanced materials [25-28]. The impetus for the growing interest in these materials partly stems up from their technological values in developing new materials with wide range of applications in electrochromic display, opto-electronic, ferromagnetic and non-linear optical devices. Indeed, these materials have been reported to display novel magnetic, electrical, optical, electro-optical, liquid crystalline, electroluminescent and catalytic properties different from those exhibited by the analogous classical metal complexes. Furthermore, modified electrode surfaces with metal-based macromolecules have lead to the discovery of new materials for a variety of electrochemical applications such as electrocatalysis, electrophotocatalysis and electrochemical sensors [29,30]. This process permits one to modulate the electrode surface characteristics to obtain the desired property. In this context, some electronic processes require materials with low metal content located at regularly inter-spaced metal sites whereas others need aggregates of metal clusters with high metal contents. Moreover, the application of metallopolymers in modeling metalloenzymes and metalloproteins are well documented [25].

An important aspect in the structure-property-function relationship of these metal-containing materials is that they combine the advantageous characteristics inherited in polymer structure (flexibility) as well as the spectroscopic, electrochemical, magnetic and catalytic properties exhibited by the embedded metal centers. Of prime significance is their ability to be easily fabricated/formulated at relatively low temperatures in different flexible forms (thin films, fibers and membranes), shapes and sizes [31-33]. The performance of these materials largely depends upon the type of metal, oxidation state, metal loading and metal-matrix mode of interaction. To this goal, a variety of metal-based macromolecular systems have been designed in which the metal centers are either σ- or π-coordinated to the polymer backbone, electrostatically bonded (ionomers) or weakly immobilized (doped) by the polymer framework [34-36]. In the first category, the metal centers can be directly coordinated, appended through a spacer or form an integral part of the polymer backbone chains as shown in Scheme 1 [34-36]. One important advantage of binding metal ions through σ or π bonds to polymer backbones is the structural versatility to obtain a monophasic homogeneous polymeric system in which the metal centers are organized at regular intervals while retaining the structural integrity and the chain packing characteristics of the backbone [34-36]. These structural advantages permit one to modulate the opto-electronic-magnetic characteristics by the proper choice of the metal, oxidation state, inter-metal distance and the coordinating matrix. Similarly, the ionomers offers an important class of metal-polymer complexes. However, in the third category the ultra-fine metal centers are non-covalently dispersed (doped) within the structural network of the polymer forming biphasic regimes [36]. In this case the metal loading can be easily controlled to meet the desired electronic-magnetic requirements. However, an important drawback of such a system is the fact that although the polymer framework immobilize the metal centers against migration and agglomeration this process cannot be avoided particularly at high temperatures.

In recent years, metal-free and metal-containing conducting polymers in which the extensive π-conjugated backbone provides a low-energy conduction pathway are widely employed in the studies of long-range electron transfer, electroluminescence and photoluminescence [36-40]. However, relatively few studies have been concerned with electron transport between metal sites coordinated to polymers in which the backbone is composed of carbon and organosilicon σ-framework skeleton [40-45]. More recently, organosilicon polymers have attracted a great deal of attention as a new class of advanced materials for numerous electronic processes [43-45].

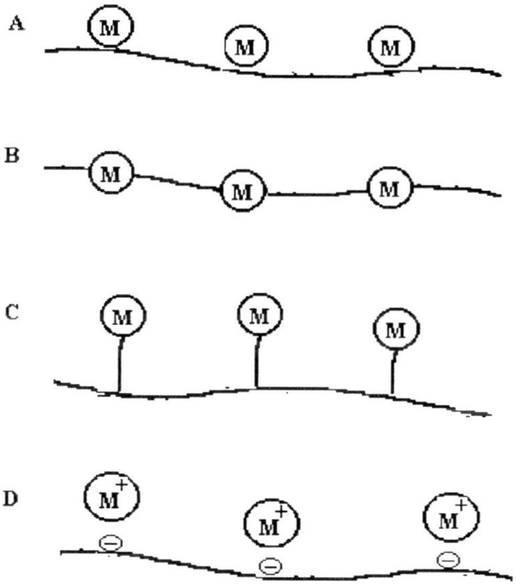

Scheme 1. Schematic representation of different classes of metallopolymers.
(A) Metal centers are covalently attached to the polymer backbone
(B) Metal centers constitute an integral part of the backbone
(C) Metal centers are appended to the backbone through a spacer
(D) Metal ions are electrostatically attached to the ionomer backbone

Numerous reports have shown that inorganic and organometallic polymers in which the main group elements constitute the polymer backbone exhibit interesting electronic, thermal and mechanical properties not found in classical organic polymers [43-53]. A variety of organosilicon (silazane, Si-N; carbosilazane, C-Si-N, Si-N-C; siloxazane, O-Si-N; carbosiloxazane, O-Si-N-C) and other inorganic polymers (phosphazane, P=N; carbophosphazane, C-P=N; borazane, B=N; sulfur nitride, S-N, oxathiazane, RS(O)=N-) have been reported [43-53].

Currently, we have directed our research efforts towards the synthesis, characterization and investigation of opto-electronic-magnetic properties of metal-based organosilicon polymers [45,54-56]. Our research work emphasizes the utilization of organosilicon polymers in which carbosilazane (Si-N-C, C-Si-N), carbosiloxazane (C-Si-O-N, -O-Si-N-C), siloxazane (O-Si-N) and carbosiloxane (O-Si-O-C) functionalities constitute an integral part of the chain backbone. A variety of organosilicon-containing aliphatic and aromatic diamine, dioxime and Schiff base have been prepared and characterized, Scheme 2 [45, 57]. Our interest in these organosilicon polymers is based on two facts: first, the high affinity of their low-dimensional backbone to coordinate a variety of transition metals at regular intervals (distances) and second, the d_{π}-p_{π} interaction in Si-O and Si-N bonds. This interaction permits regular interruption of extensive π-conjugation when the bridge (R) is aromatic and consequently produces unique dielectric and luminescent characteristics. In a series of articles, we have demonstrated that different divalent and trivalent transition metals can be readily incorporated into polyethylenediaminecarbosilazane oligomer, $[-(CH_3)_2SiNHCH_2CH_2NH-]_n$, PDMSEN **1**, Fig. 1 [54-56].

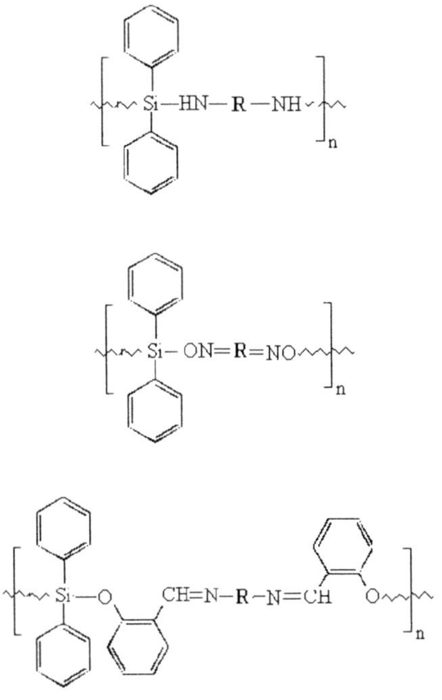

(R = aliphatic or aromatic spacer)

Scheme2. Schematic structures of the organosilicone-containing diamine, dioxime and Schiff base segments

The synthesis of this oligomer/polymer was reported several years ago [58]. Further detailed ^1H and ^{29}Si NMR studies of its backbone structure showed that it is composed of linear and cyclic segments in which the linear: cyclic ratio varies depending upon the reaction conditions, 2 [59].

In this chapter, we would like to report the formation of a series of polycarbosilazane-based homometallic (PDMSEN-Co^{II}, -Fe^{II} , -Fe^{III} and -$Fe^{II}Fe^{III}$) and heterobimetallic (PDMSEN-$Co^{II}Fe^{III}$) macromolecular chlorides complexes. In this system, the metal ions are held to the low-dimensional organosilazane chains by strong chemical bonds. Using the attractive features in these homometallic and heterometallic regimes one can construct a vast array of new electro-magneto-chromophoric materials with novel spectroscopic, magnetic and conducting properties by altering the metal centers, their oxidation states and M-M separating distances. The resulting M-PCSZ materials were characterized by electronic (UV-visible) and infrared spectroscopy (FT-IR), magnetic susceptibility, thermolysis and powder X-ray diffraction (XRD). The dc and ac electrical behavior of these materials was investigated.

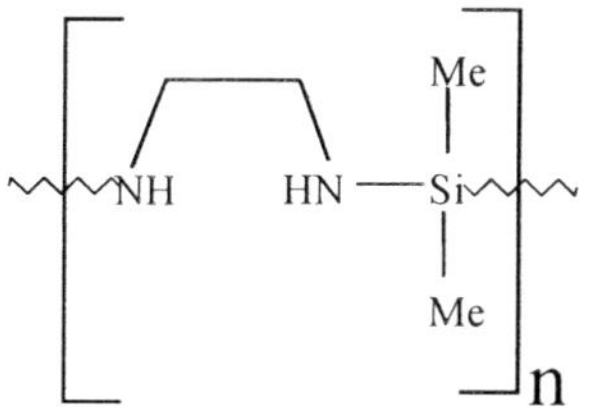

1

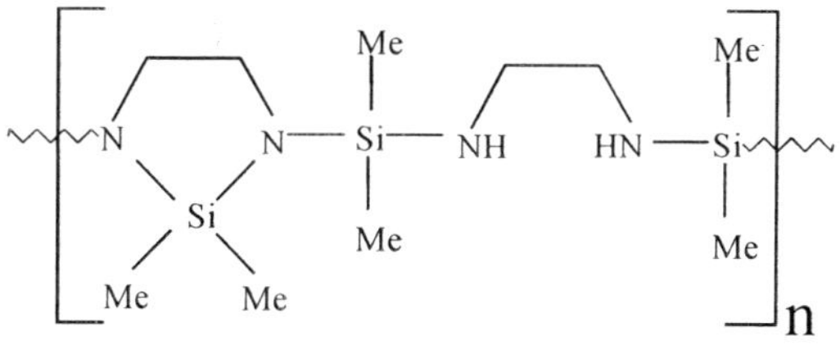

2

Figure 1. Schematic drawing showing the cyclic and linear carbosilazane segments in PDMSEN

EXPERIMENTAL

All chemicals and solvents were reagent grade and used as received without further purification unless otherwise stated [$(CH_3)_2SiCl_2$ (99%, Acros), ethylenediamine, (99%, Janssen Chimica), $CoCl_2{\cdot}xH_2O$, $FeCl_2{\cdot}xH_2O$, anhydrous $FeCl_3$ (Merck or Reidel-De Haen), triethylamine (Et_3N, Scharlau), tetrahydrofuran (THF, Scharlau) and dimethylformamide (DMF, Acros Organics)]. The synthesis of PDMSEN matrix and the incorporation of metal chlorides were carried out under nitrogen atmosphere in a sonicating bath as described later.

Infrared spectra (FT-IR) were recorded on Nicolet Impact 410 system as KBr pellets. Electronic spectral data was obtained in DMF using Shimadzu spectrophotometer. Powder X-ray diffraction (XRD) was measured on Philips PW 1729 X-ray spectrometer using Cu ($K_{\alpha 1}$ radiation, $\lambda = 1.5404$ Å) generated at 35 kV and 40 mA and 2θ angle ranging from 5-80° with 0.04-degree increments. The XRD profiles were analyzed by curve-fitting procedure using

Table Curve-2D software (Lowess Algorithm). Magnetic measurements were obtained at room temperature using a Sherwood magnetic susceptibility balance (calibrated with $HgCo(SCN)_4$). The electrical conductivity measurements were made on compressed discs. About 0.10-0.20 g of PDMSEN-M powder was compressed using a press die at 10 ton/cm^2 for 10 minutes forming a circular disc (13 mm in diameter, varied thickness 0.5-1.0 mm). A silver coating was applied to each surface of the disc then it was sandwiched between two brass circular discs to ensure good ohmic contacts with brass electrodes. The current passing through the sample was measured using an analogue electrometer (Metravo) with an accuracy of ± 0.1 nA. The potential difference across the sample was measured using a Phillips digital multimeter with an accuracy of ±10 μV.

Synthesis of [-Si(CH$_3$)$_2$NHCH$_2$CH$_2$NH-]$_n$ Matrix (PDMSEN)

The general method used to prepare $[-Si(CH_3)_2NHCH_2CH_2NH-]_n$ polymer was previously described [45, 58]. Briefly, $CH_3)_2SiCl_2$ (33.00g, 0.255 mole) dissolved in 50 ml toluene was added to a mixture of $H_2NCH_2CH_2NH_2$ (15.00g, 0.250 mole) and Et_3N (51.00g, 0.50 mole) in toluene (50 ml) at room temperature under nitrogen atmosphere. The reaction mixture was refluxed (4-5 hrs) after which it was cooled to room temperature and kept overnight under nitrogen. The white salt (Et_3NHCl) was filtered off and the product was isolated from toluene under reduced pressure. Color and state: creamy-waxy solid. M. p. (soften and swell) ≈ 50 °C. Yield: 17.50 g (60%). Average molecular weight: M_n = 2300 g/mole, n = 20 (vapor pressure osmometry, benzene, 35 °C). Characteristic FT-IR frequencies (KBr pellet, cm^{-1}): 3136 (sh) (ν_{N-H}); 1264(s) (δ_{Si-CH3}); 1085 (s), 937 (s) (ν_{N-Si-N} and ν_{Si-N}).

Synthesis of M-PCSZ Divalent and Trivalent Homometallic Macromolecules

PDMSEN-Fe^{III}: Anhydrous $FeCl_3$ powder (0.60 g, 3.72 mmole) was dissolved in THF (25 ml). The resulting dark brown solution was mixed with PDMSEN (1.50, 12.9 mmole) dissolved in minimum amount of THF (25 ml). The reaction mixture was placed in a sonicating bath at 70 °C for ≈ 2 hrs. The product was precipitated with ether, filtered off and vacuum dried. Yield: 0.65 g. Color and state: Brown powder. M.p.= 230-240 °C (decomposition). Fe%: estimated from thermolysis in air; 18.9%; Calcd for $[(CH_3)_2SiNHCH_2CH_2NH]_{1.5}$-$FeCl_3$•$H_2O$: 17.5%. Characteristic FT-IR frequencies (KBr pellet, cm^{-1}): 3372 (br) (ν_{O-H});, 3186 (s), 3010 (s) (ν_{N-H}); 1261(m) (δ_{Si-CH3}); 1035 (s), 1007 (s) (ν_{N-Si-N} and ν_{Si-N}).

PDMSEN-Fe^{II}: This macromolecular complex was formed using analogous procedure using $FeCl_2.xH_2O$ (0.63 g, 3.72 mmole) and PDMSEN (1.50 g, 12.9 mmole). Yield: 0.90 g. Color and state: Brown powder. M. p = 220-230 °C (decomposition). Fe%: estimated from thermolysis in air; 15.9%; Calcd for $[(CH_3)_2SiNHCH_2CH_2NH]_{1.5}$-$FeCl_2$•$H_2O$, 15.7%. Characteristic FT-IR frequencies (KBr pellet, cm^{-1}): 3390 (br) (ν_{O-H});, 3187 (s), 3010 (s) (ν_{N-H}); 1261(m) (δ_{Si-CH3}); 1035 (s), 1010 (s) (ν_{N-Si-N} and ν_{Si-N}).

PDMSEN-Co^{II}: In the preparation of this macromolecular complex, $CoCl_2 \cdot 6H_2O$ (0.89 g, 3.74 mmole) was dissolved in THF (10 ml), which rapidly gives a blue solution indicating the formation of $CoCl_2 \cdot xTHF$ adduct (this step was carried out to reduce the amounts of water in the reaction mixture). The solution was concentrated to ≈ 2 ml then mixed with a THF solution containing PDMSEN (1.50 g, 12.9 mmole). The reaction mixture was subjected to sonication at 70 °C for ≈ 2 hrs. Addition of ether resulted in the precipitation of blue-green PDMSEN-$CoCl_2$ macromolecular complex. Yield: 0.65 g. M. p. = M. p = 126-130 °C (decomposition). Co%: estimated from thermolysis in air; 17.7%; Calcd. for $[(CH_3)_2SiNHCH_2CH_2NH]_{1.5}$-$CoCl_2 \bullet H_2O$, 18.3%. Characteristic FT-IR frequencies (KBr pellet, cm^{-1}): 3453 (br) ($\nu_{O\text{-}H}$);, 3203 (s), 3011 (s) ($\nu_{N\text{-}H}$); 1261(m) ($\delta_{Si\text{-}CH3}$); 1034 (s), 1006 (s) ($\nu_{N\text{-}Si\text{-}N}$ and $\nu_{Si\text{-}N}$).

Synthesis of M-PCSZ Mixed Valence Homo- and Heterometallic Macromolecules

PDMSEN-$Co^{II}Fe^{III}$: This heterometallic macromolecular complex is prepared by sequential incorporation of anhydrous $FeCl_3$ into PDMSEN followed by introduction of $CoCl_2$ in THF. Anhydrous $FeCl_3$ (0.30 g, 1.85 mmole) was dissolved in 25 ml THF solution containing PDMSEN (1.50, 12.9 mmole). The reaction mixture was subjected to sonication for 30 minutes during which a dark brown solution of Fe^{III}-coordinated PDMSEN is formed. To this reaction mixture, a THF solution containing $CoCl_2$ (0.44 g, 1.85 mmole) was added and rapidly mixed. The resulting reaction mixture was placed in a sonicating bath at 70 °C for ≈ 2 hrs. The green product was precipitated with ether, filtered off and vacuum dried. Yield: 1.05 g. Color and state: dark green powder. M. p. = 190-210 °C (decomposition). (Co+Fe)%: estimated from thermolysis in air; 20.8%; Calcd. for $[(CH_3)_2SiNHCH_2CH_2NH]_2$-$CoCl_2$-$FeCl_3 \bullet xH_2O$ (x = 2-3) 20.5%. Characteristic FT-IR frequencies (KBr pellet, cm^{-1}): 3450 (br) ($\nu_{O\text{-}H}$);, 3203 (s), 3095 (s) ($\nu_{N\text{-}H}$); 1261(m) ($\delta_{Si\text{-}CH3}$); 1051 (s), 1011 (s) ($\nu_{N\text{-}Si\text{-}N}$ and $\nu_{Si\text{-}N}$).

PDMSEN-$Fe^{II}Fe^{III}$: A similar procedure as described above was used for the preparation of this mixed valence metallic macromolecular complex using anhydrous $FeCl_3$ (0.30 g, 1.85 mmole), $FeCl_2.xH_2O$ (0.32 g, 18.5 mmole) and PDMSEN (1.50, 12.9 mmole). Yield: 0.95 g. Color and state: dark brown powder. M. p = 260-275 °C (decomposition). (Fe)%: estimated from thermolysis in air; 18.8%; Calcd. for $[(CH_3)_2SiNHCH_2CH_2NH]_2$-$FeCl_2$-$FeCl_3 \bullet xH_2O$ (x = 2-3) 20.0%. Characteristic FT-IR frequencies (KBr pellet, cm^{-1}): 3390 (br) ($\nu_{O\text{-}H}$);, 3187 (s), 3009 (s) ($\nu_{N\text{-}H}$); 1262(w) ($\delta_{Si\text{-}CH3}$); 1035 (s), 1008 (s) ($\nu_{N\text{-}Si\text{-}N}$ and $\nu_{Si\text{-}N}$).

Results and Discussion

Synthesis and Structure

The preparation of PDMSEN matrix from $H_2NCH_2CH_2NH_2$ and $(CH_3)_2SiCl_2$ in an aprotic solvent containing triethylamine under inert atmosphere was reported several years ago [45,58]. Interestingly, the structure of the polycarbosilazane backbone of this matrix shows two obvious structural features. First, the backbone is composed of alternating bidentate dimethylsilylethylenediamine functionality. Second, the matrix has one-dimensional (1-D) architecture composed of cyclic and linear carbosilazane segments, Fig. 1. Furthermore, several reports have shown that Si-N bond is thermally stable but unstable towards hydrolysis [61-63]. On the other hand, the basicity and coordination characteristics of nitrogen in different silylamines have been examined using p_{π}-d_{π} interactions, negative hyperconjugation or electrostatic repulsion model indicated that the nitrogen is poorly basic and has a distorted tetrahedral geometry. Consequently, the affinity of silylamines for coordinating transition metals is less than that of the corresponding alkylamines [60-61]. These structural and chemical characteristics have encouraged us to investigate the binding affinity of this matrix towards several divalent (Cu^{II}, Co^{II}, Ni^{II} and Fe^{II}), trivalent (Cr^{III}, Fe^{III} and Co^{III}) and mixed valence ($Co^{II}Co^{III}$ and $Fe^{II}Fe^{III}$) transition metals [54-56].

In this work, the incorporation of Co^{II}, Fe^{II} and Fe^{III} chlorides into PDMSEN matrix was performed in THF under nitrogen in a sonicating bath at 70 ^{o}C. The aprotic solvent (THF) serve two main purposes, it swells the PDMSEN matrix and eventually unfold its chains thereby exposing the interior of the dimethylsilylethylenediamine functionality of the backbone to the solvated metal ions. This ensures that PDMSEN coordinating matrix uptake maximum number of metal ions uniformly. It is important to point out that sonication of the reacting mixture assist in the dissolution of the metal chloride in THF and consequently the formation of relatively homogeneous M-PCSZ macromolecules in which the metal sites are uniformly distributed in the polycarbosilazane matrix with regular inter-chain spacing. The strategy used to prepare the mixed-valence homometallic and heterometallic M-PCSZ regimes is depicted in the Scheme 3. This process has been previously described in the preparation of heterometallic magnetopolymers and heterometallic catalysts [67-69]. This method involves *in situ* sequential incorporation of the trivalent Fe^{III} into PDMSEN matrix followed by the introduction of the corresponding divalent metal ion (Co^{II} or Fe^{II}). In the first step, Fe^{III} is covalently anchored on the PDMSEN backbone. Unlike simultaneous incorporation (co-impregnation), sequential introduction of the metal ions is more likely to yield relatively uniform distribution of the trivalent and divalent metal ions.

During the complexation process two important side reactions may occur, namely, ring-opening of the strained cyclic carbosilazane segment of structure 2 and partial de-polymerization of the PDMSEN matrix as shown in Scheme 4. These two reactions are generally observed in organosilazane and carbosilazane chemistry [63-66]. In this process the cyclic carbosilazane segments were converted to linear units and therefore, the number of linear segments in the backbone of the PDMSEN matrix increases. Consequently, the bonding (chelating) sites available for metal coordination increase. Although it is difficult to draw conclusions regarding the extent of ring opening, qualitative analysis of the FT-IR spectra of these M-PCSZ materials suggests that extensive ring opening has occurred. This is evident from the comparison of the intensity of the $\nu_{N\text{-}H}$ band in the metal-free PDMSEN matrix with the same band in the metal-containing PDMSEN macromolecular complexes (see section on FT-IR). However, the de-polymerization process is expected to be slow in non-aqueous aprotic organic reaction media [63]. The insolubility of these M-PCSZ materials in organic solvents has hindered our efforts to investigate these processes quantitatively.

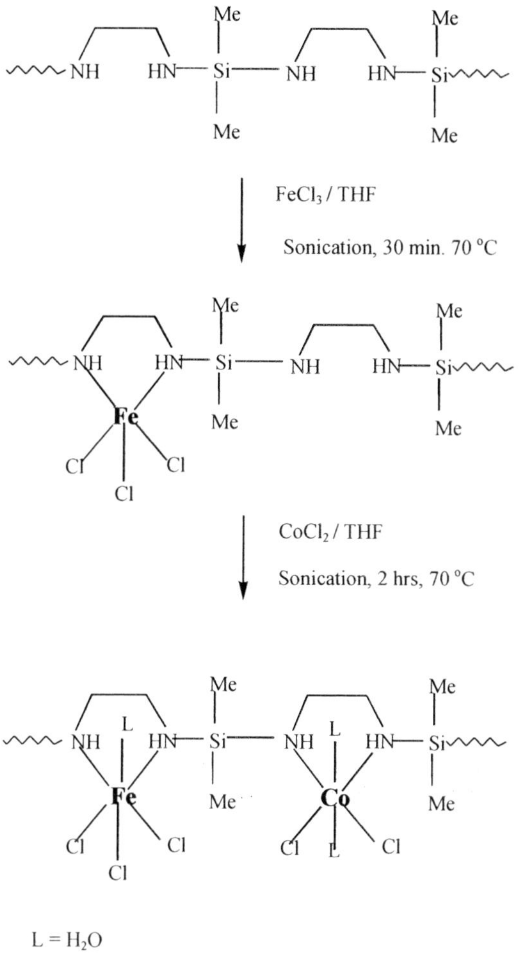

Scheme 3. Schematic diagram showing the construction of PDMSEN-$Co^{II}Fe^{III}$ macromolecule

The structural integrity of the PDMSEN backbone in the prepared homometallic and heterometallic M-PCSZ macromolecules was deduced using different techniques (XRD, FT-IR spectroscopy and thermolysis). In general, all results suggest that the studied PDMSEN matrix remained intact during the formation process of the homometallic and heterometallic M-PCSZ complexes.

This observation is consistent with earlier reports on the stability of Si-N bond [62-63]. It is reasonable to presume that the one-directional (I-D) architecture of the backbone is maintained in the homometallic and heterometallic regimes. Structural analysis of the prepared M-PCSZ using electronic spectroscopy, magnetic measurements and XRD provided strong evidence for relatively uniform distribution and covalent attachment of all the studied metal centers. This observation implies that the metal centers are homogeneous arranged at the macroscopic scale but this conclusion may not be true at the molecular level. The air-stability of these macromolecular complexes was checked over a period of several months using melting point and FT-IR spectroscopy and found to be stable when stored in a closed container. They are insoluble in common organic solvents but slightly soluble in dimethylsulfoxide (DMSO) and dimethylformamide (DMF). It is noteworthy to mention that the melting process of all the obtained M-PCSZ materials proceed via formation of gel, which swell and finally decompose over a range of about 8-20 °C. The reproducibility of the decomposition temperatures for different samples within 2-5% error suggests that the metal centers are quite uniformly distributed within the PDMSEN network. This conclusion is strongly supported by the magnetic measurements done on different samples. Furthermore, the different decomposition temperatures show that the studied metal ions influence the microstructure of the PDMSEN matrix differently.

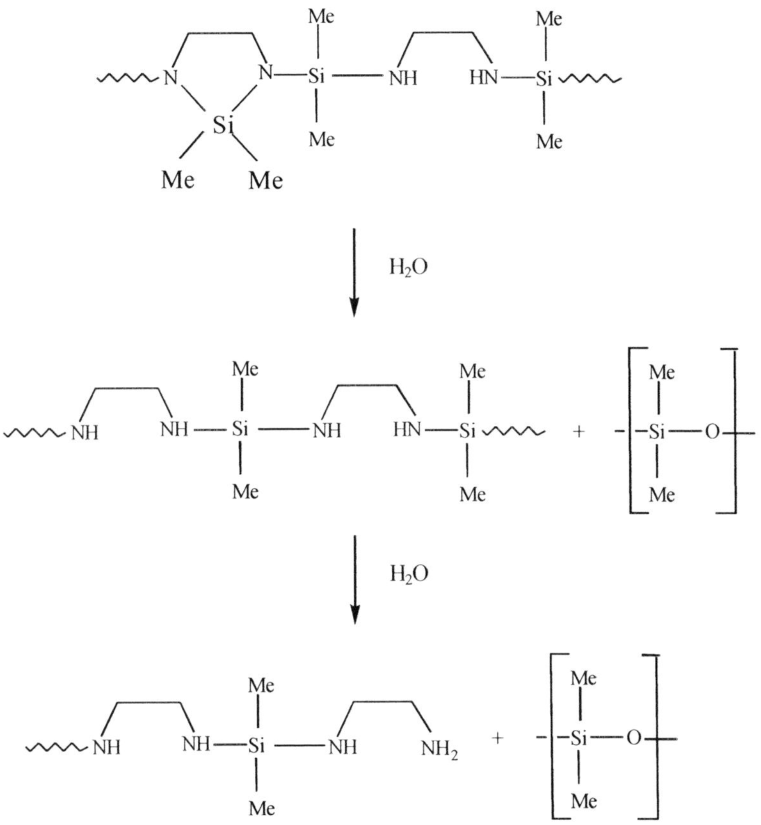

Scheme 4. Ring-opening and de-polymerization side reactions that may accompany the incorporation process of metal ions into PDMSEN matrix

Infrared Spectroscopy

The infrared absorption spectra of the metal-free and metal-containing PDMSEN M-PCSZ macromolecules were recorded in the 4000-400 cm^{-1} region as KBr pellets. The obtained FT-IR data are presented in the experimental section. FT-IR spectroscopy is particularly useful in examining the structural integrity of the polycarbosilazane backbone in metal-free and metal containing PDMSEN systems. In general, the formation of PDMSEN from the reaction of $H_2NCH_2CH_2NH_2$ and Me_2SiCl_2 is accompanied by drastic changes in the IR spectral data. Of prime importance is disappearance of asymmetric stretching frequency for the amine group ($\nu_{N\text{-}H}$) of the ethylenediamine and the appearance of new characteristic bands due to $\delta_{Si\text{-}CH3}$, $\nu_{Si\text{-}N}$ and $\nu_{N\text{-}Si\text{-}N}$. Fig. 2 displays the FT-IR spectra of the studied PDMSEN, PDMSEN-Co^{II}, -Fe^{III} and -$Co^{II}Fe^{III}$ macromolecules. The FT-IR spectrum of metal-free PDMSEN displays three important characteristic bands at 3136, 1264 and 1085 cm^{-1} regions assigned to $\nu_{N\text{-}H}$, $\delta_{Si\text{-}CH3}$ and $\nu_{Si\text{-}N}$, respectively. This tentative assignment is based on the reported IR spectral data for different monomeric and polymeric silazane and carbosilazane compounds [58,64-66]. Close inspection of the IR spectrum of the metal-free PDMSEN shows that the peak due to $\nu_{N\text{-}H}$ is broad and weak, which indicates that the nature of carbosilazane units in the backbone structure of the prepared PDMSEN is mainly cyclic rather than linear. This observation is consistent with the earlier finding by Mariam *etal* who reported that the ratio of cyclic : linear segments in this oligomer depends upon the reaction conditions [58,59].

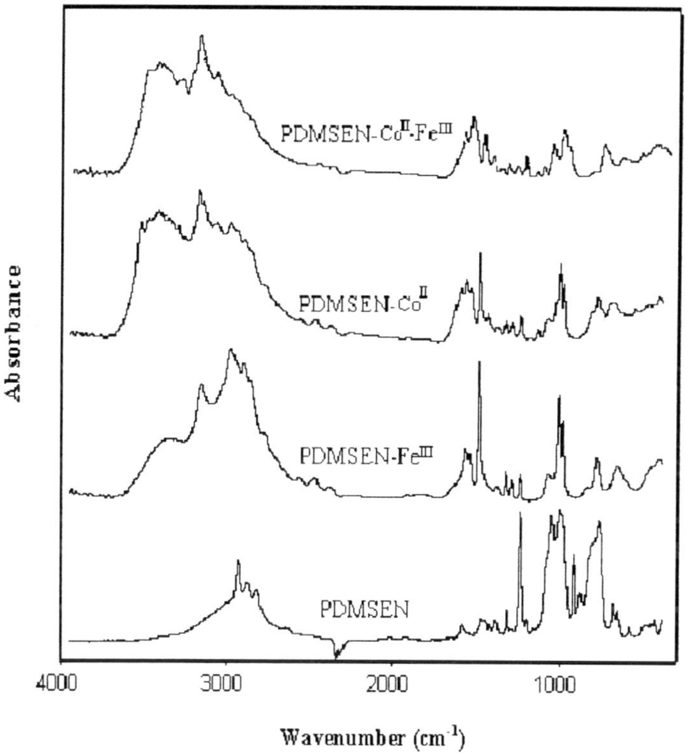

Figure 2. FT-IR spectra of PDMSEN, PDMSEN-Fe^{III}, -Co^{II} and -$Co^{II}Fe^{III}$ chlorides macromolecules

Comparison between the characteristic peaks due $\nu_{N\text{-}H}$, $\nu_{Si\text{-}CH3}$ and $\nu_{Si\text{-}N}$ functionalities in metal-free and metal-containing PDMSEN shows that the incorporation of metal ions have caused significant changes in their intensities but and positions. Of a particular importance is the band at 1264 cm^{-1} region due $\delta_{Si\text{-}CH3}$, which appeared strong in the case of PDMSEN matrix whereas it is medium-weak in the homometallic and heterometallic macromolecules. The observed decrease in the intensity may be explained based on two considerations, the number of Si–CH_3 vectors giving rise to this band and the local symmetry at silicon.

The first point suggests that the number of Si–CH_3 bonds in the metal-containing PDMSEN is less than that in the metal-free PDMSEN matrix. This observation demonstrates that the degree of polymerization decreases upon the introduction of the metal ions. In other words, the introduction of the metal ions into the matrix has lead to partial de-polymerization as shown in Scheme 4. Unfortunately, the extent of this process cannot be deduced from the intensity of this spectral band only. However, the de-polymerization process can be monitored by examining the intensity of symmetric and asymmetric $\nu_{N\text{-}H}$ bands due the formation of terminal amino functionality (–NH_2).

On the other hand, ring-opening process leads to the formation of –NH group with symmetric $\nu_{N\text{-}H}$ band. Second, the interaction between metal center and the nitrogen of the carbosilazane segment at (—$SiMe_2NHCH_2CH_2NH$—) moiety may distort the local symmetry at the silicon atoms of the silylamino groups (N-Si-CH_3) and hence cause reduction in the intensity and broadening of its band at 1264 cm^{-1}. Indeed, measurements of the bandwidth at half-height in the metal-containing PDMSEN shows that this band is slightly larger than that in the metal-free matrix. Moreover, close examination of this band in metal-containing PDMSEN shows that it is composed of two unresolved overlapping peaks. In this context, ring-opening process can also contribute to the decrease in the intensity of Si–CH_3 band. Unfortunately, no structural information about the electronic interaction between similar (-Co^{II} -Fe^{II} and Fe^{III}) and dissimilar (Co^{II} Fe^{II} and $Fe^{II}Fe^{III}$) metal sites could be obtained from the broadening and splitting features in their peaks at the 1260 cm^{-1} region. Furthermore, all the prepared homometallic and heterometallic exhibit a broad feature in 3500 cm^{-1} region attributed to $\nu_{O\text{-}H}$ of trace amounts of water that are entrapped in their crystal lattices. This conclusion is in agreement with the thermal data obtained for these M-PCSZ materials (see thermal stability section).

Electronic Spectroscopy

The electronic absorption spectra of the metal-free PDMSEN matrix and the obtained M-PCSZ macromolecular complexes were recorded in DMF at ambient conditions, Table 1. The metal-free PDMSEN matrix exhibits a single sharp band at 285 nm attributed to localized electronic transition on Si-N. This band is commonly observed in organosilicon compounds-containing Si-O and Si-N bonds due to d_π-p_π interactions [70]. This band is red-shifted to different extents in all M-PCSZ complexes except PDMSEN-Co^{II} where this band undergoes a blue shift. This variable shift may indicate that different degree of PDMSEN-metal interaction being largest in trivalent (PDMSEN-Fe^{III}, 320 nm) and mixed-valence (PDMSEN-$Co^{II}Fe^{III}$, 335 nm and PDMSEN-$Fe^{II}Fe^{III}$, 338 nm) and smallest in divalent (PDMSEN-Co^{II}, 278 nm and PDMSEN-Fe^{II}, 290 nm). The observed shift clearly indicates that the metal ions are bound to the coordinating matrix at the nitrogen of the dimethylsilylamine.

Table 1. Electronic absorption spectral data, the mass magnetic susceptibility (χ_g) and effective magnetic moments (μ_{eff}) for the studied metal-containing homometallic and heterometallic macromolecules

Macromolecules	λ_{max} (nm)[a]	χ_g x 10^6 (cgs)	μ_{eff} (BM)[b,c]
PDMSEN	285	---	---
PDMSEN-Co^{II}	673, 598, 451 (sh), 314, 278	26.5	4.53
PDMSEN-Fe^{II}	362, 315, 290	20.3	4.03
PDMSEN-Fe^{III}	475, 367, 320	13.1	3.28
PDMSEN-$Co^{II}Fe^{III}$ [d]	675, 608, 594, 452, 335	15.5	4.42
PDMSEN-$Fe^{II}Fe^{III}$ [d]	471, 338	16.7	4.92

a) Electronic absorption was recorded in DMF.
b) The reported values are the average of two or three readings with ≈ 5% error.
c) χg is the mass magnetic susceptibility per gram measured at room temperature. $\mu eff = 2.83 (\chi M \times T)^{1/2}$, $\chi M = \chi g \times$ mol. mass. χM is not corrected for diamagnetic contribution of the constituent atoms. The mol. mass of the constituting units was estimated from thermolysis data as described in the text.
d) The values represent the magnetism due to the two paramagnetic centers.

These M-PCSZ homometallic and heterometallic materials are colored and therefore exhibit electronic absorption spectra indicative of their coordination. The electronic absorption spectrum of PDMSEN-Co^{II} display three moderately weak d-d transitions at 673 nm ($^4T_{1g} \rightarrow {}^4T_{2g}$), 598 nm ($^4T_{1g}$ (F) $\rightarrow {}^4A_{2g}$) and 541nm ($^4T_{1g}$ (F) $\rightarrow {}^4T_{1g}$ (P), and an intense ligand-to-metal charge transfer (LMCT) band at 314 nm. The presence of three d-d absorption bands is suggestive of distorted octahedral field environment [70-72]. Similarly, the PDMSEN-Fe^{II} spectrum display one moderate band at 362 nm assigned to d-d ($^5T_{2g} \rightarrow {}^5E_g$) transitions and an intense LMCT band at 315 nm. The electronic spectra of PDMSEN-Fe^{III} chromophores exhibit moderate LMCT bands at 475 and 367 nm. In general, Fe^{III} exhibit LMCT band in the near UV-region with shoulders that tails off in the visible region obscuring the weak d-d transitions. These spectral features are in accordance with distorted octahedral geometry.

Comparison between the electronic transitions at the Co^{II} and Fe^{III} chromophores in the spectra of PDMSEN-Co^{II} and PDMSEN-Fe^{III} with that of PDMSEN-$Co^{II}Fe^{III}$ shows that the spectral bands due to individual Co^{II} and Fe^{III} chromophores have undergone red-shifts and blue-shift, respectively. In other words, the spectrum of PDMSEN-$Co^{II}Fe^{III}$ is not an average sum of the constituting chromophores suggesting the existence of observable electronic interaction between Co^{II} and Fe^{III} sites. This result implies that these heterometallic sites exist in close proximity. This conclusion can be similarly applied to PDMSEN-$Fe^{II}Fe^{III}$ macromolecule. Further information about the metal-metal sites interactions can be obtained from the magnetic properties of these M-PCSZ materials. In this respect, it is important to emphasize that the electronic spectra of these M-PCSZ materials are different from those of the corresponding uncoordinated transition metal chlorides. This evidence clearly supports the fact that PDMSEN matrix has a strong affinity to coordinate various transition metal ions.

Magnetic Behavior

Recent advances in magnetic materials have focused on the magnetic behavior of organic-based and metal-based magnetopolymers [73-75]. In this context, current research efforts in this field has been directed towards obtaining soluble ferromagnetic materials that can be exploited in the formulation of magnetic thin films and membranes. Although numerous studies have been concerned with classical magnetic materials, which are generally comprised of large arrays of magnetic sites inter-connected through organic ligands or aggregates of discrete magnetic complexes, relatively few studies have been concerned with magnetopolymers [73]. Currently, our work in this area focuses on the preparation of organosilicon-based magnetopolymers and other homogeneous magnetic ferrofluids employing organic sol-gel-thermolysis methodology [76].

The mass magnetic susceptibility (χ_g) and effective magnetic moment (μ_{eff}) for the studied M-PCSZ materials were investigated at room temperature using Evans method, Table 1. In all cases, the obtained magnetic data were reproducible (within 1-5% error) when different samples were used. This suggests that the corresponding metal sites are quite uniformly distributed within PDMSEN network. The magnetic moment of PDMSEN-Co^{II} was found to be 4.53 BM, which is typical for distorted octahedral high spin monomeric Co^{II} complexes ($\mu_{observed}$ = 3.88-5.2 BM). The obtained value is larger than that produced by spin-only (μ_{spin}) value of 3.87 BM for Co^{II} (d^7). This deviation from the spin-only value is ascribed to strong spin-orbit interactions between the high-spin Co^{II} sites and surrounding ligands [72]. This observation demonstrates that PDMSEN backbone offers unfavorable magnetic superexchange pathway and therefore one can infer that the Co^{II} sites are situated in a magnetically isolated environment produced by the PDMSEN matrix, Scheme 5A. However, the observed magnetic moment for divalent PDMSEN-Fe^{II} (μ_{eff} = 4.03 BM) and trivalent PDMSEN-Fe^{III} (3.28 BM) are less than what is expected for typical high spin Fe(II) (d^6, μ_{spin} = 4.90 BM) and high spin Fe(III) (d^5, μ_{spin} = 5.92 BM), respectively. In both cases the quenching of the magnetic moment may be attributed to antiferromagnetic coupling, which is more pronounced in the trivalent state. Although the actual structures of the obtained M-PCSZ macromolecules are rather complicated, one can argue that the linear PDMSEN backbone gave essentially low-dimensional magnetic materials. Since the iron sites are organized in the diamagnetic PDMSEN framework at sufficiently well separated distances, one can conclude that the superexchange mechanism between sites on the same chain is negligible and therefore, the observed magnetic quenching may be attributed to superexchange magnetic interaction between iron sites on different chains facing one another as shown in Scheme 5B. Moreover, the magnetic behavior of the mixed valence PDMSEN-$Co^{II}Fe^{III}$ and PDMSEN-$Fe^{II}Fe^{III}$ regimes suggest the existence of antiferromagnetic interactions between the homo- (Fe^{II}-Fe^{III}) and hetero- (Co^{II}-Fe^{III}) paramagnetic sites. In these cases, the different paramagnetic sites are isolated from each other and the diamagnetic PDMSEN backbone functions as an organizing medium for metal centers. In other words, the PDMSEN-$M^{II}Fe^{III}$ regimes can be thought of as a low-dimensional array of alternating different local spins. Therefore, one can infer that the PDMSEN-$M^{II}Fe^{III}$ (M^{II} is Co^{II} and Fe^{II}) macromolecules adopt structures as shown in Scheme 5B. The different observed magnetic behaviors of these M-PCSZ materials demonstrate that the different metal ions influence the overall macrostructures of PDMSEN chains differently.

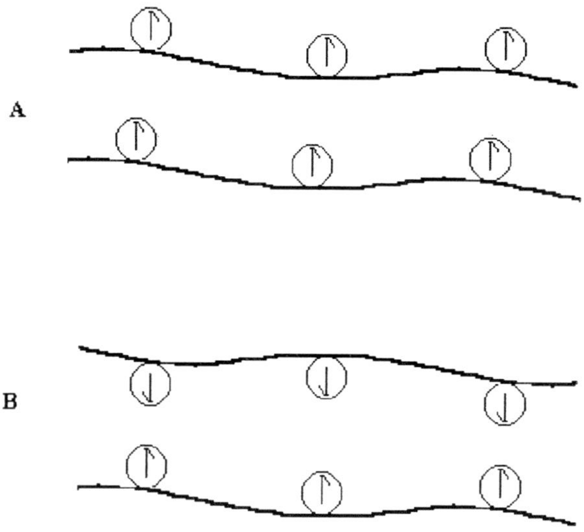

Scheme 5. Possible local spin orientation predicted from magnetic studies:
(A) Isolated metal sites with negligible magnetic interaction
(B) Isolated metal sites with weak antiferromagnetic interaction

These M-PCSZ materials provide an interesting model to study the superexchange interaction between localized spins in the magnetopolymer. Indeed, one can take the advantage of possible synergistic electronic effects and superexchange mechanism to generate new magnetopolymers. It is worth noting that the strong ferromagnetic inter-chain interaction is very useful in the preparation of magnetopolymers. In fact, upon forming inter-chain oxygen-bridged Fe^{II}-O-Fe^{III} in PDMSEN-$Fe^{II}Fe^{III}$ a strong ferrmagnetopolymer has been successfully obtained [76]. Further detailed analysis of the magnetic behavior of all these materials as a function of temperature is underway.

Powder X-Ray Diffraction

Fig. 3 depicts the XRD profiles of metal-free and all the metal-containing PDMSEN. It is clear that the XRD profile of the metal-free PDMSEN matrix contains three broad peaks of different relative intensities at 2θ of 8.0°, 14.9° and 28.5° with few sharp peaks superimposed on the middle broad peak. The positions of these broad peaks were determined by curve-fitting procedure of the experimental XRD data using Table Curve-2D software (Lowess Algorithm). This XRD pattern demonstrates that PDMSEN matrix is predominantly amorphous in nature with small crystalline regions. The diffraction patterns at 2θ of 14.9° and 28.5° can be ascribed to two different chain-chain packing modes. Using Braggs` law of diffraction ($n\lambda = 2d \sin \theta$) these broad features were found to correspond to two different chain-chain separations (d-spacing) [77,78]. In fact, polymers such as polyacetylene were reported display similar XRD diffraction pattern [78]. These two values suggest that PDMSEN may be present in two different packing forms, weakly folded chains with Van der

Waal forces and closely packed structure with an average separation distance of 5.94 and 3.14 Å, respectively.

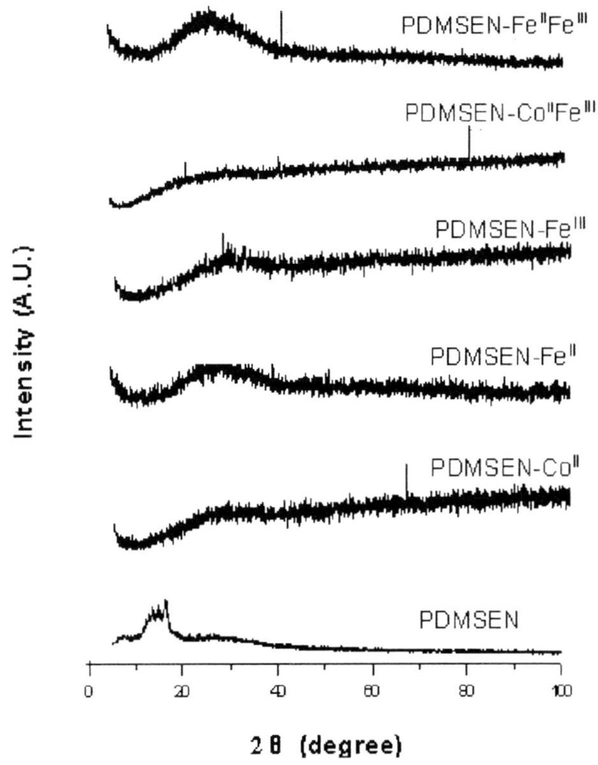

Figure 3. XRD profiles for the studied metal-free and metal-containing PDMSEN macromolecules

Close examination of the XRD patterns of the obtained M-PCSZ macromolecules show that they are composed of two diffraction sets, a set of the broad peaks due to the PDMSEN amorphous phase and a set of sharp peaks that can be indexed for the attached metal chlorides. Interestingly, the diffraction patterns displayed by the metal chlorides after incorporation into PDMSEN are totally different from those exhibited by the corresponding metal chlorides. This observation demonstrates that in all these M-PCSZ macromolecules the metal chlorides are chemically attached to the PDMSEN matrix rather than physically dispersed (doped) in the matrix framework. Furthermore, it is clear that the introduction of metal centers into PDMSEN has caused two important changes in the diffraction patterns of the coordinating matrix. First, the two broad peaks due to the weakly and closely packed structures have coalesced into a very broad feature centered at 2θ of 22.5-26.7°, Table 2. This implies that the PDMSEN chains in metal-containing macromolecules became uniformly packed with variable inter-spatial chain-chain separation as shown in Table 2. This indicates that the incorporated metal centers have influenced the microstructure and packing of the coordinating PDMSEN matrix to different extents. This observation is in agreement with earlier findings on the effect of metal ions on the morphology and chain packing of polymers

[54-55]. This inter-chain spacing has been shown to play an important role in the electric conductivity of polymers.

Table 2. XRD diffraction data (2θ) and inter-spatial chain-chain separation (d-spacing) values for the coordinating PDMSEN matrix in metal-free and metal-containing macromolecules

Macromolecules	2θ (degrees)	d-spacing (Å)
PDMSEN	14.9, 28.5	5.94, 3.14
PDMSEN-Co^{II}	26.7	3.34
PDMSEN-Fe^{II}	25.6	3.48
PDMSEN-Fe^{III}	24.8	3.59
PDMSEN-$Co^{II}Fe^{III}$	22.5	3.95
PDMSEN-$Fe^{II}Fe^{III}$	24.1	3.69

a) The 2θ values were determined by curve-fitting procedure using Table Curve-2D Software (Lowess Algorithm)

b) The d-spacing was calculated from Braggs Law ($n\lambda = 2d \sin \theta$)

It is important to point out that coalescence of these peaks was found to depend upon the percentage metal chloride loading. In other words, when metal chloride loading is in 20-30% range the PDMSEN exhibits close packing structure whereas at lower or higher loading the two packing structures are generally observed. Second, the position of the main broad diffraction peak of the PDMSEN matrix at 2θ of 28.50° is shifted to lower values as shown in Table 2. This shift permits one to deduce information about chain-chain expansion in these M-PCSZ macromolecules. This clearly suggests that in all cases the metal chlorides are incorporated between the PDMSEN chains and therefore have caused chain-chain inter-spatial separation to increase to different extents.

Moreover, comparison between the XRD profiles of the mixed-valence PDMSEN-$Co^{II}Fe^{III}$ and PDMSEN-$Fe^{II}Fe^{III}$ chlorides with the corresponding homometallic analogues permits one to deduce useful information about the relative locations of the divalent Co^{II} or Fe^{II} with respect to the trivalent Fe^{III} sites. First, the XRD profiles of the mixed-valence regimes are not the average sum of the constituting homometallic analogues. Second, the diffraction peaks width at half-height for the mixed-valence regimes are somewhat broader than those of the analogous homometallic systems. Although no quantitative information about the electronic interaction could be obtained from these XRD profiles, it can be deduced that weak electronic interaction exist between the adjacent metal sites. This conclusion is consistent with the results obtained from the magnetic studies.

Thermal Studies

The thermal behavior of metal-free and metal-containing PDMSEN was previously investigated by thermogravimetric analysis (TGA) and differential scanning calorimetry (DSC) [54-56]. In general, all thermal results showed that the metal-free and metal-containing PDMSEN decompose in multi-degradation steps between 200 and 450 °C. These studies also showed that M-PCSZ macromolecules contain trace amounts of moisture occluded into the their crystal lattices. In this study, the metal-free and metal-containing PDMSEN were heated in two steps to 100 °C then to 500 °C in air. The thermolysis data are summarized in Table 3. It is obvious that metal-free PDMSEN matrix completely degrade into volatile products. However, thermolysis of metal-containing PDMSEN showed that the coordinated PDMSEN matrix degraded almost completely during the heating process to 500 °C with the formation of the corresponding metal oxide/hydroxide and small amounts of unidentified silicon-containing ceramic residues (carbide, oxides, oxynitrides). The volatile products produced in the thermolysis of these M-PCSZ materials were chemically analyzed ($AgNO_3$/Methanol) and found to consist of chlorinated organic amines. This observation is in agreement with recent reports on the formation of chlorinated organic products during the thermolysis of organic materials in the presence of different metal chloride salts [79]. Further heating to 800 °C resulted in further small weight loss indicating that at 500 °C the degradation was almost complete.

Interestingly, the thermal data in Table 3 allow one to calculate PDMSEN : metal mole ratios, percentage mass of metal sites in these M-PCSZ macromolecules and estimate the molecular mass of the monomeric units of metal-containing PDMSEN. It is worth noting that thermolysis of the mixed-valence PDMSEN-$Co^{II}Fe^{III}$ and PDMSEN-$Fe^{II}Fe^{III}$ heterometallic materials under vacuum generated ultra-fine magnetic particles.

Table 3. Thermolysis data for metal-free and metal-containing PDMSEN macromolecules[a]

Macromolecule	Moisture (%)[b]	Organosilicon (%)[c]	Residue (%)[d]
PDMSEN[e]	traces	99.7	0.3
PDMSEN-Co^{II}	4.8	62.1	33.1
PDMSEN-Fe^{II}	3.5	66.0	30.5
PDMSEN-Fe^{III}	7.0	59.1	33.9
PDMSEN-$Co^{II}Fe^{III}$	9.0	54.6	36. 4
PDMSEN-$Fe^{II}Fe^{III}$	8.7	58.9	32.4

a) Thermolysis at 500°C in air.
b) % mass of moisture represents mass loss below 150 °C.
c) % mass of chlorinated organosilicon is mass lost between 150 and 500 °C
d) % mass of the residues is reported at 500°C.

Electrical Conductivity Measurements

The dc electrical properties of the obtained M-PCSZ macromolecules were recorded on compacted discs at variable temperatures. Fig. 4 displays the characteristic I-V curves (current-voltage) for these M-PCSZ macromolecules. Their dc electrical conductivity (σ) and ohmicity (a) were calculated from Eq. 1, where, I is the current, V is the voltage and G is the conductance of the sample ($G = A\sigma/L$, A = area, L = thickness).

$$\mathrm{Ln}(I) = a\,\mathrm{Ln}(V) + \mathrm{Ln}(G) \qquad (1)$$

The dc conductivity data in Table 4 clearly shows that the PDMSEN matrix is an insulator, whereas all the obtained M-PCSZ complexes exhibit semi-conductor behaviors. This observation indicates that the incorporation of metal ions to the PDMSEN matrix has enhanced the dc conductivity by several orders of magnitude. This observation suggests that the coordinated metal ions offer a quite facile pathway for electronic conduction to occur. However, the conductivity values obtained in this study is less than that obtained for the previously studied PDMSEN-Cu^{II}, Ni^{II} Cr^{III} by three order of magnitude whose XRD pattern exhibit the weakly packed macrostructure [54]. This difference shows that the conductivity largely depends upon the chain packing. Close examination of Fig. 4 shows that the charge transfer in these materials occur via one conduction pathway in the voltage range of 0-12 V. Furthermore, PDMSEN-Fe^{II} (a = 0.32) shows a large deviation from ohmic behavior whereas the other studied M-PCSZ macromolecules exhibits small-negligible deviations from the value of unity for ohmic (electronic) conduction. This implies that the conduction mechanism operating in PDMSEN-Fe^{II} is more complicated than in the other M-PCSZ macromolecules. Given the fact that electrical measurements are known to be sensitive to moisture, sample preparation, impurities, particle-particle distances and crystallographic orientation within the disc, the reported values were found to be reproducible within ≈ 5% [80]. The simple structure shown in Schemes 3 demonstrates that the metal centers are separated from each other by the insulating PDMSEN chain backbone. This implies that electron transfer through the bonds of the matrix spacer is hindered and therefore electron hopping through space between adjacent metal sites attached to the same chain (intra-chain electron transfer) or on different chains (inter-chain electron transfer) is feasible.

To gain more insight into the conduction mechanism in the studied M-PCSZ macromolecules, their dc electrical conductivities were investigated as a function of temperature in the temperature range of 300-340 K. Applying Arrhenius relationship, Eq. 2 (where, σ_o = constant, E_g = Energy of activation, k_B = Boltzmann Constant and T = temperature) to the conductivity data one can determine the activation energies (band gap, E_g) from the corresponding Ln(σ) *vs.* 1/T linear plot, Fig. 5. As expected, Ln(σ) increases linearly with 1/T. The variable E_g values demonstrate that the incorporated metal ions facilitate electron transfer by offering different energy states for hopping process to occur, Scheme 6. This is particularly significant in case of electron hopping between different metal sites in heterometallic regimes.

$$Ln(\sigma) = Ln(\sigma_o) - E_g/k_BT \quad (2)$$

Table 4. Electrical properties of metal-containing PDMSEN macromolecules

Macromolecules	Thickness[a] mm	Conductivity (σ) $S.m^{-1}$	Ohmicity (a)	Band Gap (eV)
PDMSEN	b	~ 10^{-14}	--	--
PDMSEN-Co^{II}	0.65	0.77×10^{-10}	0.88	0.74
PDMSEN-Fe^{II}	0.70	0.48×10^{-8}	0.32	0.97
PDMSEN-Fe^{III}	0.23	1.54×10^{-8}	1.04	0.64
PDMSEN-$Co^{II}Fe^{III}$	0.71	5.00×10^{-8}	1.01	0.78
PDMSEN-$Fe^{II}Fe^{III}$	0.84	0.10×10^{-8}	0.97	1.40

a) Area of the compacted disc is 1.27 mm^2.

b) Thin film

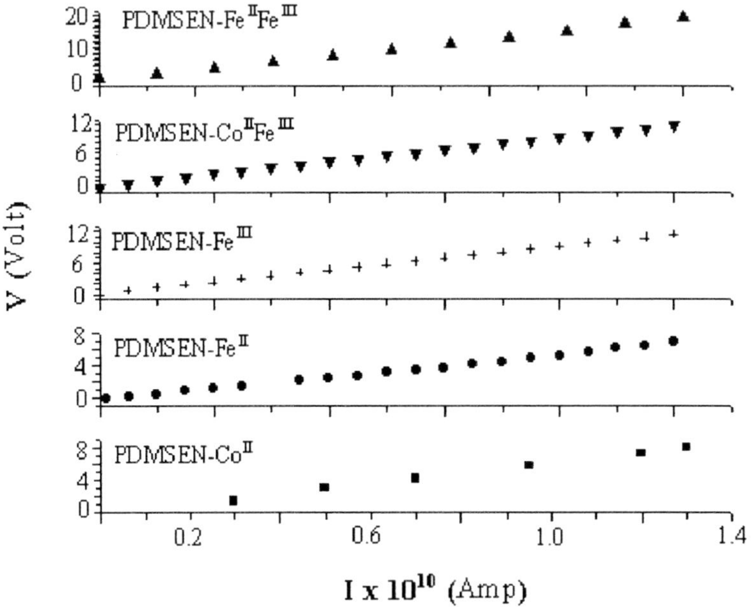

Figure 4A. I-V characteristic curves for the studied metal-containing PDMSEN macromolecules

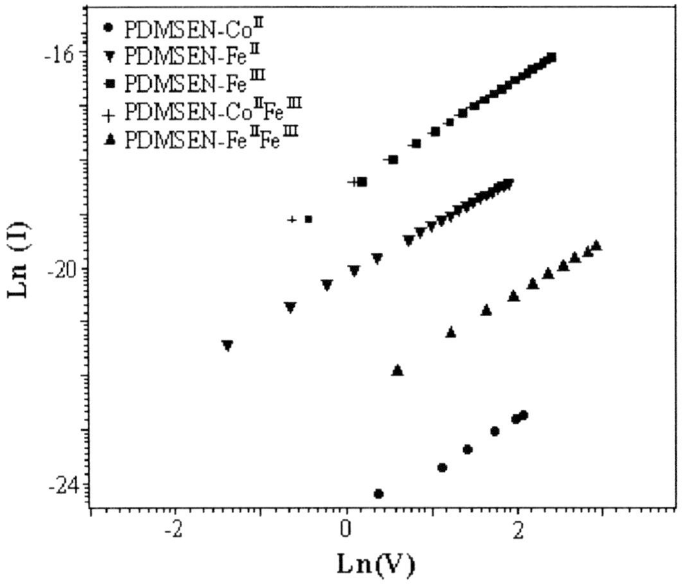

Figure 4B. Linear plots of Ln I vs. LnV for the studied metal-containing PDMSEN mcromolecules (PDMSEN-Fe^{III} and PDMSEN-$Co^{II}Fe^{III}$ overlap)

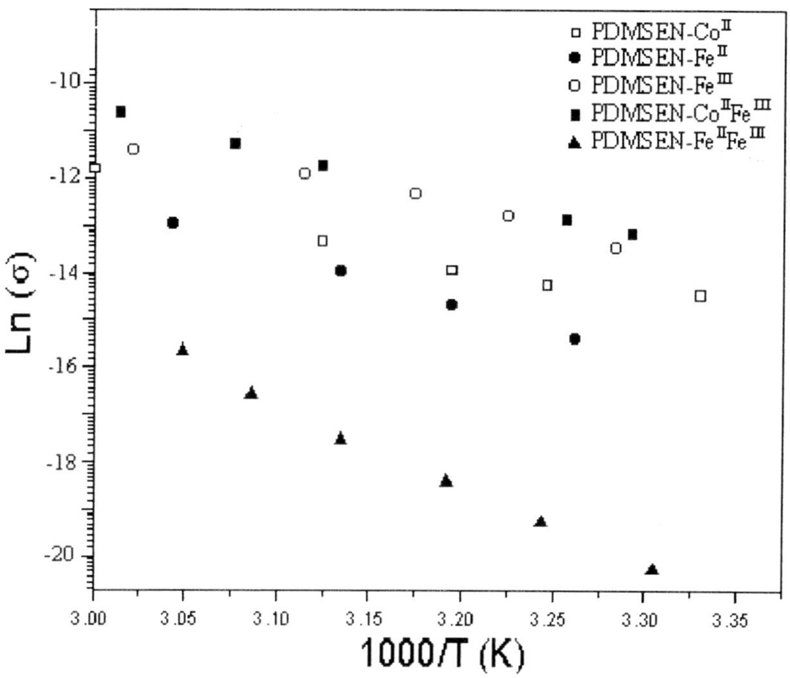

Figure 5. Plot of Ln (σ) vs. 1/T for the studied metal-containing PDMSEN systems

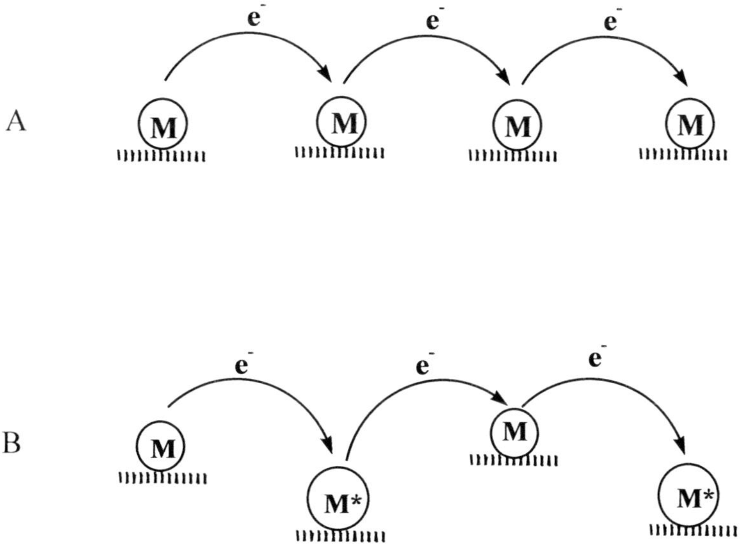

Scheme 6. Electron hopping between equivalent metal sites in divalent and trivalent homometallic systems (A), and non-equivalent metal sites in mixed-valence homo- and hetrometallic systems

In general, the distinction between electronic and ionic conduction is difficult. However, it has been suggested that electronic conduction is associated with a low energy of activation (below 0.8 eV) whereas ionic conduction that involves the transport of heavy masses (e.g., ions) requires energies larger than 0.8 eV [81]. In the case of PDMSEN-Fe^{II} (E_g = 0.97 eV) and PDMSEN-$Fe^{II}Fe^{III}$ (E_g = 1.40 eV) , the E_g values are clearly above 0.8 eV, which suggests that electronic and ionic (Cl^- migration) conductions are operative whereas in the other studied M-PCSZ regimes the E_g suggest no ionic but electronic conduction. This is probably due to the fact that chloride is bound to the metal centers and therefore, their migratory capability is restricted and hence they do not contribute to the overall conductivity. Moreover, ion transport requires appropriate vacancies (voids between the chains) to permit charge motion. Interestingly, the large difference between the band gap values for the divalent PDMSEN-Fe^{II} and trivalent PDMSEN-Fe^{III} analogue clearly demonstrate the effect of the oxidation state on charge transport processes. Although the divalent regime contains less chloride ions than the trivalent one can infer that the chloride is tightly bound to trivalent Fe(III) sites whereas it is weakly bonded to the divalent Fe(II) centers. This issue requires further detailed investigation.

The ac electrical behavior of the prepared M-PCSZ materials was examined at room temperature in the frequency range of 10 kHz to 1000 kHz. Fig. 7 depicts the frequency dependence of the relative permitivity (ε') and dielectric loss (ε'') (real and imaginary components of the dielectric constants, respectively) for the obtained M-PCSZ materials at room temperature. It is clear from the plots in Fig. 7 that these M-PCSZs exhibit significantly different dielectric properties. Since the dipolar nature of the PDMSEN backbone, chain mobility, humidity and temperature are very similar in all these M-PCSZ regimes the observed ac characteristics can be attributed to the dipolar properties of incorporated metal chlorides. It is obvious that both the type and oxidation state of the metal centers influence the dielectric behavior of these materials. The dielectric values for PDMSEN-Co^{II}, -Fe^{III}, -$Co^{II}Fe^{III}$ and -$Fe^{II}Fe^{III}$ materials decrease as a function of increasing frequency. However, the dielectric values for PDMSEN-Fe^{II} first increase to reach a maximum at ≈ 35 kHz then it decrease. This behavior is not surprising since the ohmicity for this material showed large deviation from the value of unity whereas all the other metal-containing PDMSEN regimes displayed ohmicity in the range of 0.88-1.04. Furthermore, the divalent and trivalent homometallic regimes (PDMSEN-Co^{II} -Fe^{II} and -Fe^{III}) exhibit one relaxation process in the examined frequency range whereas the mixed-valence PDMSEN-$Co^{II}Fe^{III}$ and -$Fe^{II}Fe^{III}$ exhibit two relaxation mechanisms at lower and higher frequencies. This observation indicates the presence of equivalent dipolar metal sites in the homometallic regimes and two non-equivalent dipolar metal sites in the mixed-valence systems, Scheme 6. This conclusion is consistent with our earlier findings that the metals in the mixed-valence system display weak interaction. Further detailed experiments are underway using thin sample over wide range of frequency at variable temperatures. Interestingly, modeling the electrical behavior of these M-PCSZ macromolecules showed they resemble RC electrical networks in parallel (resistance-capacitance) [82].

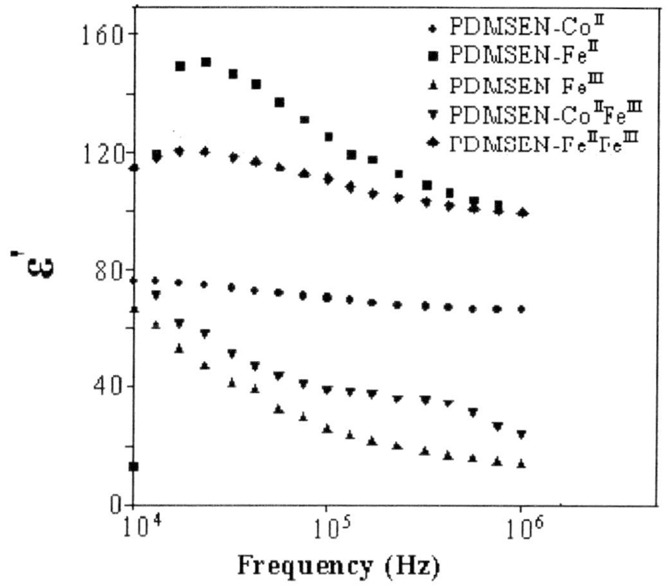

Figure 7A. Plot of the real (ε') component of the dielectric constants as a function of frequency (10 kHz-1000 kHz) for the obtained M-PCSZ materials

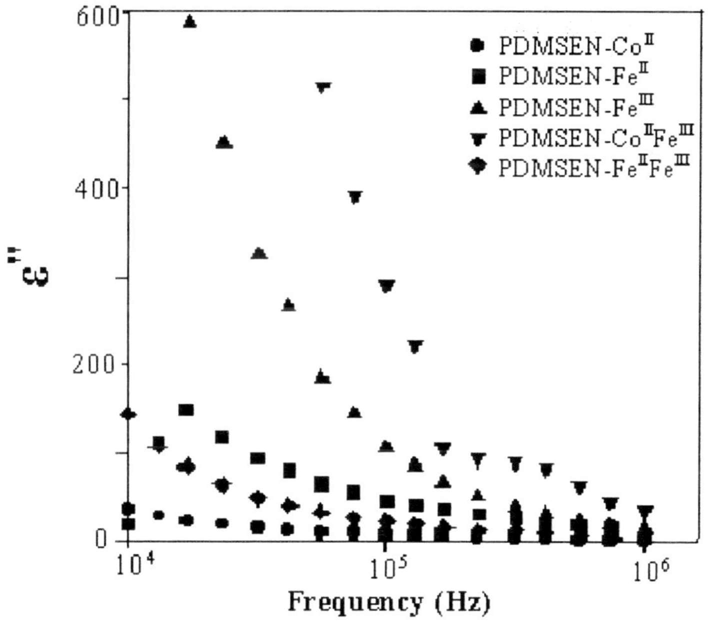

Figure 7B. Plot of the imaginary (ε'') component of the dielectric constants as a function of frequency (10 kHz-1000 kHz) for the obtained M-PCSZ materials

CONCLUSION

In previous studies, we have shown that polycarbosilazane-containing diamino groups such as PDMSEN readily coordinates various transition metals. Evidence obtained from electronic spectroscopy, X-ray diffraction and thermal measurements indicate that metal ions in are strongly held to the PDMSEN matrix. In this contribution, we have shown that the presence of transition metal in PDMSEN provide an effective pathway for electron hopping that resulted in enhancing the dc conductivity by several orders of magnitude depending upon the metal and its oxidation state. Furthermore, mixed-valence PDMSEN-$Co^{II}Fe^{III}$ and PDMSEN-$Fe^{II}Fe^{III}$ regimes show different electrochromic behavior from those of the constituting homometallic systems. Although the structural details of these homometallic and heterometallic regimes were not delineated at the molecular level, three important structural features are apparent. First, transition metals are covalently bound to PDMSEN chains. Second, the presence of different metals between the chains allows a variety of chain-chain interactions and therefore, imparts different structural and electrochromic characteristics on the PDMSEN-M-Fe frameworks. Third, the versatility of the prepared regimes is obvious as one can modify the system in an easy fashion to tune its optical, magnetic and electrical properties by replacing the nonstructural chloride with other anions without major disturbance of the structure. Indeed, interesting metallomesogens have been obtained upon metathesis of the chloride with organic (polyacrylate, polyphosphonate) and inorganic (polysilicate) polyanions.

ACKNOWLEDGEMENT

The authors wish to thank Jordan University of Science and Technology, Irbid, Jordan for financial support.

REFERENCES

[1] Kingsborough, R. P.; Swager, T.M. *J. Am. Chem. Soc.* 1999, 121, 8825-8834.
[2] Cameron, C.G.; Pickup, *P.G. J. Am. Chem. Soc.*1999, 121, 11773-79.
[3] Aggour, Y.A.; Aziz, M.S.; Youssif, M.I. *Polym. Test.* 2000,19, 919-925.
[4] Wang, Y.; Pan, C. *Eur. Polym. J.* 2001, 37, 699-704.
[5] Alexandrova, L.; Sansores, E.; Martines, E.; Rodriguez, E.; Gerasimov, G. *Polymer* 2001, 42, 273-280.
[6] Cameron, C.G.; Pittmann, T.J.; Pickup, *P.G. J. Phys. Chem.* (B) 2001,105, 8838-8844.
[7] Huber, C.; Bangerter, F.; Casert, W.R.; Weder, *C. J. Am. Chem. Soc.* 2001, 123, 3857-3863.
[8] Newkome, G.R.; He, E.; Moorefield, C.N. *Chem. Rev.* 1999, 99, 1689-1746.
[9] Cuadrado, I.; Moran, M.; Casado, C.M.; Alonso, B.; Losada, J. Coord. *Chem. Rev.* 1999, 193-195, 395-445.
[10] Grayson, S.M.; Frechet, J.M.J. *Chem. Rev.* 2001, 101, 3819-3867.
[11] Chadrdac, F.; Astruc, D. *Chem. Rev.* 2001, 101, 2991-3023.

[12] Breinbauer, R.; Jacobson, E.N. Angew. *Chem. Int. Ed.* 2000, 39, 3604-3607.

[13] Garber, S.B.; Kingsbury, J.S.; Gray, B.L.; Hoveyda, A.H. *J. Am. Chem. Soc.* 2000, 122, 8168-8179.

[14] Gomez-Elipe, P.; Resendes, R.; Macdonald, P.M.; Manners, I. *J. Am. Chem. Soc.* 1998, 120, 8348-8356.

[15] Murugavel, R.; Voigt, A.; Walawalkar, M.G.; Roesky, H.W. *Chem. Rev.* 1996, 96, 2205- 2236.

[16] Molodtsova, Yu.A.; Pozaniakova, Yu.A.; Lyssenko, K.A.; Blagodatskikh, I.V.; Katsouli, D.E.; Shehegolikhina, O.I. *J. Organomet. Chem.*1998, 571, 31-36.

[17] Zhu, S.; Swager, T.M. *J. Am. Chem. Soc.* 1997, 119, 12568-12577.

[18] Gimenez, R.; Lydon D.P.; Serrano, J.L. *Curr. Opin. Solid State Mater. Sci.* 2002, 6, 527-535.

[19] Binnemans, K.; Galyametdinov, Y.G.; Deun, R.V.; Bruce, D.W.; Collinson, S.R.; Polishchuk, A.P.; Bikchntaev, I.; Hasse, W.; Prosvirin, A.V.; Tinchurina, L.; Litvinov, I.; Gubajdullin, A.; Rakhmatullin, A.; Uytterhoeven, K.; Meervelt, L.V. *J. Am. Chem. Soc.* 2000, 122, 4335-4344.

[20] Hudson, S.A.; Maitlis, M. *Chem. Rev.* 1993, 93, 861-885.

[21] Manners, I. Angew. *Chem. Int. Ed. Engl.*1996, 35, 1602-1621.

[22] Cuadrado, I.; Moran, M.; Casado, C.M.; Alonso, B.; Losada, J. Coord. *Chem. Rev.* 1999, 193-195, 395-445.

[23] Hearshaw, M.A.; Moss, J.R. *Chem. Commun.*1999, 1-8.

[24] Nguyen, P.; Gomez-Elipe, P.; Manners, I. *Chem. Rev.* 1999, 99, 1515-1548.

[25] Kaliyappan, T.; Kannan, P. *Prog. Polym. Sci.* 2000, 25, 343-370.

[26] Sheats, I.E.; Carraher, C.E. Jr.; Pittmann, C.U. Jr.; Zeldin, M.; Currell, B. eds. *Inorganic and Metal-containing Polymeric Materials*, Plenum Press, NY, 1990.

[27] Ciardelli, F.; Tsuchida E.; Wohrle, D. eds. *Macromolecule-Metal Complexes*, 1st Ed., Springer-Verlag; Berlin, 1996.

[28] Bruce, D.W.; O'Hare, D. eds. *Inorganic Materials*, 2nd Ed. John Wiley and Sons; NY, 1999.

[29] Paula, M.M.S.; de Moraes, V.N. Jr.; Mocellin, F.; Franco, C.V. *J. Mater. Chem.* 1998, 8 (9) 2049-2054.

[30] Holstrom S.D.; Cox, J.A. *Anal. Chem.* 2000, 72, 3191-3195.

[31] MacDonald, W.A. J. *Mater. Chem.* 2004, 14, 4-10.

[32] Zoppi, R.A.; das Neves, S.; Nunes, S.P. *Polymer* 2000, 41, 5461-5470.

[33] Hijelm, J.; Constable, E.C.;Figggemeier, E.; Hagfeldt, A.; Handel, R.; Housecroft, C.E.; Mukhtar, E.; Schofield, E. *Chem. Commun.* 2002, 284-285.

[34] Pelster, R.; Simon, U. *Colloid Polym. Sci.* 1999, 277, 2-14.

[35] Wang, Y.; Pan, C. *Eur. Polym. J.* 2001, 37, 699-704.

[36] Southard, G.E.; Curtis, M.D. *Organometallics*, 2001, 20, 508-522.

[37] Cornil, J.; dos Santos, D.A.; Crispin, X.; Silbey, R.; Bredas, J.L. J. *Am. Chem. Soc.* 1998, 120,1289-1299.

[38] Setayesh, S.; Grimsdale, A.C.; Weil, T.; Enkelmann, V.; Mullen, K.; Meghdadi, F.; List, E.J.W.; Leising, G. J. *Am. Chem. Soc.* 2001, 123, 946-953.

[39] Mao, S.S.; Liu, F.; Tilley, T.D.; *J. Am. Chem Soc.* 1998, 120, 1193.

[40] Altmann, M; Bunz, UHF; *Angew Chem. Int. Ed. Engl.* 1995, 34, 569.

[41] Gad, A.M.; El-Dissouky, A.; Mansour E.M.; El-Maghraby, *A. Polym. Degrad. Stab.* 2001, 71, 267-272.

[42] Casado, C.M.; Gonzalez, B.; Guadrados, B.; Alonoso, B. Moran M. Losada, *J. Angew Chem. Int. Ed. Engl.* 2000, 39, 2135.

[43] Frey, H.; Schlenk, C. Top. *Curr. Chem.* 2000, 210, 69-129.

[44] Mark, J.E.; Allcock, H.R.; West, R. *Inorganic Polymers*, Prentice Hall, NJ, 1992.

[45] Arafa, I.M.; Al-Atrash, M. *J. Macromol. Sci Pure Appl. Chem.* 2002, A39(12),1475-1486.

[46] Yoshida, Y.; Nishihara, Y.; Ootake, R.; Fujii, A.; Ozaki, M.; Yoshino, K.; Kim, H.K.; Baek, N.S.; Choi, S.K. *J. Appl. Phys.* 2001, 92 (12) 6061-6065.

[47] Zheng, H.B.; Lu, W.; Wang, Z.Y. *Polymer* 2001, 42, 3745-3750.

[48] Blum, Y.D.; Laine, R.M. *Organometallics* 1986, 5, 2081-2088.

[49] Reiche, C.; Kleim, S.; Klingebel, U.; Noltemeyer, M.; Voit, C.; Herbst-Irmer, R.; Schmatz, S. *J. Organomet. Chem.* 2003, 667, 24-34.

[50] Y. Yoshida; Y. Nishihara; R. Ootake; A. Fujii; M. Ozaki; K. Yoshino; H.K. Kim; N.S. Baek and S.K. Choi, *J. Appl. Phys.* 2001, 92 (12) 6061-6065.

[51] Zeigler, J.M.; Fearon, F.W.G. Silicon-Based Polymer Science, Advances in Chemistry Series, 224, *Am. Chem. Soc.*Washington DC, 1990.

[52] Weinmann, M.; Haug, R.; Bill, J.; Aldinger, F.; Schuhmacher, J.; Müller, K. *J. Organomet. Chem.* 1997, 541, 345-353.

[53] Ma, K.; Scheibitz, S.; Scholz, S.; Wagner, M. J. *Organomet. Chem.* 2002, 652, 11-19.

[54] Arafa, I.M.; El-Ghanem, H.; Al-Shalabi, R. *Inorg. Organomet. Polym.* 2003, 13 (2) 69-86.

[55] Arafa, I.M.; El-Ghanem, H. *J. Macromol. Sci., Pure Appl. Chem.* 2003, A 40 (9), 975-988.

[56] El-Ghanem, H.M.; Abdul Jawad, S.; Aljundi, J.; Afaneh, F.; Arafa, I. *Polym Int.* 2003, 52 (7), 1125-1130.

[57] Arafa, I.; El-Ghanem, H.M.; Hallak, A.B.; Abdul Jawad, S. *Int. J. Polymer. Mater.* 2004, Accepted.

[58] Mariam, Y.H.; Pemawansa, K.P.W.; Abrahams, P. *Polym. Mat. Sci. Eng.* 1987, 56, 73-79.

[59] Feng, K.; Mariam, Y. H. *Macromolecules* 1991, 24, 4729-4732.

[60] Mitzel, N.W.; Angermaier, K.; Schmidbaur, H. Organometallics, 1994, 13, 1762-1766.

[61] Hu, J; Son, D.Y. *Macromolecules* 1998, 31, 8644-8646.

[62] Wynne, K.J.; Rice, R.W. *Ann. Rev. Mater. Sci.* 1984, 14, 297-334.

[63] Armitage, D.A. *In The Silicon-Hetroatom Bond*; Patai, S.; Rappoport, Z. Ed; John Wiley and Sons: NY, 1991; 365-445.

[64] Bouquey, M.; Brochon, C.; Bruzand, S.; Mingotaud, A.-F.; Schappacher, M.; Soum, A. *J. Organomet. Chem.* 1996, 521, 21-27.

[65] Bacque, E.; Pillot, J.; Birot, M.; Dunogenes, J. Bourgeois, G; Petraud, *M. J. Organomet. Chem.* 1994, 181, 167-172.

[66] Sacarescu, L.; Luchian, N.; Marcu, M.; Sacarescu, G.; Ardeleanu, R. J. *Macromol. Sci., Pure Appl. Chem.* 1997, A34 (4), 735-741.

[67] Melendrez, R.; Angel, G.D.; Bertin, V.; Valenzuela, M.A.; Barbier, J. J. Mole. Catal. A: Chemical, 2000, 157,143-149.

[68] Cernak, J.; Chomic, J.; Kappenstein, C.; Robert, F. *J. Chem. Soc.* Dalton Trans, 1997, 2981-2985.

[69] Ahmed, S.R.; Ogale, S.B.; Papaefthymiou, G.C.; Ramesh R.; Kofinas, P. *Appl. Phys. Lett.* 2002, 80 (9),1616-1618.

[70] Qiu, H.; Du, Z. J. *Polym. Sci, Part A, Polym. Chem.* 1989, 27 (9), 2861-2869.

[71] Lever, A.B.P. *Inorganic Electronic Spectroscopy*, 2nd Ed., Elsevier, 1984.

[72] Cotton, F.A.; Wilkinson, G. *Advanced Inorganic Chemistry: A Comprehensive Text*; 4th ed., John Wiley and Sons: NY, 1980, 749-798.

[73] Palacio, F.; Oliete, P.; Schubert, U.; Mijatovic, I.; Hüsing, N.; Peterlik, H. *J. Mater. Chem.* 2004, 14(12), 1873-1878.

[74] Shiho, H.; Manabe, Y.; Kawahashi, N. *J. Mater. Chem.* 2000, 10 (2), 333-336.

[75] Castro, C.; Millan, A.; Palacio, F. *J. Mater. Chem.* 2000, 10 (8), 1945-1947.

[76] Hammam, Y.A.; Arafa, I.M.; El-Ghanem, H.; Said, M.R.; Abu-Aljarayesh, I. Submitted.

[77] Reidinger, F.; Murthy, N.S.; Correale, S.T. In *A Guide to Materials Characterization and Chemical Analysis;* Sibilia, J.P. Ed; VCH, NY, 1988, 115-135.

[78] Bodor, G. *Structural Investigation of Polymers*, Engl. Ed., Harwood, England, 1991, 295-322.

[79] Conesa, J.A.; Fullana, A.; Font, R.*J. Anal. Appl. Pyrol.* 2001, 58, 553-568.

[80] El-Shekeil, A.; Al-Maydama, H.; Al-Karbooly, A.; Khalid, M.A. *Polymer* 1999, 40, 2879-2887.

[81] Jonscher, A.K. *Thin Solid Films*, 1967, 1 (3), 213-234.

[82] El-Ghanem, H; Arafa, I.; Nemrat, S. In preparation.

In: Recent Developments in Polymer Research
Editor: Antony V. Hopper, pp. 29-51
ISBN 1-60021-346-4

Chapter 2

SYNTHESIS AND CHARACTERIZATION OF POLY(DIBROMOPHENYLENE OXIDE)S THROUGH THERMAL DECOMPOSITION OF VARIOUS TRANSITION METAL COMPLEXES IN SOLID STATE

Pınar Çamurlu, and Duygu Kısakürek*
Middle East Technical University, Department of Chemistry,
06531 Ankara, Turkey

ABSTRACT

The synthesis of poly(dibromophenylene oxide)s by thermal decomposition of Bis(2,4,6-tribromophenolato)di(N-methyl imidazole)Cu(II) (C1), Bis(2,4,6-tribromophenolato)-di(N-methyl imidazole)Co(II) (C2) and Bis(2,4,6-tribromophenolato)di(N-methyl imida-zole) Ni(II) (C3) complexes were achieved by solid state polymerization. The characterizations of the complexes were performed by Fourier Transform Infrared Spectroscopy (FTIR), Differential Scanning Calorimeter (DSC), Mass Spectroscopy (MS), Magnetic Susceptibility and CHN elemental analysis. The effects of temperature and time (at the highest conversion) on percent conversion and intrinsic viscosity of the polymers were examined. Among the synthesized polymers, the highest percent conversion and intrinsic viscosity was achieved by the decomposition of $Cu(TBP)_2(NMIZ)_2$ complex, whereas $Ni(TBP)_2(NMIZ)_2$ yielded the lowest. Varying the decomposition temperature during 3 h scan showed that the $Cu(TBP)_2(NMIZ)_2$ complex revealed lowest and $Ni(TBP)_2(NMIZ)_2$ complex revealed the highest decomposition temperature. The $Ni(TBP)_2(NMIZ)_2$ (having the highest minimum decomposition temperature) complex was more stable than either $Co(TBP)_2(NMIZ)_2$ and $Cu(TBP)_2(NMIZ)_2$ complexes. Synthesized polymers were characterized by Fourier Transform Infrared Spectroscopy (FTIR), Differential Scanning Calorimeter (DSC), ^{1}H-Nuclear Magnetic Resonance (^{1}H-NMR), ^{13}C-Nuclear Magnetic Resonance (^{13}C-NMR) , Scanning Electron Microscope (SEM) and viscometric measurements. The poly(dibromophenylene oxides)s, synthesized from Cu, Co and Ni transition metal complexes, having nonchelating ligand N-methyl imidazole, displayed selectivity in the

* Correspondence to: Duygu Kısakürek (E-mail: duygu@metu.edu.tr) Tel: +903122103236, Fax: +903122101280

favor of 1-2 and 1-4 additions, taking place at equal rates irrespective of the type of the metal used. All the polymers having T_g values in between 211 ^{0}C to 190 ^{0}C from Cu to Ni, were rigid and have branched structures.

Keywords: Poly(dibromophenylene oxide), transition metal complexes, thermal decomposition, solid state polymerization.

INTRODUCTION

Poly(dihalophenylene oxide)s had been synthesized in various forms since 1916 because of their mechanical, chemical and thermal properties. These polymers are found to be extremely fire resistant, and found use in paints [1]. Poly(dihalophenylene oxide)s can be linear or branched, amorphous or crystalline and have high or low molecular weight depending on the type and the position of the substituent on the starting phenol. Ether linkage is either at ortho-or para-position or both. When it is at para-position the polymer is called poly(1,4-phenylene oxide) (Figure 1a) and when it is at ortho-position it is called (1,2-phenylene oxide) (Figure 1b), and 1,4 and 1,2 addition may take place at the same monomeric unit (Figure 1c), where(X : Halogen or Hydrocarbon): (a)1,4-addition (b) 1,2-addition and (c) both 1,4-and 1,2-addition.

The original synthesis was achieved by Hunter and his colleagues through the ethyl iodide –induced decomposition of anhydrous silver 2,4,6 trihalophenoxide [2]. This synthesis was achieved from the decomposition of the silver salts of trihalogenated phenols in the presence of ethyl iodide in hot benzene. The products were silver halide and a white amorphous polymerized residue with softening point above 300°C. The proposed reaction was: n $C_6H_2X_3OAg \rightarrow$ n AgX + $(C_6H_2X_2O)_n$

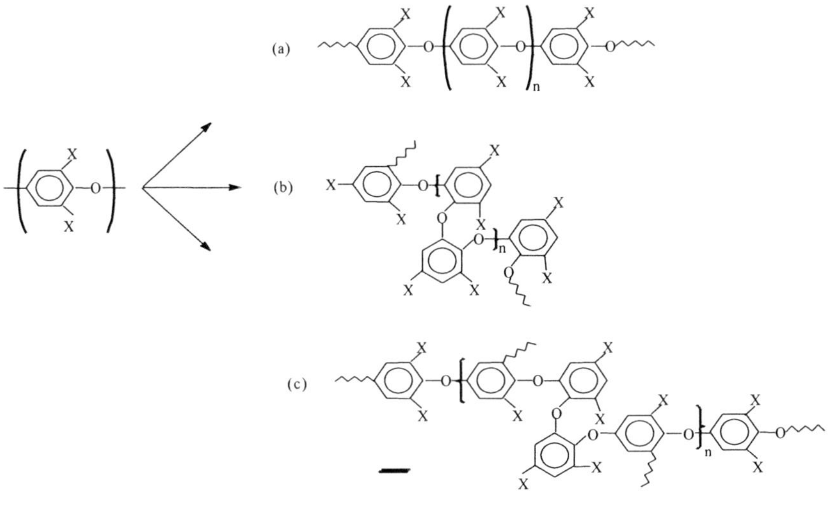

Figure 1. Possible structures of poly(halogenated phenylene oxide)s, (a)1,4-addition (b) 1,2-addition and (c) both 1,4-and 1,2-addition

Further work had been devoted to investigate the effect of the position of the halogen during polymerization [3]. Initially for this purpose silver salts of symmetrically substituted phenols was used. Results implied that the para position reacted more readily than ortho. Consecutive studies revealed halogen selectivity of the displaced phenols in the order of I> Br>Cl and only the halogen in the hydroxyl-bearing nucleus is removed during the decomposition [4,5].

In 1962 Blanchard et.al. [6] showed that 4-halo-2,6-dimethylphenols can be polymerized to high molecular weight poly(2,6-dimethylphenylene ether)s by the action of cupric ions, providing pyridine was also present under a wide variety of reaction conditions. They observed that this reaction requires a stoichiometric amount of cupric ions and proceeds through the pyridine complexes of cupric phenolates. In case of synthesis of poly(dihalophenylene oxide)s, very low intrinsic viscosities led the authors to conclude that the polymers were all significantly branched. Later Harrod et. al.[34-36,27,46] and Kısakürek et. al. [8,27,35,39,40] revealed much progress in understanding the mechanism of the reaction. Recently polymerization of halogenated phenols has been studied by the decomposition of transition and inner transition metal complexes with various amine ligands through various decomposition techniques which are electroinitiation in solution [7-21] thermal decomposition in solid state [22-29] and in solution [30-33]. It was found that besides decomposition method, type of amine ligand and the type of halogen substituted on the phenol ring were found to be effective on the structure, molecular weight and the conversion of the polymer.

In 1973, Carr et. al. proposed the mechanism of the decomposition of phenoxocopper(II) complexes [34]. The proposed mechanism was involved the several key steps. Firstly there is a fast induced decomposition of bis(2,4,6-trichlorophenoxo)bis(pyridine)copper(II) via a radical to give Cu(I) complex and radical substituted 2,4,6-trichlorocyclohexadienone. Secondly 2,4,6-trichlorocyclohexadienone slowly dissociate into initiator and 2,4,6-trichlorophenoxy radical forms, which could to be the reason of chain branching. Thirdly abstraction of a chlorine atom from 2,4,6-trichlorocyclohexadienone occurs where Cu(I) to give Cu(II) and a substituted phenoxy radical.

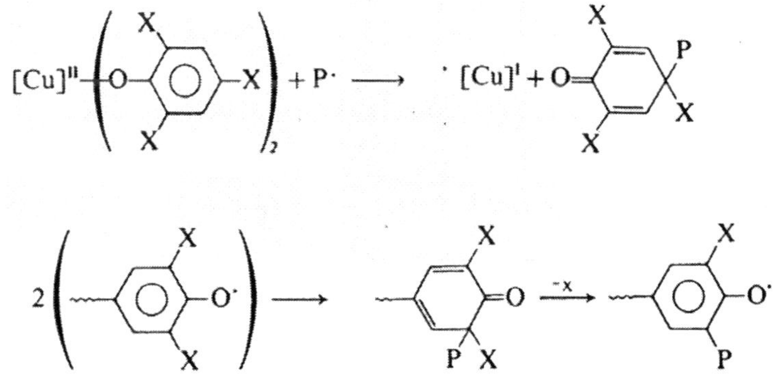

Figure 2. Proposed mechanism of poly(dihalophenylene oxide)s formation by Carr et.al [34]

In 1981 Harrod et. al. [35] modified his previously proposed mechanism for the polymerization reaction by performing further studies with ^{13}C NMR. ^{13}C NMR revealed that there was a high density of monomeric branches in the structure poly(phenylene oxide)s , where this observation could not be explained on the basis of a mechanism. According to the results they proposed ortho substitution occurring by an intramolecular process that involves two adjacent phenoxy ligands at a single copper center.

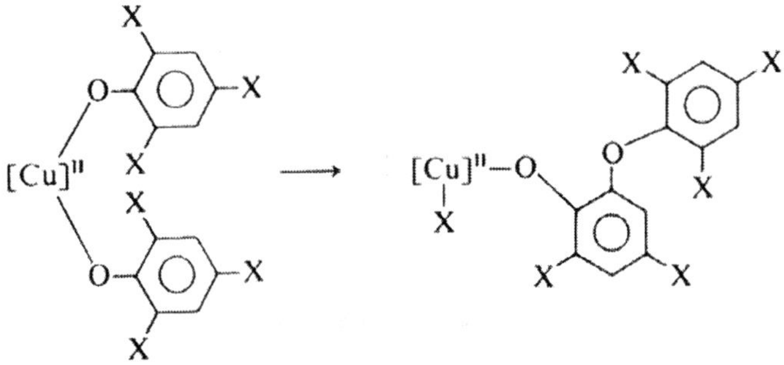

Figure 3. Proposed mechanism of poly(dihalophenylene oxide)s formation by Harrod et.al [35]

To test this hypothesis a sample of poly(dibromophenylene oxide) was prepared by free radical induced decomposition of alkaline tribromophenol. Spectrum of this polymer was compared with that of a polymer derived from the decomposition of copper tribromophenoxide. It was quite remarkable that these two spectrums were very much similar [36].

Results of the intensive studies of Kısakürek et. al. revealed that polymers with lowest molecular weight via electroinitiated polymerization and the highest via thermal decomposition in solution, whereas the solid state polymerization was close to solution polymerization with lower values. Through out theses studies, effect of the type of the amine ligand has also been studied. Thus it was observed that the amine ligand increase the oxidizing power of metal and decreased thermal stability of the complexes. The phenoxo complexes with chelating ligand were exceptionally resistant to thermal decomposition compared to non-chelating ligand increasing the induction periods of polymer formation [21, 24, 25, 39]. The structure of the polymer obtained also affected by the nature of the phenoxo ligand. Bromine atom in the para-position of 2,4,6-trihalophenoxides results generally in polymers with higher linearity [11,15,19] independent of the method of initiation and the type of the ligand substituted [11,15,19,24]. In the case of trichlorophenoxide, 1,2- and 1,4-addition is taking place at equal rates with non-chelating ligand pyridine in thermal polymerization, either in solution or in solid state or in electroinitiation with ligands pyridine and ethylenediamine [10,17,25,28,40,41]. However, 1,2- addition was favored over 1,4-addition in the case of chelating ligands, TMED and ethylenediamine, in thermal polymerization in solution and in electroinitiation.

The effect of the type of the transition and inner transition metal was also examined. Copper, cobalt, nickel, zinc, manganese and cerium were studied as metals. It was found that in solid state Cu [28,31,35] complexes decompose much more easier than Co [29,32,33], Ni [31,37,38], Mn [41-43], Zn [44] and Ce [46,47] complexes. Hence, the type of metal was found to be effective on the structure of the polymer and on the percent conversion.

As summarized above, a detailed study on the field of synthesis of poly(dihalophenylene) oxides through decomposition of transition and inner transition metal complexes revealed that; the decomposition method, type of the amine ligand, the type of the halogen substituted on the phenol ring and type of the metal used effects the structure, molecular weight and the conversion of the polymeric product. Although there has been several attempts to synthesize of poly(chlorophenylene oxide)s utilizing several different amine ligands, success was achieved only via solid state decomposition of $Cu(DCP)_2(NMIZ)_2$ [32]. Thus authors considered the use of N-methylimidazole ligand as a powerful approach to archive better percent conversions.

As a part of a broader study on the field of polymerisation of trihalogenated phenols, the effect of NMIZ and type of the transition metal on structure, molecular weight and percent conversion and glass transition temperatures (T_g) of the poly(dibromophenylene oxide)s synthesised through the thermal decomposition of Bis(2,4,6-tribromophenolato) di(N-methyl imidazole)Cu(II) (C1), Bis(2,4,6-tribromophenolato)di(N-methylimidazole)Co(II) (C2) and Bis(2,4,6-tribromophenolato) di(N-methyl imidazole) Ni(II) (C3) complexes were investigated.

EXPERIMENTAL

Materials

Analytical grade 2.4.6-tribromophenol (Aldrich Chemical Co. Ltd.) was used without further purification. The reagent grade N-Methyl imidazole (Merck) was the ligand in the preparation of complexes. Commercially available reagent grade copper sulphate pentahydrate; ($CuSO_4.5H_2O$), cobalt sulphate heptahydrate; ($CoSO_4.7H_2O$) and nickel sulphate hexahydrate; ($NiSO_4.6H_2O$) (Merck) were used metals for preparation of the complexes.

Synthesis of Complexes

Stoichiometric amount of tribromophenol (TBP) was dissolved in minimum amount of distilled water with sodium hydroxide (solution 1). $CuSO_4.5H_2O$, $Co(SO_4).7H_2O$ and $Ni(SO_4).6H_2O$ were dissolved respectively in minimum amount of distilled water containing the stoichiometric amount of N-methylimidazole (NMIZ) was added to the solution (solution 2). Solution 1 was added to solution 2 dropwise with vigorous stirring at room temperature. A slurry solution formed was stirred for two hours. Then precipitated complex was filtered, washed with distilled water and dried until constant weight under vacuum.

Complex Characterization

The complexes were characterized by elemental analyzer using Leco 932 CHNSO elemental analyzer. The elemental analysis of the complexes are (experimental errors ±0.3): $Cu(TBP)_2(NMIZ)_2$ (C1): Calc. C 27.04, H 1.81,N 6.31; Exp.C 26.87,H 1.65, N 6.10; $Co(TBP)_2(NMIZ)_2$ (C2): Calc. C 27.18, H 1.87,N 6.41; Exp. C 27.01, H 1.67, N 6.21; $Ni(TBP)_2(NMIZ)_2$(C3): Calc. C 27.19, H 1.81,N 6.34; Exp. C 26.95, H 1.79, N 6.26.Oxidation state of the metal in complex was readily assigned by Sherwood Scientific

Magnetic Susceptibility Balance. Results were (in Bohr Magneton, BM): $Cu(TBP)_2(NMIZ)_2$-(C1) 1.87, $Co(TBP)_2(NMIZ)_2$ -(C2) 4.66 and $Ni(TBP)_2(NMIZ)_2$ -(C3) 3.89. C1 and C2 were subjected to Mass Spectroscopy (MS) (LC/MC Platform II Micromass UK (Electron Impact) mass spectrometer) to characterize the weight loss during solid state polymerization of the complexes (Table 1) Thermal behaviour of the complexes were investigated by differential scanning calorimeter with a scanning rate of 10 °C/min as shown in Figure 4 a,b and c.

Table 1. MS data for $Ni(TBP)_2(NMIZ)_2$ and $Cu(TBP)_2(NMIZ)_2$ (The acceptable error limit in data evaluation is ± 2 units)

$Ni(TBP)_2(NMIZ)_2$			$Cu(TBP)_2(NMIZ)_2$		
Relative intensity %	M/Z	Possible origin of the peak	Relative intensity %	M/Z	Possible origin of the peak
100	90	TBP	100	91	TBP
23.37	81	NMIZ	22.4	81	NMIZ
18.49	327	TBP	16.5	327	TBP
12.62	141	TBP	10.8	41	NMIZ
11.83	41	NMIZ	10.7	42	NMIZ
10.80	42	NMIZ	8.1	249	TBP
10.50	40	NMIZ	3.6	220	TBP
2.51	249	TBP	5.5	116	TBP
2.30	220	TBP	13.9	141	TBP

Polymer Synthesis

For the synthesis of polymers about 4g of the complex was placed in to a glass sample holder (3.5 cm in height and 3 cm in diameter) and decomposed under pre-determined temperature and time condition. Decomposed product was allowed to cool to room temperature in a desicator and then weighed to determine the weight loss during the thermal decomposition. The decomposed complex was dissolved in toluene and after removing undisolved by-product, $(NMIZ)_2MBr_2$, poured into large amount of ethyl alcohol containing few drops of concentrated HCI. Then the precipitated polymer was filtered, washed with ethyl alcohol and dried till constant weight under vacuum.

Polymer Characterization

Automatic Viscometer: Viscosity of the polymers measured in toluene at 30°C by using Schott Gerate AVS 400 model automatic viscometer equipped with Schott Gerate CT 1150 model thermostat.

Differential Scanning Calorimeter (DSC): Thermal behaviour of the complexes and glass transition temperatures of the polymers were obtained by Dupont thermal analyst 2000 DSC 910S model differential scanning calorimeter with 10 °C/min for 7 mg sample. Treatment of the polymer and complex samples were performed under nitrogen and air atmosphere respectively.

Fourier-Transform Infrared Spectrometer (FTIR): Infrared spectra of the complexes and the polymers were recorded on a Mattson 1000 model FTIR spectrometer by dispersing the sample in KBr pellets.

Nuclear Magnetic Resonance Spectrometer (NMR): ^{1}H and decoupled ^{13}C-NMR spectra of samples were recorded by using BRUCKER Instruments, Avance Series (DPX 400), 400 MHz High Performance Digital FT-NMR Spectrometer. Deuterated chloroform was used as solvent and TMS as an internal reference.

Scanning Electron Microscopy (SEM): Surface morphologies of the polymer was investigated by JEOL JSM-6400 scanning electron microscope.

RESULTS AND DISCUSSION

Synthesized amorphous complexes were subjected to elemental analysis to characterize the complex. Elemental analysis results of all complexes correlate well with the given stoichiometries. Thus, in all complexes, two moles of amine ligand and two moles of phenolate coordinate with one mole of metal ion, revealing the chemical structure in Scheme 1. There were no sign of neither crystalline water nor trapped phenol in the structure of complexes. This observation was also supported by DSC thermograms of the complexes (Figure 4 a,b and c).

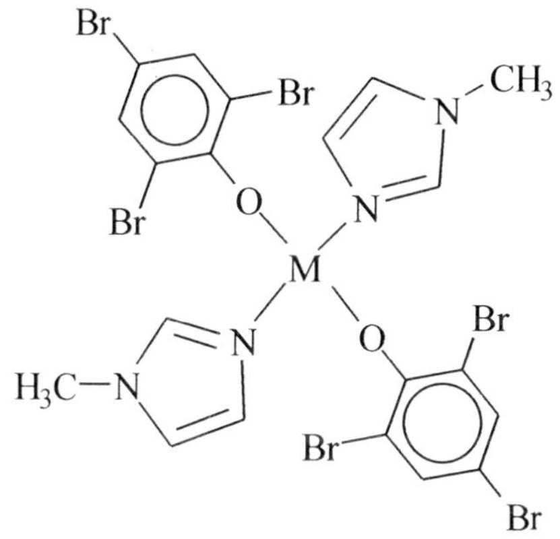

Scheme1. Chemical diagram of the synthesized complexes, where M stands for Cu, Co and Ni for C1,C2 and C3, respectively

FTIR spectra of the complexes 1,2 and 3 are similar to those of complexes given in literature [49]. Phenols are characterized by O-H vibrations around 3500 cm^{-1}, aromatic C-H stretching at 3070-3090 cm^{-1}, C=C ring stretching at 1460-1580 cm^{-1} and C-O stretching at 1165-1295 cm^{-1} (Figure 4-a). Amine ligand was characterized by aromatic C-H stretching at 3000-3100 $cm^{-1}$and C-N stretching at 1250-1350 cm^{-1}. Synthesized complexes revealed all the above mentioned bands except the band around 3500 cm^{-1} which is due to the O-H stretching of phenolic group. This band disappears as a result of complexes formation because H atom in the phenol is replaced by the transition metal during the reaction (Figure 4-a).

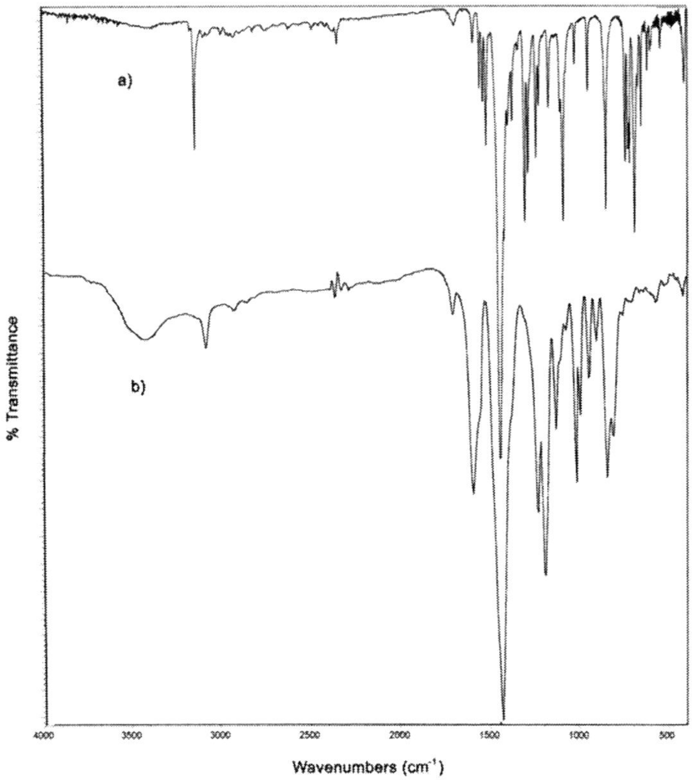

Figure 4. FTIR Spectrum of the a) $Cu(TBP)_2(NMIZ)_2$ complex and b) polymer obtained through thermal decomposition

The magnitude of interaction of a coordination compound with a magnetic field frequently allows the determination of the number of unpaired electrons in the complex. Thus, the oxidation state of a metal in a complex may be readily assigned on the basis of its number of unpaired electrons. For this purpose magnetic susceptibilities of the complexes were measured at room temperature and magnetic moments (μ_{eff}) in Bohr Magneton, BM. According to magnetic susceptibility measurements all complexes μ_{eff} values at room temperature were within the limit values expected for +2 oxidation state of transition metal complexes [49]. Measurements also reveal that there was no significant magnetic interaction between the metal ions which confirmed the monomeric nature of the complexes.

In order to investigate the real structure of the synthesized complexes authors tried to crystallize the resultant amorphous complexes. Although there had been several attempts with various procedures, only cobalt complex yielded in single crystals. Single crystal X-ray studies [50] revealed that the Co(II) ion had a distorted tetrahedron geometry involving two O atoms from phenol and two N atoms from the imidazole rings of the Bis(2,4,6-tribromophenolato)di(N-methyl imidazole)Co(II), (C2). In the coordination sphere, the Co-O and Co-N bond lengths were [1.920-1.922] Å and [2.014-2.078] Å, respectively. Results showed that there was a large deviation from ideal tetrahedral angle value of 109.5°, which could be due to steric effects. Incase of $Cu(TBP)_2(NMIZ)_2$ complex powder X-ray diffraction pattern was used for crystallographic identification and revealed the following results: a = 8.586 Å, b= 13.128 Å, c= 8.579 Å, α=107.027 °, β= 102.828 °, γ = 77.414 °, V= 889.34 $Å^3$.

Thermal behaviors of the complexes were investigated by DSC measurements under atmospheric condition for all complexes in order to achieve the normal polymerization conditions. A sharp exothermic peak at 116.88 °C and two endothermic peaks at 171.95 °C and 205.83 °C were observed in DSC thermogram of Cu(TBP)$_2$(NMIZ)$_2$ (Figure 5-a). Co(TBP)$_2$(NMIZ)$_2$ complexes (Figure 5-b) showed a sharp exothermic peak at 132.5 °C and an endothermic peak at 184.53 °C ,whereas Ni(TBP)$_2$(NMIZ)$_2$ (Figure 5-c) showed a single sharp exothermic peak at 143.93 °C. The exothermic peaks could be attributed to the new bond formation during polymerization. This proposed idea was also supported by DSC-IR study [32]. The complex was subjected to DSC by a heating rate of 10 °C/min and just after the exothermic peak, the sample was taken off from the instrument and KBr pellet of the resultant product was prepared. The pure complexes do not reflect any peaks (etheric bonds) in 900-1030 cm^{-1} region. However, the appearance of the peak at the resultant product in this region was considered as a reflection of the polymer formation. This fact was accompanied by the increase in the % conversions in all complexes during decomposition beyond the exothermic peak temperatures. In order to investigate the origin of the endothermic peaks authors investigated the DCS thermograms of the by-products formed during the decomposition (Figure 6). It was observed that the by-products achieved from decomposition of Cu(TBP)$_2$(NMIZ)$_2$ and Co(TBP)$_2$(NMIZ)$_2$ complexes revealed endotherms that were close to the values reflected in thermograms of the original complexes (Figure 5). Whereas, by-product of the Ni(TBP)$_2$(NMIZ)$_2$ complex did not display any endotherm in accordance with the DSC thermogram of the original complex. Thus the endothermic peaks were attributed to the phase transformation (melting) of the by-products. The absence of endothermic peak near 85°C and 100 °C indicates that there was neither trapped phenol nor crystalline water in the complexes.

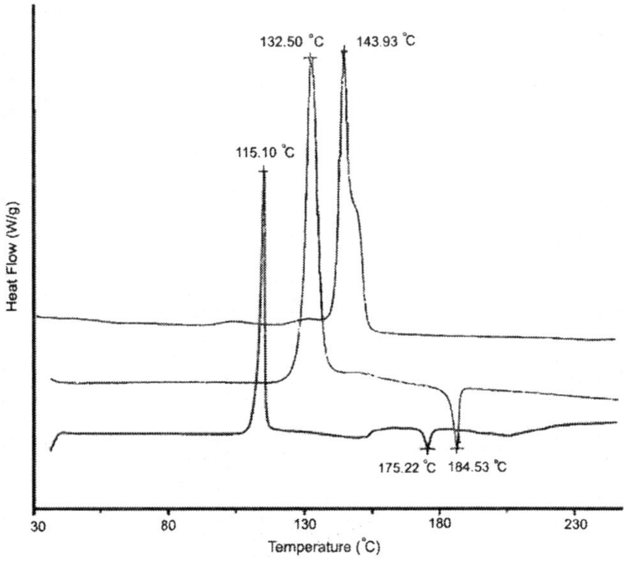

Figure 5. DSC Thermograms of a) Cu(TBP)$_2$(NMIz)$_2$ -(C1), b) Co(TBP)$_2$(NMIz)$_2$ -(C2) and c) Ni(TBP)$_2$(NMIz)$_2$ -(C3) by heating rate of 10 ^{0}C/ min

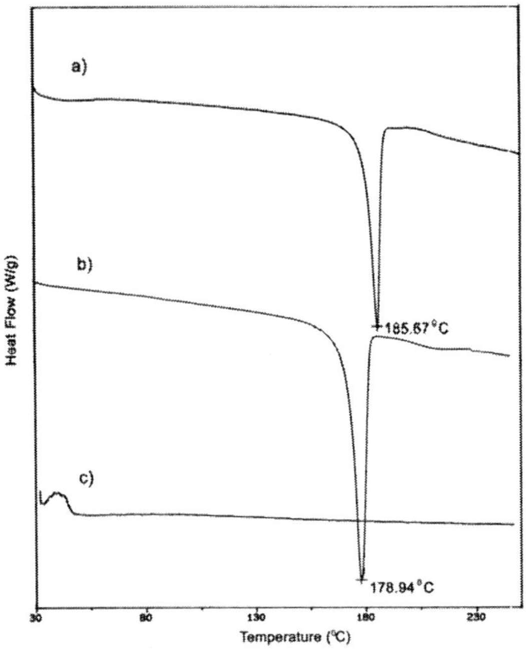

Figure 6. DCS thermograms of the by-product produced from the decomposition of a) $Co(TBP)_2(NMIZ)_2$,b) $Cu(TBP)_2(NMIZ)_2$ and c) $Ni(TBP)_2(NMIZ)_2$ complexes

Synthesis of the polymers through the thermal decomposition of the complexes results in a weight loss of the initial complex due to the formation of gaseous products during the polymerization. Hence, complexes were subjected to mass spectroscopy to investigate the origin of the gaseous products. Within the light of the previous study [51], results of the mass spectroscopy (Table 1) implied that the gaseous products originated from the loss of mainly halogenated phenol and to some extend from the neutral amine ligand. These results would be useful in understanding the relation between the weight loss and % conversion.

Solid state thermal polymerization of $Cu(TBP)_2(NMIZ)_2$, $Co(TBP)_2(NMIZ)_2$ and $Ni(TBP)_2(NMIZ)_2$ complexes were performed in two different conditions. In the first condition, time kept constant at 3 hours and temperature varied between lowest decomposition temperature of the complex to 250 ^{0}C. In the second condition, temperature kept constant at maximum conversion temperature and time varied from 3 to 48 hours. The calculated yields for the polymerizations were based on the initial weight of the complexes; since the ligand and copper did not incorporate in to the polymer, the results may appear to be low in value. For all syntheses, the percent conversion and the weight loss were calculated as follows;

Weight of the polymeric product
%Conversion=x100
Initial weight of the complex

Weight loss = Initial wt of complex –wt of the products after decomposition

Weight loss
% Weight loss= x100
Initial weight of the complex

In this study FTIR used as an effective approach to reveal the formation of poly(dibromophenylene oxide)s through thermal decomposition of the complexes. Polymer formation was followed by the appearance of the peaks around 950- 1030 cm^{-1} because of the etheric bond formed during the polymerization (Figure 4-b). The polymers were characterized by the C=C stretching at 1380, 1440, 1560 and 1610 cm^{-1} ; C-O-C stretching at 950, 1006 and 1033 cm^{-1}; C-O stretching at 1030-1245 cm^{-1} ; out of plane C-H bending at 820 and 850 cm^{-1} and C-H stretching of benzene ring at 3079 cm^{-1}. The peak around 3500 cm^{-1} showed phenolic end group.

For $Cu(TBP)_2(NMIZ)_2$ it was observed that, as the temperature increased, percent conversion increased and reached a maximum value of 30.6 % at 200 °C, followed by a slight decrease (Figure 7). This result could be related with the increase in the weight losses. As previously discussed, MS results revealed the loss of halogenated phenol during thermal decomposition. Hence, MS results could be considered as the underlying reason of the decline in the % yield during polymerizations at high temperatures. When the second condition was applied, temperature was kept constant at maximum conversion temperature of 200°C and time varied from 3 to 48 hours; it was observed that, both percent conversion and weight loss increased slightly (Figure 8).

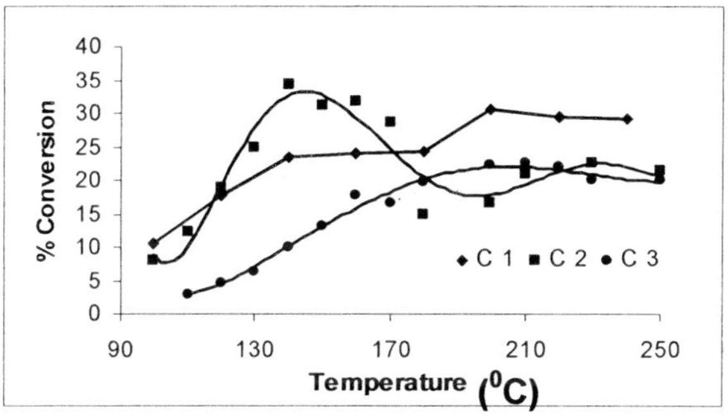

Figure 7. %Conversion vs temperature graph of $Cu(TBP)_2(NMIZ)_2$–(C1), $Co(TBP)_2(NMIZ)_2$–(C2) and $Ni(TBP)_2(NMIZ)_2$–(C3) complex at constant decomposition time (3hour)

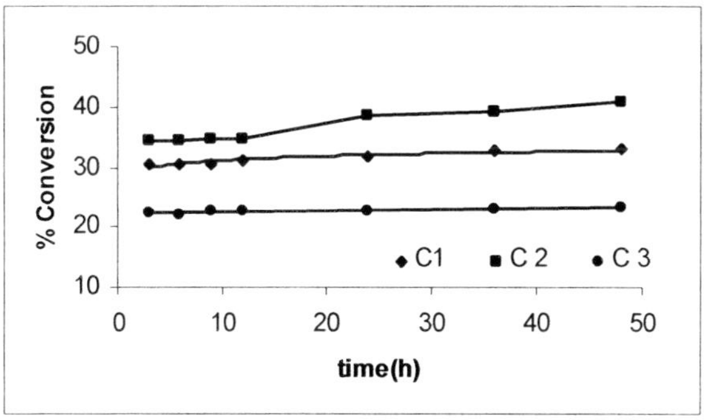

Figure 8. %Conversion vs time graph of $Cu(TBP)_2(NMIZ)_2$, $Co(TBP)_2(NMIZ)_2$ and $Ni(TBP)_2(NMIZ)_2$ complex at constant temperature (200°C, 140°C and 210°C, respectively)

Maximum conversion value of 34.4 % was achieved at 140 °C for $Co(TBP)_2(NMIZ)_2$ complex. As the temperature increased, percent conversion increased up to 140 °C and than decreased to % 21.3, till to the melting point (184.53 °C, corresponding the melting of the by-products, confirmed by DSC). Increase in temperature beyond the maximum percent conversion temperature (140 °C) decreased the percent conversion in C2, which could be attributed to the hindrance of the diffusion of the monomers due to both the formation of a large amount of by-product, $(NMIZ)_2MBr_2$ and the increase in percent weight loss. Beyond the melting point of the by-product percent conversion gradually increased slightly. This inspection could be related to the ease of diffusion in the molten state. At 140°C, different decomposition times between 3 and 48 h was performed and it was observed that longer than 12 h decomposition time, resulted in an increase in percent conversion (Figure 8).

In case of $Ni(TBP)_2(NMIZ)_2$ there was a sharp increase in percent conversion up to 210°C followed by a slight decrease to %19.9 while approaching to 250°C. This observation could again attribute to the weight losses, which mainly originated from halogenated phenol itself. The maximum conversion value of 22.4% was achieved at 210°C (Figure 7). Percent weight loss increased gradually as the decomposition temperature approached to 250°C. At 210°C, increase in decomposition time had no effect on percent conversion of C3 (Figure 8).

Generally all the synthesized polymers revealed low intrinsic viscosity, indicating the high degree of branching. This fact was also supported by the ^{1}H-NMR and ^{13}C-NMR studies. The polymers synthesized from the decomposition of C1 had the highest whereas C3 had the lowest value of intrinsic viscosity (Figure 9). This observation could be related to the fact that the complex with higher thermal stability resulted in lower intrinsic viscosity. The increase in intrinsic viscosities of the polymers synthesized by the decomposition of C1 and C2 above their melting point of their by-products, could be due to the ease of diffusion of the complex in molten state. As confirmed by DSC, this effect did not appear in the C3 complex. Hence lower % conversion and lower molecular weight of nickel complex relative to copper and cobalt complexes showed a significant difference among them, due to difference in the structure and the oxidizing powers of these two metals.C1 and C3 seem to be less affected than C2, with the variation of decomposition time (Figure 10).

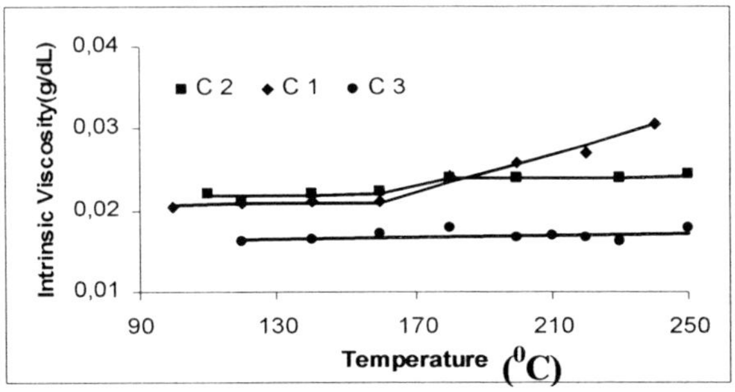

Figure 9. [η] vs Temperature graph of polymer obtained from $Cu(TBP)_2(NMIZ)_2$–(C1), $Co(TBP)_2(NMIZ)_2$–(C2) and $Ni(TBP)_2(NMIZ)_2$–(C3) complex at constant decomposition time (3hour)

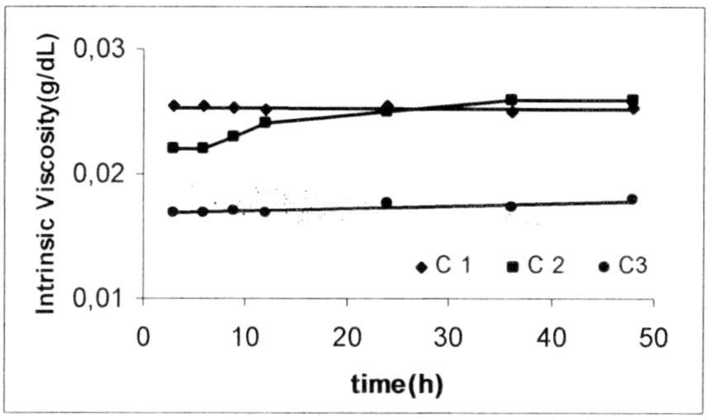

Figure 10. [η] vs time graph of polymer obtained from $Cu(TBP)_2(NMIZ)_2$–(C1), $Co(TBP)_2(NMIZ)_2$–(C2) and $Ni(TBP)_2(NMIZ)_2$–(C3) complex at constant temperature (200°C, 140°C and 210°C, respectively)

In the previous [32] study, the trichlorophenol counterparts of the C1 and C2 were synthesised .The comparison on the basis of their minimum decomposition temperature, it was observed that $Cu(TCP)_2(NMIZ)_2$ and $Co(TCP)_2(NMIZ)_2$ displayed higher minimum decomposition temperatures than C1 and C2, which could be related to the fact that the complexes with 2,4,6-trichlorophenol are more stable than 2,4,6-tribromophenol complexes (Table 2). Although the $Cu(TCP)_2(NMIZ)_2$ had higher maximum % percent conversion than C1, $Co(TCP)_2(NMIZ)_2$ had the inverse relation with C2 complexes, which could be due to the difference in the structures of the complexes.

Table 2. The structures, maximum polymerization yields and temperatures, T_g (°C) values, and [η] (g/dL) of various poly(dihalophenylene oxide)s

Complex	Ref	Structure	Tg (°C)	VİSH [η] S g/dL)	Max. polymer conversion (%)	Max. polymer conversion T (°C)
$Cu(NMIZ)_2(TCP)_2$	32	1,2 add.	175	0.043	50.0	140
$Cu(NMIZ)_2(TBP)_2$		equal	210	0.026	30.5	200
$Co(py)_2(TCP)_2$	33	1,2 add.	189	0.036	21.2	250
$Co(py)_2(TBP)_2$	33	equal	211	0.019	42.7	105
$Co(NMIZ)_2(TBP)_2$	32	equal	197	0.022	34.4	140
$Co(NMIZ)_2(TCP)_2$		1,2 add.	--	--	1.1	170
$Ni(py)_2(TCP)_2$	31	1,2 add.	190	0.039	34.3	195
$Ni(py)_2(TBP)_2$	31	1,2 add	208	0.021	41.9	160
$Ni(NMIZ)_2(TBP)_2$		equal	190	0.017	22.4	210

* Present study

In the previous studies [33,31] of the thermal decomposition of Bis(2,4,6-tribromophenolato)di(pyridine)Co(II) and Bis(2,4,6-tribromophenolato)di(pyridine)Ni(II) complexes resulted in higher percent conversion than their NMIZ counterpart which were C2 and C3 respectively. This observation could be explained on the basis of MS results of the previous studies, where the gaseous product found to be mainly of ligand pyridine, not leading to a reduction in the phenol content. In the solid state, prepared in the presence of the neutral ligand pyridine, using the same tribromophenol and metals, there was always higher percent conversion values along the same time and temperature ranges in comparison to the synthesized complexes in the current study, indicating the effect of the structure of the complexes for ease of decomposition and the percent conversion of the polymers ^{13}C-NMR decoupled spectra of the polymers obtained from complexes were displayed in Figures 11 a, b and c. The theoretical ^{13}C-NMR chemical shift data for the main five possible products were calculated by using appropriate correlation tables [48] and binary relations between observed and calculated data for all polymers are listed in Figures 12, 13 and 14. Within the five possible structures of synthesized poly(dibromophenylene oxide)s from all complexes ^{13}C-NMR shift data showed that the structure c (1,2 and 1,4 addition) was the most predominant one with the highest correlation of the calculated and the observed data.

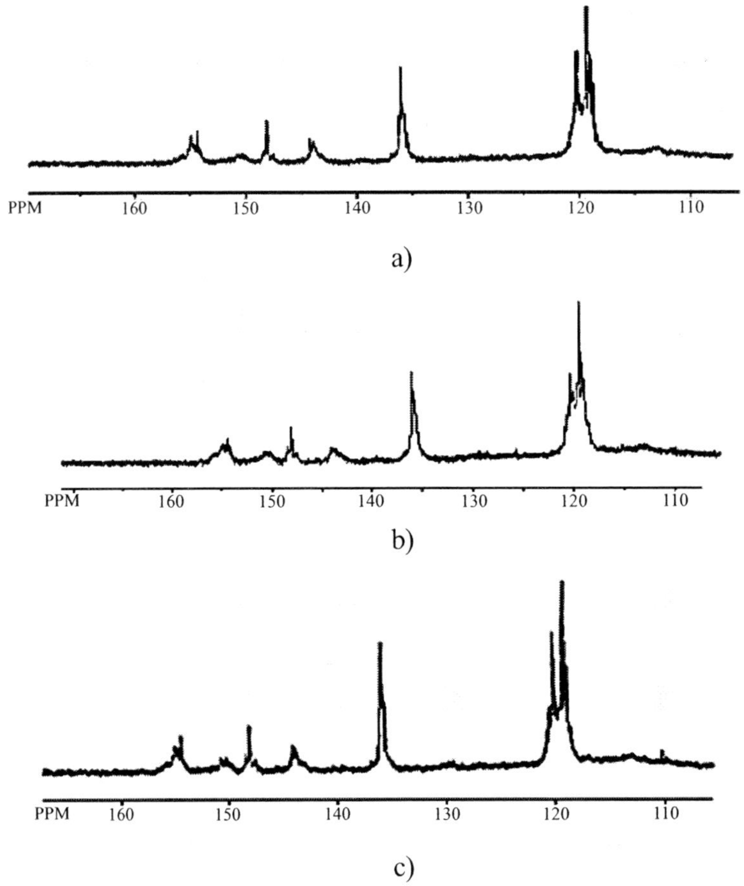

Figure 11. ^{13}C-NMR Spectrum of the polymer obtained from the decomposition of a) C1, b) C2 and c) C3

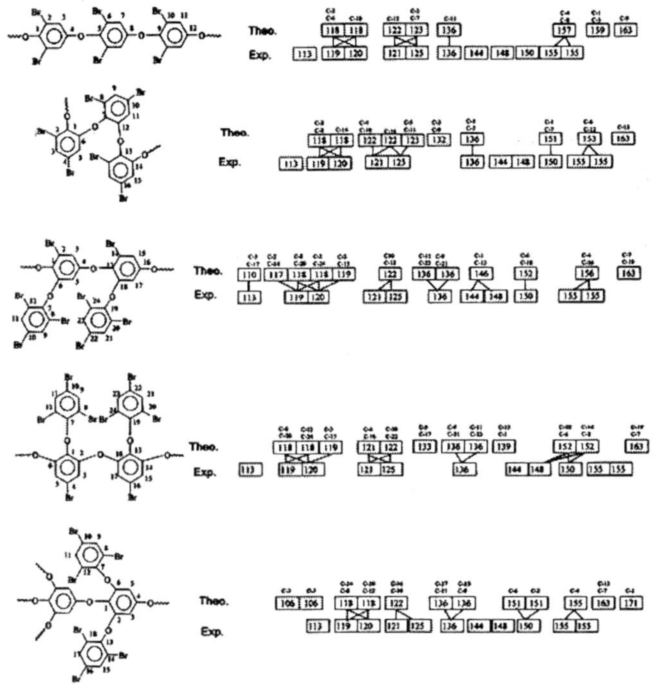

Figure12. ^{13}C-NMR shift data of the polymer obtained from $Cu(TBP)_2(NMIZ)_2$ complex

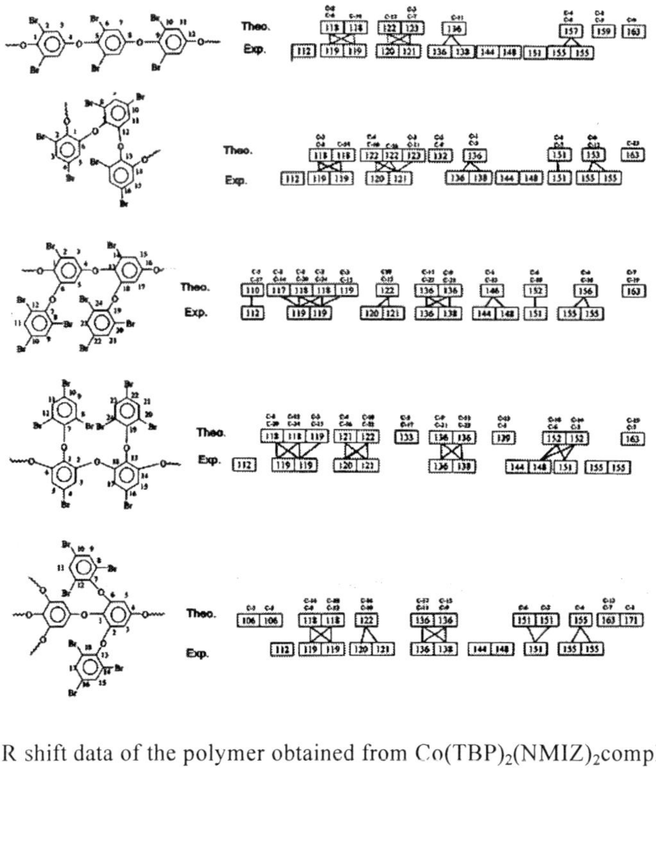

Figure13. ^{13}C-NMR shift data of the polymer obtained from $Co(TBP)_2(NMIZ)_2$complex

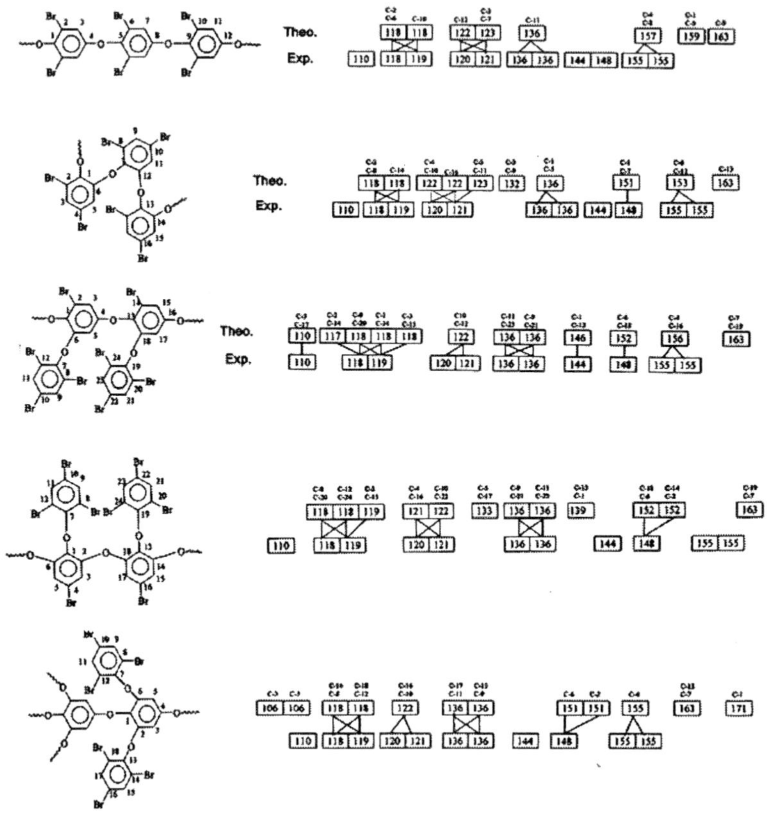

Figure 14. ^{13}C-NMR shift data of the polymer obtained from $Ni(TBP)_2(NMIZ)_2$complex

^{1}H-NMR spectra of polymers obtained from decomposition of the complexes were shown in Figures 15 a,b and c. ^{1}H-NMR spectrum of the polymers obtained from decomposition of complexes revealed that the peaks between δ = 7.85-7.65 and δ =7.19-7.26 ppm were due to the fifth and third protons of 2,4-dibromo-1,6-phenylene oxide units (1,2- addition) respectively; peaks between δ = 6.78-7.1 were due to the protons of 2,6-dibromo-1,4-phenylene oxide units (1,4-addition) and the broader peaks at higher field indicated the presence of both 1,2- and 1,4-addition at the same monomeric unit, which were well correlated with ^{13}C-NMR data. Thus, results implied that polymerization proceeds through both 1,2-addition and 1,4-addition at equal rates for all of the synthesized polymers without regarding the type of the transition metal in the complex, leading to branched structure.

DSC thermograms of the of the polymer synthesized through solid state thermal decomposition of the complexes $Cu(TBP)_2(NMIZ)_2$, $Co(TBP)_2(NMIZ)_2$ and $Ni(TBP)_2(NMIZ)_2$ complexes reveallled glass transition temperatures of 210.7, 197.7 and 189.9°C respectively. High T_g values indicated the rigidity of the synthesized polymers.

Scanning electron microscopy was employed to investigate the morphology of the synthesized polymer. Micrographs (Figure 16) implied granular structure of the polymer which was different than poly(dichlorophenylene oxide) [52].

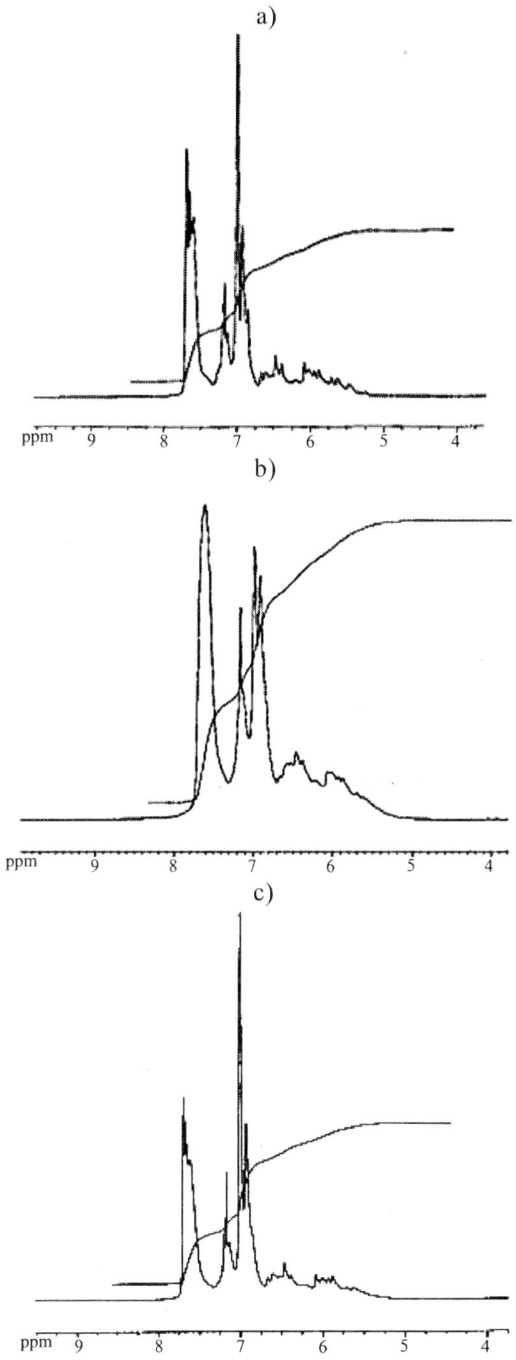

Figure 15. ^{1}H-NMR Spectrum of the polymer obtained from the decomposition of a) C1, b) C2 and c) C3

Figure 16. SEM micrograph of the poly(dibromophenylene oxide)

CONCLUSION

In this study Bis(2,4,6-tribromophenolato)di(N-methyl imidazole) Cu(II), Bis(2,4,6-tribromophenolato)di(N-methyl imidazole) Co(II) and Bis(2,4,6-tribromophenolato)di(N-methyl imidazole) Ni(II) complexes were synthesized and polymerized through thermal decomposition in solid state for the first time. The polymer syntheses were achieved either the temperature range between 90 °C to 250 °C at the end of 3 h to investigate the effect of temperature or at maximum conversion temperature at 200 °C, 140 °C and 210 °C for $Cu(TBP)_2(NMIZ)_2$, $Co(TBP)_2(NMIZ)_2$ and $Ni(TBP)_2(NMIZ)_2$ complexes respectively, within the time range between 3 to 48 h to investigate the time effect on decomposition.

The optimum conditions for the synthesis of poly(dibromophenylene oxide)s through thermal decomposition of $Cu(TBP)_2(NMIZ)_2$,$Co(TBP)_2(NMIZ)_2$ and $Ni(TBP)_2(NMIZ)_2$ complexes were 200 °C, 140 °C and 210 °C; revealing 30.6 , 34.4 and 22.4 percent conversions at decomposition time of 3 h, respectively. Among the synthesized polymers, polymers synthesized through the decomposition of $Ni(TBP)_2(NMIZ)_2$ complexes yielded lowest intrinsic viscosity and percent conversion. This fact could be related with the difference in the structure of the complexes and the oxidizing power of metal. Effect of time on both percent conversion and intrinsic viscosity was also investigated. The intrinsic viscosity of the polymer synthesized via decomposition of $Cu(TBP)_2(NMIZ)_2$ and $Co(TBP)_2(NMIZ)_2$ increased slightly as the time elongated , whereas $Ni(TBP)_2(NMIZ)_2$ was not affected.

In this study mass spectroscopy used as an effective tool to understand the origin of the gaseous products evolved during decomposition of the complexes and its effect on the percent conversion during polymer synthesis. Results revealed that, gaseous products originated from the loss of mainly halogenated phenol and to some extend from the neutral amine ligand. Thus, decline in the percent conversions of polymeric product synthesized via decomposition of complex C1 and C2, at elevated temperatures, were considered to be due to loss of phenol during polymerization.

An extensive work has been devoted to understand the effect of by-product formation on percent conversion and intrinsic viscosities for the first time. For this purpose DSC thermograms of the by-product, $(NMIZ)_2MBr_2$, was investigated and results revealed that the melting points of the by-products in accordance with the endotherms observed in DSC thermograms of the original complexes. An increase in both percent conversion and intrinsic viscosity of the polymeric products was observed beyond the melting point of the by-products for both complexes C1 and C2. This observation was concluded to be due to ease of diffusion in solid state which was an important factor in achieving high percent conversions. Hence, authors concluded that thermal behavior of the by-product and the MS results can enlighten the underlying reason of the variations at percent conversion and intrinsic viscosity values of the resultant polymeric products.

In comparison to previous studies, it was observed that pyridine, as a neutral amine ligand, was more powerful than NMIZ ligand, irrespective of the type of the metal used in terms of both percent conversion and intrinsic viscosities of the resultant polymer. In another previous study although same amine ligand and metals were used, $Cu(TCP)_2(NMIZ)_2$ had higher percent conversions than $Cu(TBP)_2(NMIZ)_2$, but inverse relation between $Co(TCP)_2(NMIZ)_2$ and $Co(TBP)_2(NMIZ)_2$ was observed. Thus it was concluded that, the adverse relation between the counter part complexes could be due to the difference in the structure of the complexes.

In this study, ^{1}H-NMR and ^{13}C-NMR results suggest that polymerization proceeds though both 1,2- and 1,4- addition at equal rates, without respecting the type of the metal used, whereas in previous $Cu(TCP)_2(NMIZ)_2$ and $Co(TCP)_2(NMIZ)_2$ revealed 1,2 addition products, which could be considered as the effect of the type of the halogenated phenol. All the polymeric products have high T_g values indicating that they are rigid polymers.

ACKNOWLEDGEMENTS

Partial financial support by METU research fund is gratefully acknowledged.

REFERENCES

[1] Gündüz, G; Kısakürek, D; Kayadan, S. Flame Retardant Alkyd Paint. *Polym Degrad Stabil,* 1999, 64, 501-504.

[2] Hunter, WH; Olson, AO; Daniels, EA, Decomposition of Certain Phenol Silver Salts. *J. Am. Chem. Soc,* 1916, 38, 1761-1771.

[3] Hunter, WH; Joyce, FE. A Catalitic Decomposition of Certain Phenol Silver Salts II. *J. Am. Chem. Soc.,* 1917, 39, 2640-2646.

[4] Kısakürek, D; Poly(dichlorophenylene oxides)s by Pyridine/Copper Complex, Polymeric Materials Encyclopedia. Salamone, *J C. CRC Press*. Vol. 8, 5731, 1996.

[5] Hunter, W; Dahlen, MA. A Catalytic Decomposition of Certain Phenol Silver Salts. VI. The Synthesis of 4-Hydroxy-3,5-Dibromo-2',6'-Diiodo-4'-Chlorodiphenyl Ether and the Decomposition of Its Metallic Salts. *J. Am. Chem. Soc., 1*932, 54, 2456-2463.

[6] Blanchard, HS; Finkbeiner, HL; Russell, GA. Polymerization by Oxidative Coupling. IV. Polymerization of 4-Bromo- and 4-Chloro-2,6dimethylphenol and Preperation and Decomposition of the Silver and Copper Salts of Certain Other Phenols. *Journal of Polym. Sci.*, 1962, 58, 469-490.

[7] Türker, L; Kısakürek, D; Şen, Ş; Toppare, L; Akbulut, U.Electroinitiated Polymerization of Halophenols - Electroinitiation Potential Lumo Relations. *J Polym Sci Pol Phys,* 1988, 26 , 2485-2498.

[8] Saçak, M; Akbulut, U; Kısakürek, D; Türker, L; Toppare, L. Electroinitiated Polymerization of Bis(Trichlorophenoxo)-Tetramethylethylene Diamine Copper(II). *J Polym Sci Pol Chem,* 1989 ,27,1599-1608.

[9] Şen, Ş; Kısakürek, D; Türker, L; Toppare, L; Akbulut, U. Electroinitiated polymerization of (4-chloro-2,6-dibromophenol). *New Polymeric Materials,* 1989, 1, 177-88.

[10] Yiğit, S; Kısakürek, D; Türker, L; Toppare, L; Akbulut, U. Polymerization of Bis(Trichlorophenoxo) Ethylenediamine Copper (II) Complex by Electro-Oxidation. *Polymer,* 1989 ,30 , 348-352 1989.

[11] Saçak, M; Akbulut, U; Kısakürek, D; Toppare, L. Electroinitiated Polymerization of Bis(4-Bromo-2,6-Dichlorophenoxo)N,N,N',N'-Tetramethylethylenediamine Copper(II) Complex. *Polymer,* 1989 ,30, 928-932.

[12] Akbulut, U; Saçak, M; Kısakürek, D; Toppare, L. Electroinitiated Polymerization of The Bis(2,6-Dibromo-4-Chlorophenoxo)-N,N,N',N'-Tetramethylethylenediamine Copper (II) Complex. *J Macromol Sci Chem,* 1989, A26, 1623-1635.

[13] Akbulut, U; Saçak, M; Kısakürek, D; Toppare, L. Electroinitiated Polymerization of Bis(Tribromophenoxo)-N,N,N',N'-Tetramethylethylenediamine Copper(II) Complex. *Brit Polym J* ,1990, 22 ,65-71.

[14] Toppare, L; Türker, L, Yiğit, S; Kısakürek, D; Akbulut, U. Electro-Initiated Polymerization of Bis-(2,4,6-Tribromophenoxo) Ethylenediamine Copper(II) Complex in DMF and Its Kinetic-Study by Cyclic Voltammetry.*Eur Polym J,* 1990, 26 ,255-259.

[15] Kısakürek, D; Yiğit, S. Polymerization of Bis(4-Bromo-2,6-Dichlorophenoxo) Ethylenediamine Copper(II) Complex By Electro-Initiation.*Eur Polym J,*1991 , 27 , 955-958.

[16] Şen, Ş; Kısakürek, D; Toppare, L. Polymerization of Bis(Pyridine)Bis-(2,4,6-Tribromophenoxo)Copper(II) Complex ($Py_2Cu(TBP)_2$) by Electrooxidation and Structural Characterization By H^{-1}-Nmr, C^{-13}-Nmr, and FTIR Spectroscopies.*J Macromol Sci Pure*, 1993 , A30 , 481-492.

[17] Şen, Ş; Kısakürek, D. Synthesis and Characterization of Poly(Dichlorophenylene Oxide)s Based On The Electrooxidation of Bis(2,4,6-Trichlorophenoxo)Bis(Pyridine)Copper(II) Complex. *Polymer*, 1993 , 34 4146-4149.

[18] Pulat, M; Önal, A M; Kısakürek, D. Electro-initiated polymerization of bis(4-chloro-2,6-dibromophenoxo)ethylenediamine copper(II) complex. Kinetic study by cyclic voltammetry and UV spectroscopy. *New Polymeric Materials* ,1994, 4, 111-18.

[19] Aras, L; Şen, Ş; Kısakürek, D; Synthesis of Mainly Linear Poly(Dichlorophenylene Oxide) Based On The Electrooxidation of Bis(Pyridine)Bis(4-Bromo-2,6-Dichlorophenoxo)Copper(II) Complex in Dimethylformamide. *Polymer,* 1995 ,36 , 3013-3018.

[20] Kısakürek, D; Şen, Ş; Aras, L; Türker, L; Toppare, L. Polymerization of Bis(2-Bromo-4,6-Dichlorophenoxo)Bis(Pyridine) Copper (II) Complex in DMF By Electroinitiation. *Polymer,* 1991 ,32, 1323-1328.

[21] Kısakürek, D; Aras, L; Şanlı, O. Synthesis and Characterization of Different Poly(dihalophenylene oxides). *METU J. Pure and Appl. Sci.,* 1990, 23, 49-56.

[22] Pulat, M; Şanlı, O; Kısakürek, D. Synthesis and Structural Characterization of Poly(Dihalophenylene-Oxide)s Based on The Thermal-Decomposition of Bis(4-Chloro-2-6-Dibromophenoxo) Ethylenediamine Copper(II) Complex in Toluene.*Pol J Chem,* 1994, 68 , 453-458.

[23] Kısakürek, D; Akbaş, M; Bilir, N. Thermal Polymerization of Bis(4-Fluoro-2,6-Dibromophenolato)Bis(Pyridine)Copper(II) Complex in Solution and in Solid State. *Macromol Chem Physic,* 1998 , 199 , 169-173.

[24] Şanlı, O; Kısakürek, D. Synthesis of Mainly Linear Poly(Dichlorophenylene Oxide)s By Thermal Polymerization in Solution. *Makromol Chem,* 1992 ,193, 619-625.

[25] Kısakürek, D; Şanlı, O. Thermal Polymerization of Bis(Trihalophenoxo)Copper(II) Complexes in Excess Dimethylformamide and Dimethyl-Sulfoxide as Ligands. *Makromol Chem,* 1989 ,190 , 1843-1853.

[26] Kısakürek, D; Türker, L. Further Experimental Considerations On Poly(Dihalophenylene Oxide). *Brit Polym J* ,1989 ,21 , 367-368.

[27] Kısakürek, D; Binboğa, N; Harrod, J F Characterization Of Poly(Dihalophenylene Oxides) In Solution. *Polymer,* 1987, 28 ,1767-1771.

[28] Baştürkmen, M; Kısakürek, D. Synthesis And Characterization Of Poly(Dichlorophenylene Oxide)s Through Solid-State Thermal-Decomposition of Bis(Pyridine)Bis(Trichlorophenoxo)Copper(II) Complexes.*Polymer,* 1993, 34, 625-629.

[29] Baştürkmen, M; İşçi, H; Kısakürek, D. Solid-State Polymerization of Bis(Trichlorophenoxo) Cobalt(II) Complex to Give Poly(Dichlorophenylene Oxide). *Polym Int,* 1993 ,30, 387-392.

[30] Şanlı, O; Pulat, M; Kısakürek, D. Synthesis of Poly(Dihalophenylene Oxide)s By Solid-State Thermal-Decomposition of Bis(4-Chloro-2,6-Dibromo Phenoxo) Ethylenediamine Copper(II) Complex. *Eur Polym J,* 1995 ,31, 1255-1259.

[31] Kısakürek, D; Aslan, A. Polymerisation of Bis(Trichlorophenolato)Tri(Pyridine)Nickel and Bis (Trihalophenolato)Di(Pyridine)Nickel (II) Complexes in Solid State. *Polymer,* 2000 41 ,5149-5155.

[32] Baştürkmen, M; Kısakürek, D. Effect of N-Methyl Imidazole and Dimethyl Pyrazole Ligands on Polymerization of di(2,4,6-Trichlorophenolato)Cu(II)/Co(II) Complexes in Solid State. *J. Macmol. Sci. Part A—Pure and Applied Chemistry,* 2004, 41, 1071–1086.

[33] Kesici, N; Kisakurek, D. Polymerization and Characterization of Various Di(Pyridine)Bis(Trihalophenolato)Cobalt(II) Complexes in Solid and Melt States.*Polym Int,* 2001, 50, 1061-1067.

[34] Carr, B; Harrod, J F. Chemistry of Phenoxo Complexes 1. Mechanism of Decomposition of Phenoxocopper(II) Complexes. *J Am Chem Soc,*1973,95 , 5707-5712.

[35] Harrod, J.F; Van Gheluwe, P; Kısakürek, D; Shaver, A., Further-Studies Of The Structure Of Poly(Dihalophenylene Oxides) Using Carbon Nuclear Magnetic-Resonance Spectroscopy. *Macromolecules,* 1981,14, 565-570.
[36] Harrod, J.F; Shaver, A. Comparison of The Structures of Poly(Dibromophenylene Oxides) Produced By Free-Radical Initiation and By Decomposition of Copper Tribromophenoxide. *Macromolecules* , 1982, 15 , 676-678.
[37] Karabey, L.M; Kisakurek, D. Synthesis and Characterization of Poly(Dihalophenylene Oxide)s Through Bis(Trihalophenolato) (Ethylene-Diamine)(Pyridine)Nickel(II) Complexes in Solution in The Presence of Iodine.*Macromol Chem Physic,* 2001, 202 , 2999-3007.
[38] Molu, LK; Kisakurek, D. Polymerization And Characterization Of Bis(Trichlorophenolato)Di(Pyridine) Nickel(II) And Bis(Tribromophenolato)Di(Pyridine) Nickel(II) Dihydrate Complexes in Solution in The Presence Of Iodine. *J Appl Polym Sci,* 2002, 86, 2232-2239.
[39] Ünal, Hİ; Sanli, O; Kisakure, D. Synthesis and Characterization of Poly(Dihalophenylene Oxides) by Thermal-Decomposition of Copper-Complexes with Tetramethylethylenediamine and Ethylenediamine Ligands.*Polymer,* 1989, 30, 344-347.
[40] Türker, L; Kısakürek, D. Theoretical and Experimental Considerations On Poly(Dihalophenylene Oxide)s. *Brit Polym J,* 1988, 20, 131-141.
[41] Pulat, M; Sanli, O. Synthesis of Poly(Dihalophenylene Oxide)s by Solid-State Thermal Decomposition of Manganese Phenolate Complexes With Ethylenediamine Ligand. *Macromol Chem Physic,* 1996, 197 ,811-819.
[42] Pulat , M; Aksu, B. Synthesis and Characterization of Poly(Dihalophenylene Oxide)s By Solid-State Thermal Decomposition of Manganese Phenolate Complexes With Pyridine. *React Funct Polym,* 1998,36 ,91-98.
[43] Pulat, M; Çalışkan, G. Synthesis and Characterization of Poly(Dihalophenylene Oxide)s by Solid-State Thermal-Decomposition of Manganese Phenolate Complexes with Tetramethylethylenediamine. *Polym Int,* 1999 ,48 ,228-234.
[44] Pulat, M; Çakır E. Synthesis and Characterization of Poly(Dihalophenylene Oxide)s by Solid-State Thermal Decomposition of Zinc Phenolate Complexes with Ethylenediamine and Pyridine. *React Funct Polym* , 2000,46,117-126.
[45] Harrod, J F. Phenoxo Complexes of Copper (II). *Can J Chem,* 1969,47,637-647.
[46] Şenturk, O; Sultan, W; Kisakurek, D. Polymerization and Characterization of Various Tetrakis(Trihalophenolato)Cerium (IV) Complexes in The Solid State and in The Melt. *React Funct Polym,* 2002 ,53 ,73-81.
[47] Şenturk, O; Kisakurek, D. Polymerization and Characterization of Trihalophenolato Cerium(IV) Complexes in The Solid State and in The Melt.*Polym-Plast Tecnol,* 2003, 42 ,373-387.
[48] Silverstein, K.M ;Bassler, G.C; Morrill, T.T. Spectrometric Identification of Organic Compounds,5th ed.,John Wiley and Sons,New York,1991.
[49] West, A R. Basic Solid State Chemistry, John Wiley and Sons, New York, 1988.
[50] Camurlu, P; Yilmaz, A; Tatar, L; Kisakurek, D; Ülkü, D. Synthesis , Crystal Structures and Characterization of New Transition Metal Compounds of Bromophenols: Bis(2,4,6-Tribromophenolaato)Di(N-Methylimidazole)M(II) (M= Co,Cu). *Crystal Research and Tecnology* (in press)

[51] Fares, MM; Hacaloğlu, J; Süzer, S. Characterization Of Degradation Products Of Polyethylene Oxide By Pyrolysis Mass-Spectrometry. *Eur Polym J* ,1994, 30 ,845-850

[52] Çakmak, O; Baştürkmen, M; Kısakürek, D, Synthesis and Characterization of Conducting and Non-Conducting Polymers of Sodium 2,4,6,-trichlorophenolate by Microwave Initiation. *Polymer* (in press)

In: Recent Developments in Polymer Research
Editor: Antony V. Hopper, pp. 53-71
ISBN 1-60021-346-4

Chapter 3

MULTIPLE MELTING BEHAVIOR IN POLY(BUTYLENE TEREPHTHALATE): CALORIMETRIC AND MORPHOLOGICAL ANALYSES

***M. L. Di Lorenzo*[1] *and M. C. Righetti*[2]**

[1] Istituto di Chimica e Tecnologia dei Polimeri (CNR)
c/o Comprensorio Olivetti - Via Campi Flegrei, 34 - 80078 Pozzuoli (NA)
[2] Istituto per i Processi Chimico-Fisici (CNR)
Area della Ricerca - Via G. Moruzzi, 1 56124 Pisa

ABSTRACT

The melting behavior of PBT was investigated after isothermal crystallization from the melt at a fixed temperature. Analyses conducted by differential scanning calorimetry (DSC) revealed a multiple melting behavior. Temperature-modulated differential scanning calorimetry investigations (TMDSC), performed at various modulation periods, showed that the frequency of modulation largely affects the apparent reversing specific heat capacity (c_p) measured during melting, but not the total c_p. In addition, a detailed examination of the modulated heat-flow-rate curves revealed that the melting process of PBT is the sum of simultaneous events of melting of small and defective crystals and their immediate recrystallization into more stable structures.

Quasi-isothermal TMDSC analyses, conducted for extended times at different temperatures in the whole melting range, revealed different degrees of reversible melting in PBT. The total latent heat involved in the fusion/recrystallization/crystal-perfection processes was found to decrease with time for the quasi-isothermal analyses performed at temperatures below 500 K. At higher temperatures, instead, recrystallization resulted prevailing with respect to fusion, with a increase of the apparent reversing heat capacity as a function of time.

1. Introduction

Poly(butylene terephthalate) (PBT) is a thermoplastic engineering material with a very high crystallization rate [1-2]. Morphology and crystal structure of PBT have been studied by a number of authors. Data available in the literature mainly concern analysis of the bulk crystallization rate of PBT, measured both in isothermal and non-isothermal conditions, and only limited data exist on analysis of growth rate (G) of PBT spherulites [1,3-10]. Growth rate data measured with the traditional isothermal method, i.e. by monitoring at a fixed temperature the increase in dimensions of the spherulite radius as a function of time, have been reported in the literature only within temperature ranges of a few degrees [4,10]. Difficulties in the determination of G arise from the very high nucleation density of PBT. The use of digital methods, which permit to record the growth of spherulite radii at very short time intervals before crystallite impingement, allow measurements of crystal growth rate even in case of very fast crystallization kinetics [11]. These methods have been used to measure growth rate of PBT spherulites, and results detailed in Ref. 12.

Fusion of PBT has also been the subject of several studies reported in the literature, due to multiple melting behavior. The occurrence of multiple melting endotherms is a common phenomenon for synthetic macromolecules, being observed for several other polymers, including poly(ethylene terephthalate) (PET) [13], poly(trimethylene terephthalate) [14], aliphatic nylons [15-16], poly(L-lactide) [17], poly(p-phenylene sulfide) (PPS) [18]. However, a general explanation of such a complex process has not been achieved yet. This behavior can, in principle, be ascribed to a number of factors, like the presence of more than one crystal modification, molecular weight segregation that accompanies crystallization, different crystal morphologies, orientation effects, or reorganization processes that occur during the DSC heating scan [19].

This chapter reviews the studies conducted by our research group on melting behavior of PBT [20-22]. Main analyses have been conducted by conventional and temperature-modulated calorimetry, and completed by morphological investigations conducted by optical microscopy.

2. Experimental Part

2.1. Material

Poly(butylene terephthalate) (PBT) of molar mass M_n=44,000 g/mol and M_w=81,000 g/mol, was purchased from Sigma-Aldrich Corp and used as received.

2.2. Preparation of Compression-Molded Samples

PBT sample chips were compression-molded with a Carver Laboratory Press at a temperature of 523 K for 3 minutes, without any applied pressure, to allow complete melting. After this period, a load of about 0.5 ton was applied for 2 min. Successively the sample was quickly cooled to room temperature by means of cold water circulating in the plates of the press.

2.3. Calorimetry

DSC and TMDSC measurements were performed with a Perkin-Elmer Dynamic Differential Scanning Calorimeter (DDSC). The instrument was calibrated in temperature with high purity standards (indium, naphthalene and zinc) at 2 K/min, according to the procedure for standard DSC. The heat-flow-rate was initially calibrated with the heat of fusion of indium, then refined with a baseline run of two empty aluminum pans, and a calibration run with sapphire as a standard [23]. Dry nitrogen was used as purge gas at a rate of 30 ml/min. In order to reduce temperature gradients, the sample mass was kept small, approximately equal to 4 mg.

The following standard test procedure was used: each sample was heated up to 523 K, maintained at this temperature for 3 min to cancel any previous thermal history, then quickly cooled down to 473 K. At this temperature, PBT crystallization is fast, and the leveling off of the DSC trace occurs within 5 min. Nevertheless, the samples were kept at 473 K for 30 min to improve crystal perfection. Subsequent analyses were conducted from 473 to 508 K using the dynamic program that is obtained through a sawtooth temperature modulation in the Perkin-Elmer DDSC. The underlying heating rate was 0.5 K/min, the temperature amplitude was 0.2 K, and the modulation periods (p) ranged from 48 to 120 s. The sawtooth was obtained either with alternation of heating and cooling segments ($p = 48$ and 72 s), or with both heating steps at different rates ($p = 120$ s).

Standard DSC measurements were also conducted at the same heating rate of 0.5 K/min after isothermal crystallization performed with the same conditions (30 min at 473 K that followed fusion at 523 K for 3 min).

Quasi-isothermal TMDSC analyses on PBT were conducted for extended times, using the following temperature program. After isothermal crystallization at 473 K for 30 min, the PBT sample was heated from 473 to the chosen base temperature for quasi-isothermal modulation (T_o), using the dynamic temperature program that is obtained through a sawtooth modulation in the Perkin-Elmer DDSC. The underlying heating rate was 0.5 K/min, the temperature amplitude 0.2 K, and the modulation period was 120 s. When the temperature reached T_o, quasi-isothermal analyses were started with $A_T = 0.2$ K $p = 120$ s. Each quasi-isothermal measurement was extended for 6 hours (in a few cases, indicated below in the text, for 7 hours). After the quasi-isothermal analysis, the sample was heated from T_o to 512 K at a linear scanning rate of 3 K/min, without temperature modulation.

2.4. Optical Microscopy

Optical microscopy was used to investigate melting kinetics and morphology of PBT spherulites, using a Zeiss polarizing microscope equipped with a Linkam TMHS 600 hot stage. A small piece of PBT was squeezed between two microscope slides, then inserted in the hot stage. The thickness of the squeezed sample was less than 10 μm. Each PBT sample was heated from 303 to 523 K, kept at 523 K for 3 min to allow complete melting, then quickly cooled to 473 K and allowed to crystallize for 30 minutes. After isothermal crystallization, the sample was heated at a rate of 0.5 K/min until fusion occurred. Morphology was recorded during the final heating at 0.5 K/min.

To determine melting kinetics, an automatic exposure apparatus, mounted on the optical microscope, was used to measure the intensity of light transmitted between crossed polars. The time to exposure (t_e), needed to impress a film of a given sensitivity in a MC-100 camera, was detected by an electronic digital display, and recorded as a function of temperature or time. This was performed after isothermal crystallization at 473 K for 30 min, followed by a heating scan at a rate of 0.5 K/min (as before). When the temperature reached 503 K, quasi-isothermal sawtooth modulation started, with temperature oscillating between 502 and 503 K, using heating and cooling rates of 1 K/min (A_T = 0.5 K; p = 120 s).

3. Results and Discussion

3.1. Analysis of PBT Melting During Heating

A conventional DSC analysis of the melting behavior of PBT is illustrated in Figure 1. This measurement was performed upon heating at a constant rate of 0.5 K/min, after isothermal crystallization at 473 K for 30 min. In the figure, the experimental c_p data are compared to thermodynamic c_p values as taken from ATHAS data bank [24]. It is worth pointing out that the plot presented in Figure 1 is qualitatively similar to several other DSC thermograms reported in the literature that demonstrate the complex fusion behavior of PBT [25-28]. The apparent specific heat capacity has contributions from the baseline c_p, i.e. the heat necessary to change the temperature of a sample in absence of latent heat effects, and from enthalpy changes due to melting, crystallization or annealing processes [19]. From Figure 1 several processes involving latent heat exchanges can be evidenced. Soon after beginning of the heating scan, the experimental c_p values are higher than the thermodynamic c_p data, because of the occurrence of a broad endothermic process ("Peak 1"). This weak transition precedes a sharper endothermic event ("Peak 2"), which seems to be followed by an exotherm ("Peak 3") and by a final large endothermic peak ("Peak 4").

The multiple melting behavior of PBT, shown in Figure 1, has been sometimes correlated to variations in spherulitic morphology during the heating scan that leads to fusion [29-30], because of the observation of different types of spherulites after primary crystallization: depending on crystallization conditions, PBT forms "unusual" spherulites with optical axes laying at an angle of about 45° to the spherulite radius, as well as "usual" spherulites with optical axes either along or perpendicular to the spherulite radius and a maltese cross along the polar direction when analyzed under a polarizing optical microscopy [29]. In order to test the validity of this hypothesis, the effect of fusion on PBT spherulites was analyzed by optical microscopy during heating at a constant rate. Analyses were conducted on a very thin PBT film, in order to obtain quite large spherulites. Figure 2-a reports the optical micrograph of a typical PBT sample isothermally crystallized at 473 K for 30 min, showing a mixed-type spherulitic superstructure [30], with extinction patterns in multiple directions, due to superposition of extinction patterns at 0 and 45°.

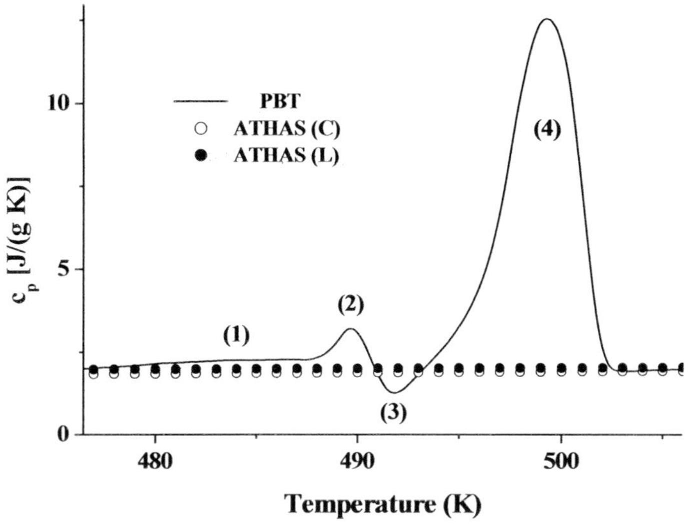

Figure 1. Apparent specific heat capacity of PBT crystallized isothermally at 473 K as a function of temperature. The thick line represents the experimental curve by standard DSC (heating rate: 0.5 K/min). The solid and open symbols are the specific heat capacity of the liquid and crystalline phases, respectively, as reported in Ref. 24. Numbers highlight major transitions, as detailed in the text

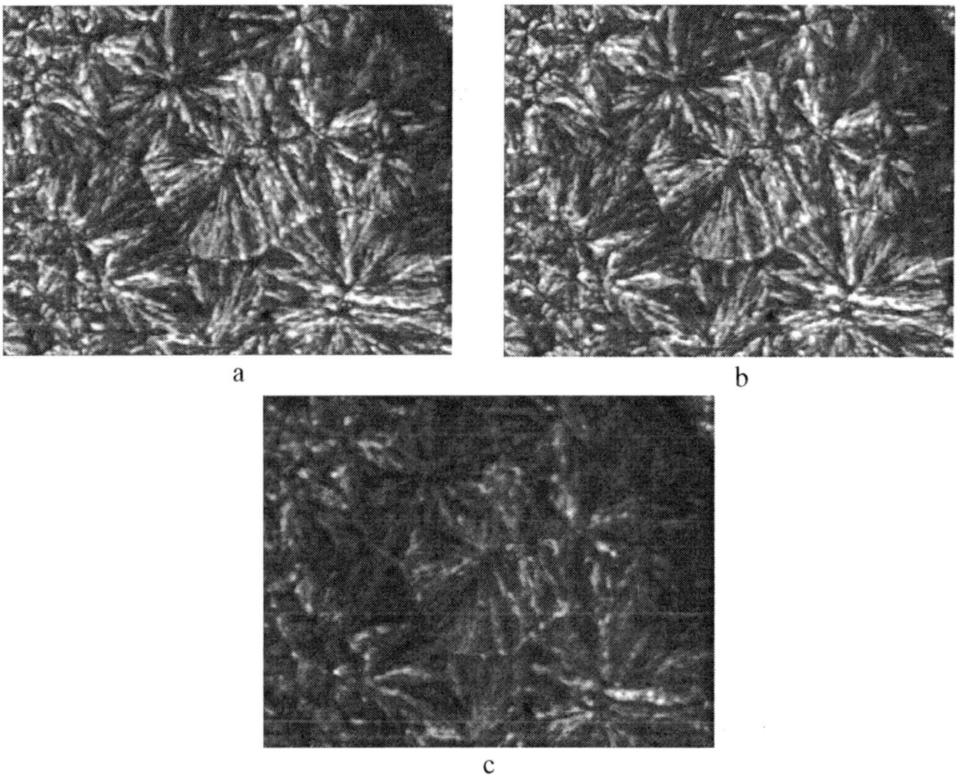

Figure 2. Optical micrographs (crossed polars) of PBT isothermally crystallized at 473 K for 30 min, then heated at 0.5 K/min. Micrographs were taken at various temperatures during the heating at 0.5 K/min: (a) 473 K; (b) 493 K; (c) 502 K

A few unusual spherulites can also be distinguished [30]. This PBT sample was heated at a rate of 0.5 K/min, and optical micrographs were taken at various temperatures during the heating scan, as illustrated in Figure 2-b and 2-c. Analysis of the optical micrographs at various temperatures shows that the multiple extinction patterns of PBT spherulites are unaffected by the raising temperature until complete fusion, and only a variation in the intensity of depolarized light that follows the temperature changes can be observed. This proves that the double melting behavior shown in Figure 1 is independent on the initial type of spherulitic morphology developed upon isothermal crystallization, ending the controversy arisen from contrasting literature reports [29-31].

Integration of the apparent c_p curve of Figure 1, according to the method proposed by Mathot et al. [32], provides the enthalpy-based temperature variation of the crystalline fraction (w_c), as illustrated in Figure 3. Initial crystallinity of the PBT sample, after isothermal crystallization at 473 K for 30 min is around 32%. Again, Figure 3 evidences a slow decrease in crystallinity of PBT, which begins together with the onset of the heating scan, followed by a shallow minimum at about 490 K, a slight increase in w_c and finally a drastic decrease of w_c down to a value of zero crystallinity when all the material under analysis becomes liquid.

The results of the analysis performed with the Mathot's method on DSC data was compared with data from optical microscopy, obtained by monitoring the variation in the intensity of light transmitted between the crossed polars by the birefringent PBT spherulites as a function of temperature. The use of depolarized light transmission is often used as a tool for the study of crystallization kinetics [33-35], or to detect melting temperatures of crystallized polymers [36], but to our knowledge no report exists on the use of depolarized light intensity to follow melting kinetics.

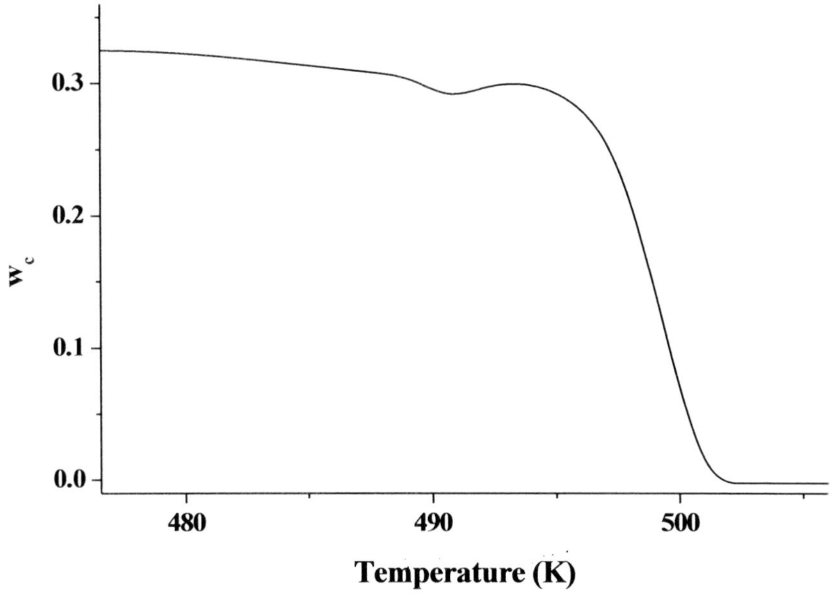

Figure 3. Enthalpy-based crystallinity of PBT, as a function of temperature, calculated from the experimental data of Figure 1

It is generally assumed that, at any time, the intensity of depolarized light increases with the fraction of crystallized material (w_c) [37], which allows to determine phase change kinetics as a function of temperature and/or time. Unfortunately, the detected light intensity depends not only on the light source, the glass slide, the optics, and the detection system, all parameters that can be easily kept constant during a series of experiments, but also on sample thickness, which is a very difficult parameter to control in very thin samples (generally below 10 μm). For these reasons it is not possible to obtain quantitative evaluations of melting kinetics by analysis of the depolarized light transmission intensity [37], but only to derive qualitative information that can complete the DSC analysis reported in Figures 1 and 3.

Figure 4 illustrates the reciprocal of the exposure time necessary for a photomicrograph ($1/t_e$), which depends on the crystalline fraction of the material (w_c), as it varies during the heating scan at 0.5 K/min. The trend of $1/t_e$ with temperature (T) is very similar to the w_c vs. T plot obtained by DSC measurements, shown in Figure 3. The very good agreement between the quantitative data of Figure 3 with the qualitative data illustrated in Figure 4, shows that the analysis of depolarized light intensity can be successfully applied, not only for the study of crystallization, but also to follow melting kinetics of polymers. Due to its reliability, this method was also used to monitor the variation in crystallinity of PBT during prolonged permanence at high temperatures, as detailed in the following section.

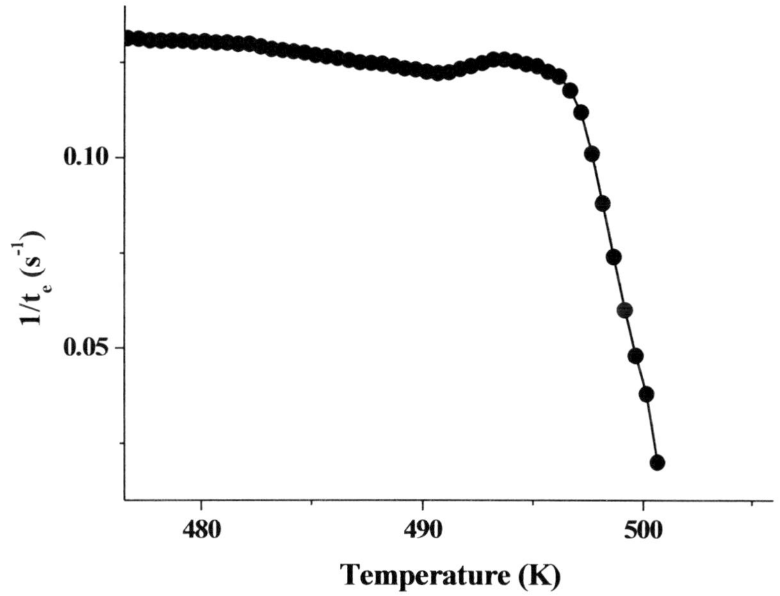

Figure 4. Reciprocal of the exposure time ($1/t_e$) as a function of temperature, recorded during heating at 0.5 K/min of a PBT sample isothermally crystallized at 473 K for 30 min

To better clarify the nature of the processes involved during fusion of PBT crystallites, TMDSC analyses were also conducted in the melting region. TMDSC measurements were performed with an underlying heating rate of 0.5 K/min, and at three different frequencies, as illustrated in Figure 5, where the measured modulated heat-flow-rate is reported as a function of the underlying temperature.

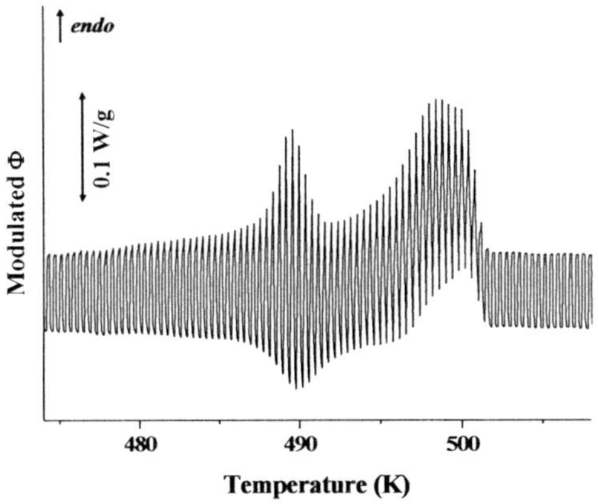

a

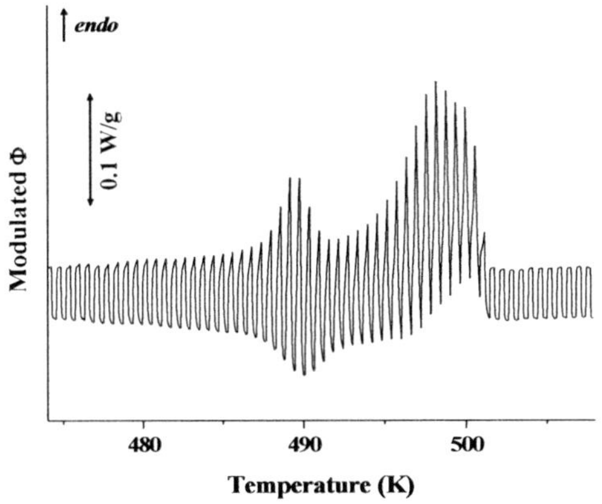

b

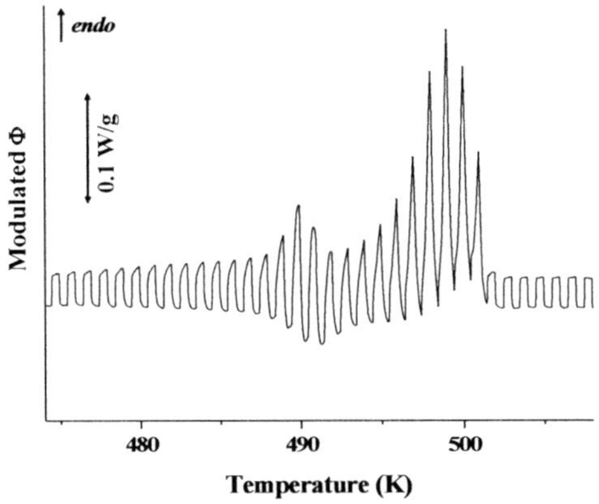

c

Figure 5. Modulated heat-flow-rates measured at different modulation periods: (a) $p = 48$ s; (b) $p = 72$ s; (c) $p = 120$ s. $A_T = 0.2$ K, $q = 0.5$ K/min

From the three TMDSC curves, the total heat-flow-rate was determined as the sliding average over the modulation period. The resulting total heat capacity displays no frequency dependence and practically overlaps, within experimental errors, with the c_p curve of Figure 1, that was obtained in absence of modulation and with the same underlying heating rate.

Large variations with modulation period were instead observed for the apparent reversing specific heat capacity, shown in Figure 6, where a strong frequency dependence appears, similarly to results reported in the literature for other semicrystalline polymers [38].

When polymer melting is analyzed during a TMDSC heating scan, crystal reorganization generally occurs, involving exothermic and endothermic events. Both development and absorption of latent heat produce an increase in the modulation amplitude of the heat-flow-rate, independently of the direction of heat exchange [39]. An apparent reversing specific heat capacity higher than the baseline value indicates that thermal processes occur under the chosen experimental conditions, without, however, distinguishing whether the phenomena are truly reversible or irreversible. In other words, when TMDSC measurements of polymer melting are conducted with a non-zero underlying heating rate, TMDSC analysis with Fourier series cannot discriminate between exothermic and endothermic processes [39-40], and more detailed information can be gained from the modulated heat-flow-rate profile, rather than from the calculated reversing contributions.

A close examination of the modulated heat-flow-rate curves of Figure 5 highlights the following features. In the temperature range 473-488 K (corresponding to Peak 1 in Figure 1) a slight distortion in the modulated curve is observable in its endothermic parts, with a smaller exothermic effect during the subsequent semi-periods. As a consequence, the modulated heat-flow-rate amplitude is slightly higher than it would be in absence of transitions, as observable in Figure 6, where an apparent reversing specific heat capacity higher than the baseline value is evident in this temperature range. At higher temperatures (about 488-491 K, Peak 2), the modulated heat-flow-rate profiles of Figure 5 reveal the occurrence of both endothermic and exothermic events within the same modulation period. The modulated heat-flow-rate curve is largely deformed, due to the ongoing phase transformation. Large endothermic events take place during the heating segments, followed, during the subsequent half-cycle, by exotherms that have almost the same size of the preceding endotherms. In this temperature range a well marked peak appears in the apparent reversing specific heat capacity of Figure 6. In particular, with regards to the measurement conducted at $p = 120$ s, one can observe that, around 490 K, both endothermic and exothermic processes occur during both the two semi-periods, as evidenced in the enlargement of Figure 5-c presented in Figure 7. In the areas indicated by the arrows, during the first half-cycle, characterized by a higher heating rate, the initial increase in heat-flow-rate, caused by the switch to a different heating rate overlapping partial melting, is followed by exothermal effects, as revealed by comparison with corresponding semi-periods of modulation in the melt, at temperatures above 503 K, where no latent heat is exchanged. Similarly, during the second half-cycle, that has a lower heating rate, the starting decrease in the heat-flow-rate is followed by an endothermic increase. This reveals that fusion and re-crystallization take place at the same time, and the long semi-period (60s) allows to evidence both the processes occurring within the same modulation segment.

In the narrow temperature range of the exothermic dip of Figure 1 (Peak 3), the raw modulated data of Figure 5 also underline the concurrent occurrence of both endothermic and exothermic processes, as outlined by the apparent reversing c_p of Figure 6, that is higher than the baseline value, and even during the final melting (Peak 4) some recrystallization occurs.

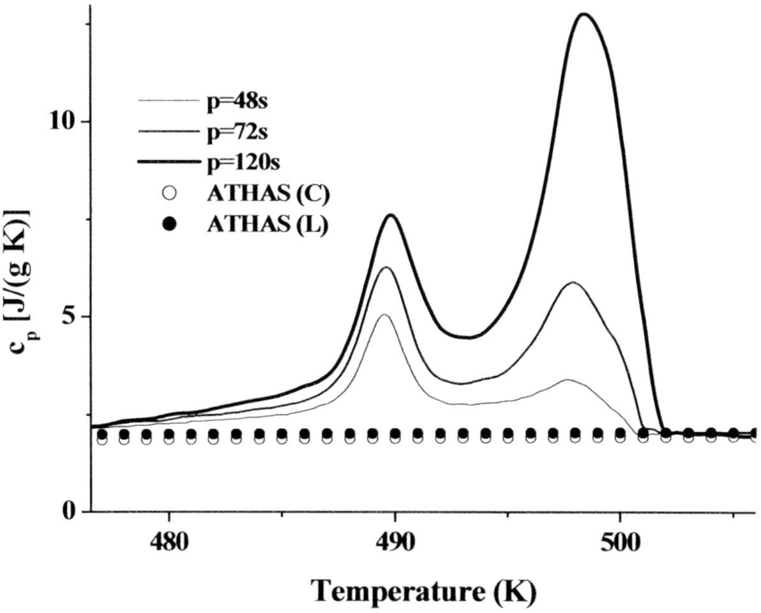

Figure 6. Apparent reversing specific c_p at various frequencies of modulation. The solid and open symbols are the specific heat capacity of the liquid and crystalline phases, respectively, as reported in Ref. 24

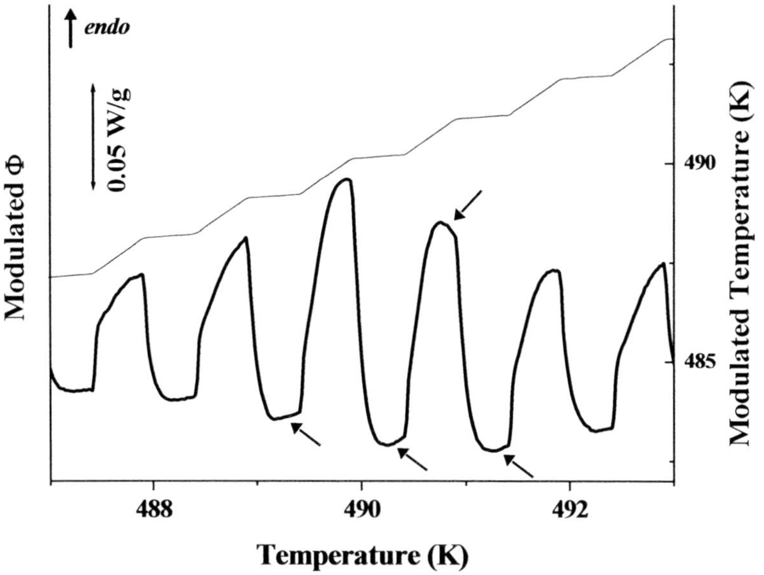

Figure 7. Enlargement of Figure 5-c in the temperature range 487 – 493 K. The thick line represents the modulated heat-flow-rate, the thin line refers to the modulated temperature. Arrows highlight the area of discussion

When endothermic and exothermic events take place simultaneously, standard DSC reveals only the net excess heat-flow-rate, i.e. in Figure 1 the recrystallization exotherms in correspondence of Peaks 1, 2 and 4 are hidden by the simultaneous more intense melting. This obscuring is overcome in TMDSC analysis, where the continuous switches between heating and cooling segments allow to evidence the occurrence of multiple phenomena within the same modulation cycle. It needs to be underlined that the successive melting-recrystallization processes in TMDSC scan are not induced by the temperature modulation, i.e. by the alternation of heating and cooling steps, as they are well evident also in the modulated heat-flow-rate of Figure 5-c ($p = 120$ s), where modulation was obtained with heating-only steps ($p = 120$s). In other words, the modulated heat-flow-rate profiles of Figure 5 prove that recrystallization in PBT during heating takes place immediately after fusion of portions of the polymer chains.

The large reversing apparent c_p displayed in Figure 6 is due to the large latent heat contributions arising from the continuous and repeated immediate regrowth of the just melted crystals, which are not reversible processes, as the thermal stability of the newly developed structures is higher than that of the original crystals. The influence of modulation period detected in the apparent reversing c_p curves of Figure 6 is generally explained by considering that, with decreasing the period, the amount of crystalline material that can follow the modulation within one period reduces, since the temperature range covered by each modulation cycle is progressively smaller. This way, the irreversible latent heat effect per cycle becomes smaller at higher frequencies. The intensity of the high-temperature peak seems to decrease faster with frequency with respect to the peak at lower temperatures. The overall amount of latent heat exchanged in the final endotherm is much higher, which results in a larger fraction of latent heat evolved and/or absorbed by the sample during each modulation period. This in turn produces an erroneously high modulation amplitude within each period and a larger frequency dependence of the apparent reversing c_p, being the final melting essentially irreversible. In addition, with respect to more perfect crystallites, poor crystals are characterized by a smaller difference between fusion and crystallization temperatures, since the folding of the chains is subjected to less restrictions [19,38]. As a consequence, melting and recrystallization of poor crystals occur more easily within the same modulation period and the frequency dependence of the apparent reversing c_p comes out smaller.

3.2. Quasi-Isothermal Analyses

To complete the investigation on melting behavior of PBT, detailed quasi-isothermal analyses were performed. Initial measurements were conducted by modulating around $T_o = 489.6$ and 498.4 K, the temperatures corresponding to the two maxima of the reversing specific heat capacity peaks from the TMDSC scan shown in Figure 6.

The time evolution of the apparent reversing specific heat capacity at 489.6 and 498.4 K, calculated with eq. (1), is displayed in Figure 8. The amplitude of temperature oscillations remains constant for the whole time of analysis, as expected for a power compensation calorimeter. The heat flow rate amplitude, instead, gradually decreases with time, as a consequence of the decreasing amount of latent heat that is evolved and/or absorbed by the sample during the quasi-isothermal measurement.

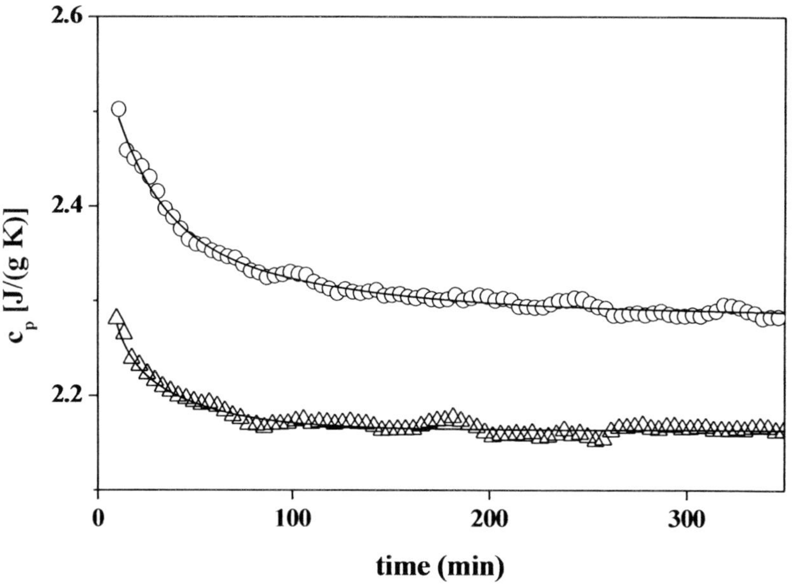

Figure 8. Time dependence of the apparent reversing heat capacity of PBT during quasi-isothermal analyses at $T_o = 489.6$ K (△) and T_o=498.4 K (○). $p = 120$ s, $A_T = 0.2$ K

This results in a decrease of the apparent reversing c_p with time, as usually observed for semicrystalline polymers of sufficiently high molar mass [19]. The trend is typically a double-exponential decay, with the relative equation shown in the upper right corner of Figure 8. The two relaxation times τ_1 and τ_2, that are commonly attributed to irreversible processes occurring at the interface between the metastable crystals and the surrounding melt during the analysis [19], are found to be of the order of 10 ÷ 12 min and 60 ÷ 100 min, respectively, for the two analyzed temperatures.

The kinetics of recrystallization and crystal perfection during the quasi-isothermal TMDSC experiments was followed by a close inspection of the modulated heat-flow-rate profiles, as presented in Figure 9 for the analysis conducted at $T_o = 498.4$ K. At the initial stages of the quasi-isothermal treatment, both endothermic and exothermic events take place within the same modulation period (as highlighted by the arrows in the left half of the figure), whereas at longer times recrystallization predominates. This finding supports the interpretation of the two relaxation times τ_1 and τ_2 (Figure 8), both related to the events that progressively remove sample portions from the reversing melting/crystallization-process: the contribution to the apparent reversing specific heat capacity connected to the relaxation time τ_1 is associated with partial melting and recrystallization/crystal perfection, whereas the contribution to τ_2 comes only from recrystallization/crystal perfection processes. Similar results were found also for the quasi-isothermal measurement conducted at $T_o = 489.6$ K.

The kinetics of reorganization processes of the crystal phase was also followed by monitoring the melting behavior after increasing times of the quasi-isothermal treatment. From integration of the DSC curves, the crystalline fraction at the end of the quasi-isothermal analyses ($w_{c,qi}$) was calculated, using for the fusion enthalpy of 100% crystalline PBT the literature value of 145 J/g [24]. The results, reported in Figure 10 for the quasi-isothermal modulations around 498.4 K, reveal that both the melting-endotherm temperature (T_m) and the crystalline-fraction increase with time of analysis, proving the occurrence of recrystallization and crystal perfection processes during temperature modulations. These processes are very fast at the initial stages of the quasi-isothermal treatment, then proceed more slowly. It is worth noting that the considerable increase in T_m parallels the corresponding variation in $w_{c,qi}$. In fact, the T_m vs. $w_{c,qi}$ plot obtained for the data of Figure 10 is almost linear.

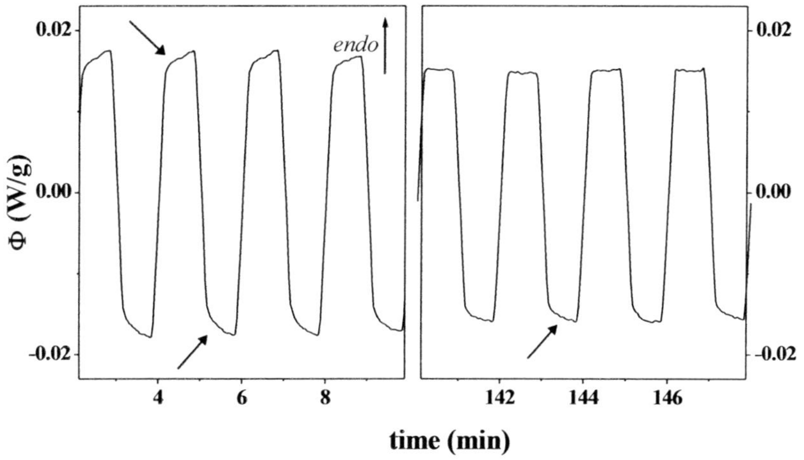

Figure 9. Modulated heat flow rate profile for PBT at different times of the quasi-isothermal measurement at T_o=498.4 K. The arrows highlight endothermic and exothermic events

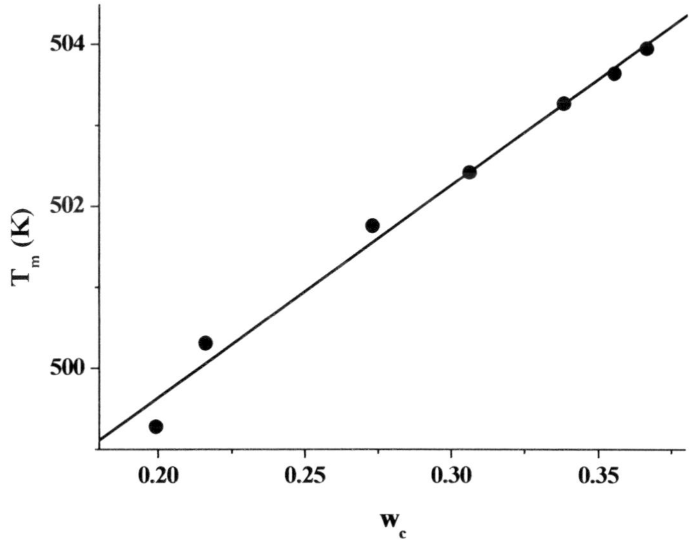

Figure 10. Experimental melting temperature vs. crystal fraction of a PBT sample isothermally crystallized at 473 K, heated at q = 0.5 K/min, then quasi-isothermally modulated around T_o = 498.0 K for different times: 0 min, 3 min, 30 min, 60 min, 180 min, 300 min, 360 min

This clarifies the type of structural reorganization occurring during the quasi-isothermal measurement. The increase in crystallinity with time has to be mainly ascribed to crystallization with progressive perfection of the crystals, being fusion characterized by higher and higher melting temperature.

The effect of the small temperature oscillations on the recrystallization/crystal perfection/annealing processes was evaluated by comparison with simple isothermal treatments conducted in the conventional DSC mode. The crystal fraction, calculated after quasi-isothermal analyses of different duration at 498.4 K was always found perfectly coincident with the crystallinity measured after isothermal treatments of the same time, but without temperature modulation. Thus it was inferred that the small temperature oscillations around T_o (± 0.2 K) does not influence the extent of the total recrystallization process of PBT, in agreement with TMDSC results obtained with an underlying heating rate of 0.5 K/min, discussed above.

Besides the experiments conducted at the temperatures corresponding to the two maxima of the reversing heat capacity of Figure 6, further quasi-isothermal TMDSC analyses were performed at increments of 1.0 K, starting from 487.4 K. The time of the measurements was 6 hours, excluding those in the T_o range 501.4-503.4 K, whose duration was raised to 7 hours, because of the slower reorganization kinetics. Results of these analyses are summarized in Figure 11. They indicate that the structural reorganizations of the crystal phase in PBT are strongly affected by temperature. Below 500 K, the apparent reversing specific heat capacity plotted vs. time displays the commonly observed double exponential decay [38]. When PBT is analyzed at 500.4 K, the apparent reversing c_p remains practically constant at a value of 2.21 J/(g·K) during the whole measurement. This value is higher than the thermodynamic c_p of liquid PBT, determined in absence of thermal events, similarly to the trend reported in the literature for a poly(ether-amide) segmented copolymer [41]. For quasi-isothermal analyses carried out at higher temperatures, the apparent reversing c_p shows a quite unusual trend, increasing with measurement time.

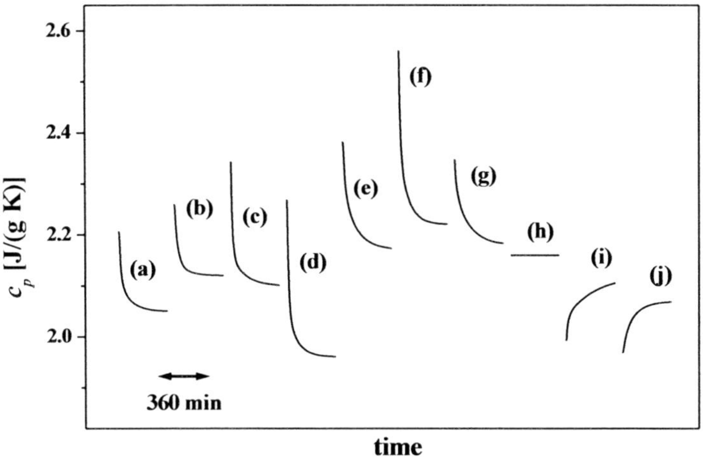

Figure 11. Time dependence of the apparent reversing specific heat capacity of PBT during quasi-isothermal measurements at: (a) 488.4 K, (b) 490.4 K, (c) 491.4 K, (d) 493.4 K, (e) 495.4 K, (f) 498.4 K, (g) 499.4 K, (h) 500.4 K, (i) 501.4 K, (j) 503.4 K. Data from separate measurements are collected in a single graph, in order to compare the reversing c_p trends for different T_o's

As evidenced above, a complex process is involved in the decrease of the apparent reversing heat capacity with time, including both fusion and recrystallization and/or crystal perfection. A fully melted molecule or molecular segment longer than a critical length is removed from the reversing melting if the supercooling required for the recrystallization is not achieved during the modulation cycle [38]. Similarly, since the thermal stability of a recrystallized structure is higher than that of the original crystals, each more perfect crystal has a higher melting temperature (see Figure 10) and is thus removed from the successive melting cycle, so that the crystal fraction that participates in the melting process reduces progressively. This results in a corresponding decrease of latent heat contributions to the apparent reversing c_p.

Different is the case of quasi-isothermal analyses performed at temperatures ranging from 501.4 to 503.4 K. Measurements conducted at these temperatures start with an apparently completely melted (or almost completely melted) PBT sample, that recrystallizes during the quasi-isothermal modulation around T_o. Recrystallization at high temperatures, starting from a melted sample, was also observed in poly(ethylene terephthalate) [42]. Recrystallization of PBT immediately occurs at the beginning of the quasi-isothermal analysis, continuously adding new crystals that participate in the reversing melting process: the effect is a continuous increase in the measured reversing c_p. At intermediate temperatures (500.4 K), fusion, recrystallization and crystal perfection processes are counterbalanced during the 6 hours of experimental measurement, which results in a constant apparent reversing c_p.

The crystalline fractions before and after the quasi-isothermal treatments ($w_{c,o}$ and $w_{c,qi}$ respectively) are presented in Table 1. As already observed for the measurements at 498.4 K (Figure 10), the increase of the crystalline fraction is considerable for all the quasi-isothermal treatments performed in the whole melting range. The large increase in crystallinity might be favored by a lower molecular mass resulting from chain degradation, especially at high T_o: a low molecular mass increases chain mobility, which, in turn, may favor formation of crystals in the presence of proper nuclei. Literature data indicate that PBT degrades upon prolonged exposure at temperatures of 513.2 K and above [43]. Degradation involves chain scission via elimination of butadiene, and, after 6 hours of permanence at 513.2 K, the average molecular weight approximately reduces to a half. Unfortunately, no literature data were found on degradation of PBT at the temperatures where quasi-isothermal measurements were conducted in this study. Ester interchange reactions at high T_O were also found to favor recrystallization and crystal perfection in the lower homologue of PBT, poly(ethylene terephthalate): due to the break of main-chain bonds outside the crystals, the mobility of the polymer segments is enhanced and the effects of the constraints, such as entanglements and tie chains, reduced [44-45].

In order to test if the above chemical reactions can modify the reorganization process of PBT crystals monitored by the quasi-isothermal measurements, a comparison was made at T_o= 495.4 K between a fresh sample and a sample previously maintained for a prolonged time (6 hours) at 508.2 K. The apparent reversing c_p values were found perfectly coincident for both samples during the whole time of analysis, thus the effect of degradation on the heat capacity increase at high T_o was considered minor.

Table 1. Crystalline fraction before and after the quasi-isothermal treatments ($w_{c,o}$ and $w_{c,qi}$ respectively), determined at the indicated temperatures of quasi-isothermal analyses (T_o)

T_o (K)	$w_{c,o}$*	$w_{c,qi}$
487.4	0.307	0.351
488.4	0.305	0.359
489.6	0.298	0.367
490.4	0.292	0.380
491.4	0.292	0.385
492.4	0.299	0.384
493.4	0.300	0.383
494.4	0.298	0.380
495.4	0.288	0.368
498.0	0.212	0.360
498.4	0.188	0.350
499.4	0.117	0.318
500.4	0.049	0.266
501.4	0.008	0.133
502.4	0	0.087
503.4	0	0.057

* Calculated from the data reported in Figure 3

Recrystallization of PBT during quasi-isothermal TMDSC experiments at high temperatures, detailed in Figure 11 and Table 1, was confirmed by analysis of depolarized light intensity, using the method described above. Results are illustrated in Figure 12, for a quasi-isothermal modulation around 502-503 K. At the beginning of the quasi-isothermal analysis, the PBT sample is in the melt state, although some residual birefringence remains. Crystallinity rapidly increases as soon as the heating scan stops and temperature starts to modulate between 502 and 503 K, indicating large and immediate recrystallization of the almost melted sample. The insert in Figure 12 shows that recrystallization proceeds during both heating and cooling steps between 502 and 503 K, but with different kinetics: recrystallization is rapid during the cooling cycles, and continues more slowly during the heating segments of temperature modulation.

The recrystallization process of PBT at high temperature was also followed by taking optical micrographs of PBT spherulites at various times of permanence at 503 K. Results are exhibited in Figure 13, and refer to the same PBT sample shown in Figure 2. Birefringence of PBT spherulites rapidly increases with time, following the occurrence of large recrystallization. Besides the increase in birefringence, no other significant variation in the spherulitic morphology can be observed, and the spherulitic microstructure of the recrystallized material results identical to that of the original sample crystallized at 473 K, differing only in the intensity of depolarized light. This confirms that memory effects play a fundamental role in recrystallization of the melted PBT chains. The newly formed PBT spherulites illustrated in Figure 13 grow exactly in the same position, and with the same appearance of the original spherulites of Figure 2, which refer to PBT before and during the fusion process. The recrystallization process of PBT spherulites is thus facilitated by the small chain aggregates that, remaining in the liquid phase, act as precursors for crystal growth.

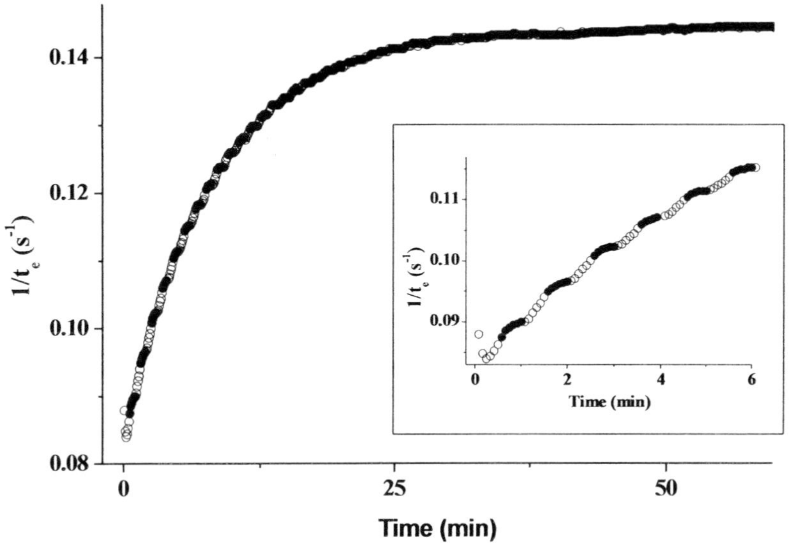

Figure 12. Reciprocal of the exposure time ($1/t_e$) as a function of time during quasi-isothermal modulation between 502 and 503 K of a PBT sample isothermally crystallized at 473 K for 30 min, then heated at 0.5 K/min to the quasi-isothermal temperature; $A_T = 0.5$ K; $p = 120$ min

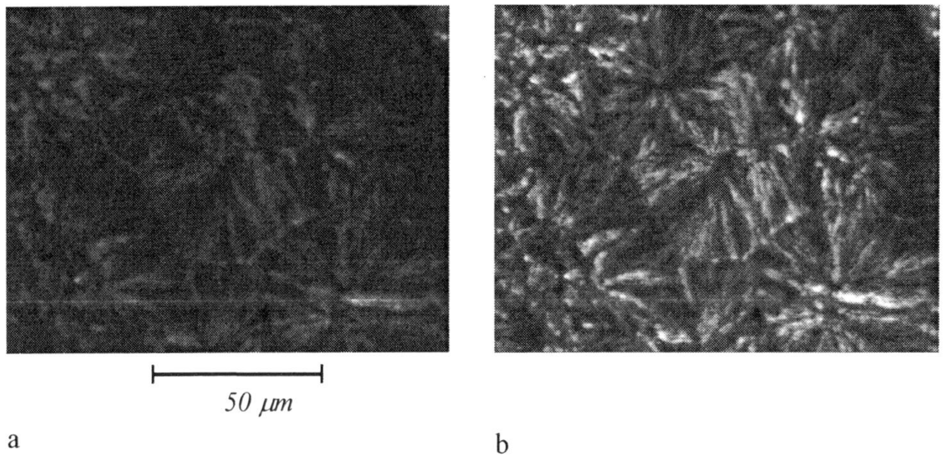

Figure 13. Optical micrographs (crossed polars) of PBT isothermally crystallized at 473 K for 30 min, then heated to 503 K at a rate of 0.5 K/min. Micrographs illustrate recrystallization at 503 K: (a) after 30 s; (b) after 90 min

REFERENCES

[1] Runt, J.; Miley, D. M.; Zhang, X.; Gallagher, K. P.; McFesters, K.; Fishburn, J. *Macromolecules* 1992, *25*, 1929.

[2] Righetti, M. C. *Thermochim Acta* 1999, *330*, 131.

[3] Stein, R. S.; Misra, A. *J Polym Sci, Polym Phys Ed* 1980, *18*, 327.

[4] Gilbert, M.; Hybart, F.J. *Polymer* 1972, *13*, 327.

[5] Pratt, C. F.; Hobbs, S. Y. *Polymer* 1976, *17*, 12.
[6] Righetti, M. C.; Munari, A. *Macromol Chem Phys* 1997, *198*, 363.
[7] Chisholm, B. J.; Zimmer, J. G. *J Appl Polym Sci* 2000, *76*, 1296.
[8] Pillin, I.; Pimbert, S.; Feller, J. F.; Levesque, G. *Polym Eng Sci*, 2001, *41*, 178.
[9] Kalkar, A. K.; Desphande, A. A. *Polym Eng Sci* 2001, *41*, 1597.
[10] Pracella, M.; Chiellini, E.; Donelli, D. *Makromol Chem* 1989, *190*, 175.
[11] Di Lorenzo, M. L. *Polymer* 2001, *42*, 9441.
[12] Di Lorenzo, M. L.; Righetti, M. C. *Polym Eng Sci* 2003, *43*, 1889.
[13] Mehta, A.; Wunderlich, B. *J Polym Sci, Polym Phys Ed* 1978, *16*, 289.
[14] Pyda, M.; Wunderlich, B. *J Polym Sci: Part B: Polym Phys* 2000, *38*, 622.
[15] Todoki, M.; Kawaguchi, T. *J Polym Sci, Polym Phys Ed* 1977, *15*, 1067.
[16] Xenopulos, A.; Wunderlich, B. *J Polym Sci: Part B: Polym Phys* 1990, *28*, 2271.
[17] Celli, A.; Scandola, M. *Polymer* 1992, *33*, 2699.
[18] Cebe, P.; Chung, S. *Polym Compos* 1990, *11*, 265.
[19] Wunderlich, B. *Macromolecular Physics, Vol. 3: Crystal Melting*; Academic Press: New York, 1980.
[20] Righetti, M. C.; Di Lorenzo, M. L. *J Polym Sci: Part B: Polym Phys* 2004, *42*, 2191.
[21] Righetti, M. C.; Di Lorenzo, M. L.; Angiuli, M.; Tombari, E. *Macromolecules* 2004, *37*, 9027.
[22] Di Lorenzo, M. L.; Righetti, M. C. *Polym Bull* 2004, 53, 53.
[23] Archer, D. G. *J Phys Chem Ref Data* 1993, *22*, 1441.
[24] ATHAS Data Bank, http://web.utk.edu/~athas/databank/welcome-db.html. Ed. M. Pyda (1994).
[25] Cheng, S. Z. D.; Pan, R.; Wunderlich, B. *Makromol Chem* 1988, *189*, 2443.
[26] Kim, J.; Nicols, M. E.; Robertson, R. E. *J Polym Sci: Part B: Polym Phys* 1994, *32*, 887.
[27] Kim, H. G.; Robertson, R. E. *J Polym Sci: Part B: Polym Phys* 1998, *36*, 133.
[28] Yasuniva, M.; Tsubakihara, S.; Ohoshita, K.; Tokudome, S. *J Polym Sci: Part B: Polym Phys* 2001, *39*, 2005.
[29] Stein R. S. , Misra A. *J Polym Sci: Polym Phys Ed* 1980, 18, 327.
[30] Ludwig, H. J.; Eyerer, P. *Polym Eng Sci* 1988, *28*, 143.
[31] Yeh, J.T.; Runt J. *J Polym Sci: Part B: Polym Phys* 1989, 27, 1543.
[32] Mathot, V. B. F.; Pijpers, M. F. J. *Thermochim Acta* 1989, *151*, 241.
[33] Hock, C.W.; Arbogast, J. F. *Anal Chem* 1961, *33*, 462.
[34] Magill, J.H. *Polymer* 1961, *2*, 221.
[35] Magill, J.H. *Polymer* 1962, *3*, 219.
[36] Kishore, K.; Vasanthakumari, R.; Pennings, A.J. *J Polym Sci: Polym Phys Ed* 1984, 22, 537.
[37] Ghanem, A. *J Polym Sci: Part B: Polym Phys* 1999, 37, 997.
[38] Wunderlich, B. *Progr Polym Sci* 2003, *28*, 383.
[39] Di Lorenzo, M. L.; Wunderlich, B. *Thermochim Acta*, 2003, *405*, 255.
[40] Wurm, A.; Schick, C. *Coll Polym Sci* 2003, *281*, 113.
[41] Di Lorenzo, M. L.; Pyda, M.; Wunderlich, B *J Polym Sci: Part B: Polym Phys* 2001, *39*, 2969.
[42] Zachmann, H. G.; Stuart H. A. *Makromol Chem* 1960, *41*, 148.

[43] Passalacqua, V.;Pilati, F.; Zamboni, V.; Fortunato, B.; Manaresi, P. *Polymer* 1976, *17*, 1044.

[44] Miyagi, A.; Wunderlich, B. *J Polym Sci, Polym Phys Ed* 1972, *10*, 2073. ibid. 2085.

[45] Wang, Z.G., Hsiao, B.S.; Sauer, B.B.; Kampert, W.G. *Polymer* 1999, *40*, 4615.

In: Recent Developments in Polymer Research
Editor: Antony V. Hopper, pp. 73-94
ISBN 1-60021-346-4

Chapter 4

UV-CURABLE POLYDIMETHYLSILOXANE EPOXY ACRYLATE

Weizhi Wang*
Fudan University, Institute of Advanced Materals,
220 Handan Rd, Shanghai, 200433 Peoples R China

ABSTRACT

In this article, UV-curable polydimethylsiloxane epoxy acrylate (PSEA) was synthesized by hydrosilylation of allyl glycidyl ether with hydrogen-containing polydimethylsiloxane to give polydimethylsiloxane-type epoxy resin which modified with acrylic acid. The influences of the synthetic process were discussed in detail. The properties of the cured film were also investigated by thermogravimetric analyzer, dynamical thermal mechanical analysis, and etc. In order to improve the photo-oxidative stability of photocured PSEA film, photocured PSEA without and with added photostabilizers have been investigated by measuring Fourier transform infrared photoacoustic spectroscopy and X-ray photoelectron spectroscopy. The hindered amine stabilizers based on 2,2,6,6-teteramethyl-piperdine, largely decrease the rate of surface oxidation of photocured PSEA. As an application sample, PSEA, polyethylene glycol urethane acrylate (PGUA) and trimethylolpropane triacrylate (TMPTA) were mixed in different proportions to modify the properties of blends to prepare UV-curable optical fiber coating. Formulation was optimized by the orthogonal table. When the proportion was PGUA: PSEA:TMPTA = 10:4:3, it was used for a single coating (OFC-6). However, the proportion PGUA: PSEA:TMPTA = 8:3:2 was used for a primary buffer coating. The rheological behavior and the selection of photoinitiators for OFC-6 were also discussed. The photopolymerization of OFC-6 was inspected by the FT-IR spectrum. Results show OFC-6 film is completely cured after irradiated 5 s. The thermal behaviors of the cured film were also investigated by TGA, DMA, and etc.

Keywords: polydimethylsiloxane epoxy acrylate; synthesis; characterization; UV-curing; properties; photostabilization; hindered amine light stabilizer; optical fiber coating.

---

* Corresponding author. Fax: +86-021-65644180. E-mail address: weizhiwang@fudan.edu.cn

1. INTRODUCTION

Polydimethylsiloxanes(PDMSs), which are usually known as "silicone"or "silicone elastomers", have received widespread attention as a special family of polymers since their commercial introduction in the 1940's and are by far the most important inorganic backbone polymers [1-3]. Their own unique properties, such as, low glass transition temperature (T_g), low surface tension and surface energy, low solubility, low dielectric constant, transparent to visible and UV light, very resistant to ozone, and stable against atomic oxygen even oxygen plasmas. In recent years, PDMSs with different organofunctional substituents are synthesized and used as starting compounds in the preparation of polymeric materials with those properties. Photocrosslinking of functional PDMSs is one of the most efficient methods in rapidly producing siloxane based polymeric networks. Polydimethylsiloxane epoxy acrylate (PSEA) containing acrylic ester groups, linked to the siloxane chain as pendant units, gives the PSEA reactivity, absorptivity, coupling properties, etc. This compound combines the properties of silicones with the fast crosslinking ability of acrylates. Siloxane-based networks have been used for metal coating and for the protection of glass or optical fibers[4,5] owing to their transparent nature.

In this article, UV-curable polydimethylsiloxane epoxy acrylate (PSEA) was synthesized by hydrosilylation of allyl glycidyl ether with hydrogen-containing polydimethylsiloxane to give polydimethylsiloxane-type epoxy resin which modified with acrylic acid. The influences of the synthetic process were discussed in detail. The properties of the cured film were also investigated by thermogravimetric analyzer, dynamical thermal mechanical analysis, and etc. In order to improve the photo-oxidative stability of photocured PSEA film, photostabilizers in photocured PSEA film can largely decrease the rate of surface oxidation of photocured PSEA. As an application sample, PSEA, polyethylene glycol urethane acrylate (PGUA) and trimethylolpropane triacrylate (TMPTA) were mixed in different proportions to modify the properties of blends to prepare UV-curable optical fiber coating.

2. SYNTHESIS OF UV-CURABLE POLYDIMETHYLSILOXANE EPOXY ACRYLATE (PSEA)

2.1. Synthesis of Polydimethylsiloxane Type Epoxy Resin(PSE)

Two methods for preparing PSE are generally used, they are, copolymerization of cyclosiloxanes by base-catalyzed equilibration and hydrosilylation of allyl glycidyl ether with hydrogen-containing polydimethylsiloxane [6]. However, the conversion percentage of the former one is relatively low. Furthermore, the epoxy rings are easily opened in acidic or basic solution. On the contrary, the high conversion of the later one at similar conditions with fewer by-products can be obtained. Therefore, the later method is selected for preparing PSE in this study.

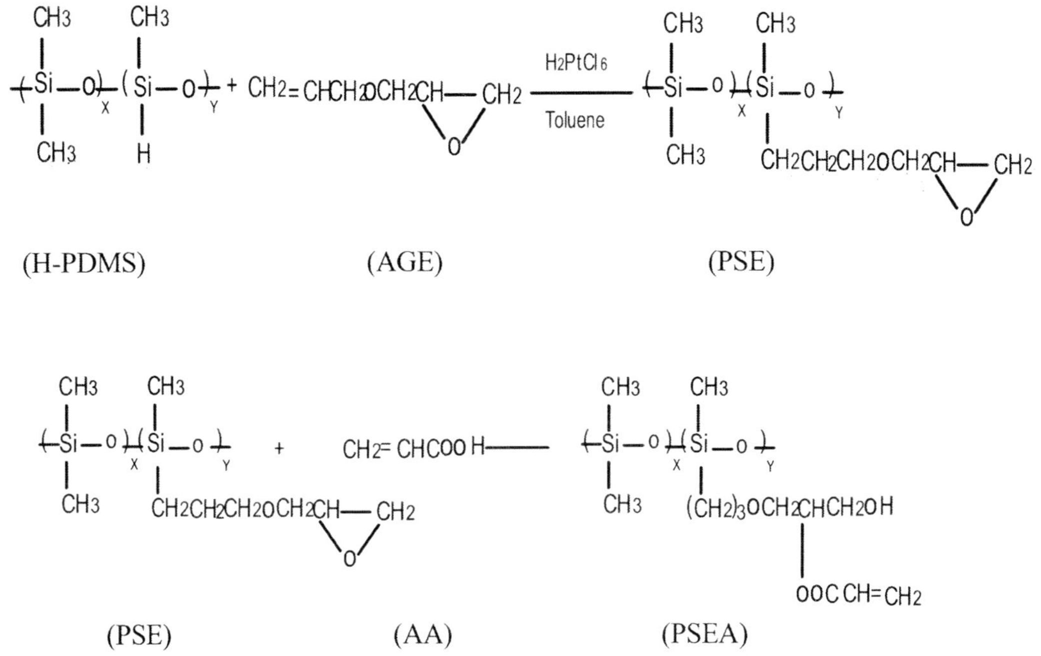

Scheme 1

2.2. The Influence in Preparing PSEA

In order to obtain UV curable PSEA, the PSE is modified by AA, which provides acrylated PES with unsaturated double bonds. The reaction rate is affected by the catalyst, reaction temperature and the concentration of reactants.

2.2.1. Effects of Catalyst

Figure 1 shows relationship between the catalyst concentration and the yield of PESA when PSE reacts with AA at 110 °C. It can be seen that the increasing of catalyst has obvious positive influence on the conversion of PSEA. The yield of PSEA is about 65 % after reacting for 6 hr without catalyst. The yield increases obviously with catalyst increasing, especially at initial stage. The optimistical catalyst concentration is 0.8 wt% of reactant, because the further increasing of catalyst can not effectively enhance the yield of PSEA.

2.2.2. Effects of Temperature

Figure 2 shows the effect of reaction temperature on the hydrosilylation of AGE with H-PDMS. It clearly reflects that polydimethylsiloxane with unsaturated epoxy gives yield almost 100% at 130 °C after 2.5 hr. It can be proved that Si-H characteristic peak (2140cm^{-1}) in FT-IR spectra disappears. However, the conversion can not reach over 80% when the temperature is below 130 °C, even if enhances the time of reaction. For example, the system still can detect active hydrogen after 10 hr at 110 °C. On the other hand, excessive high temperature results in the formation of crosslinked structure and even gel. The same results are obtained by chemically titration.

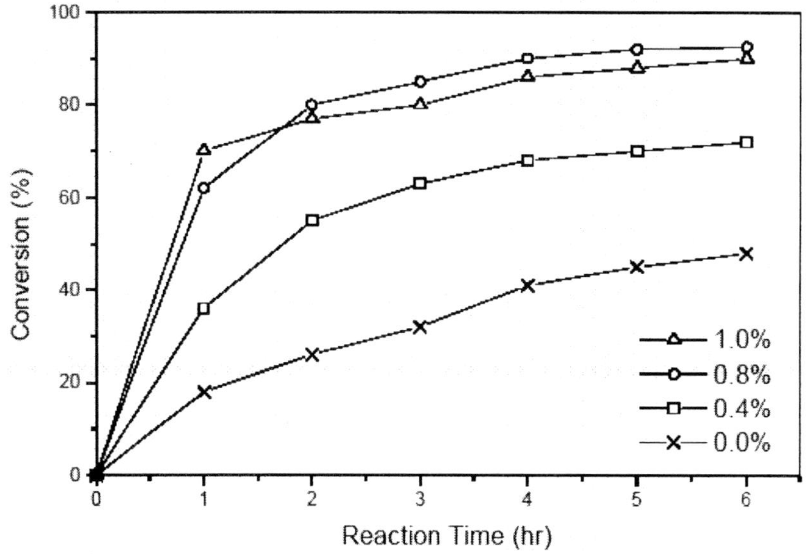

Figure 1. The concentration of catalyst Vs epoxy conversion in hydrosilylation reaction of PSE with AA measured at 110 °C

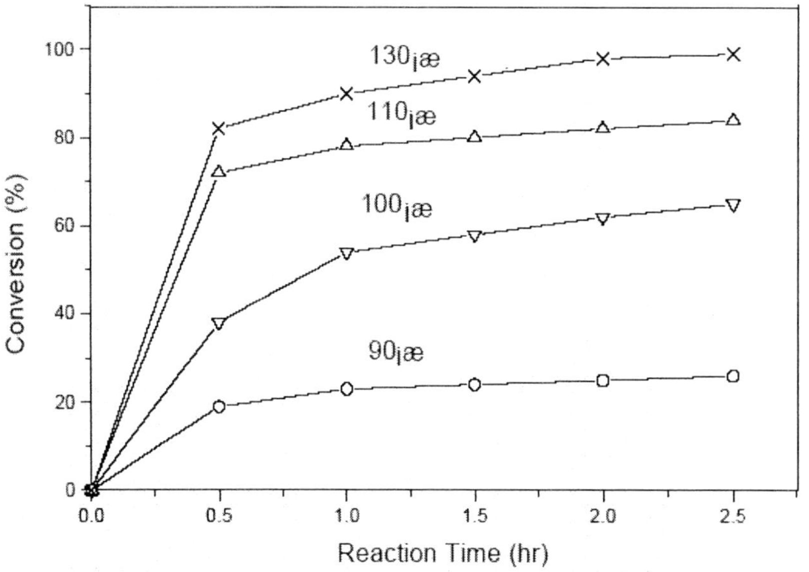

Figure 2. The effect of reaction temperature on hydrosilylation reaction of AGE with H-PDMS

Figure 3 shows the different yields of PSEA at different reaction temperatures. The reaction rate and the yield are greatly affected by temperature. The yield at 110 °C increases more than 10 % than that of at 100 °C. It is proved that improving temperature has positive influence on reaction rate. However, exorbitant temperature will result in gelation because of the reaction between double bonds, such as the reaction at 130 °C.

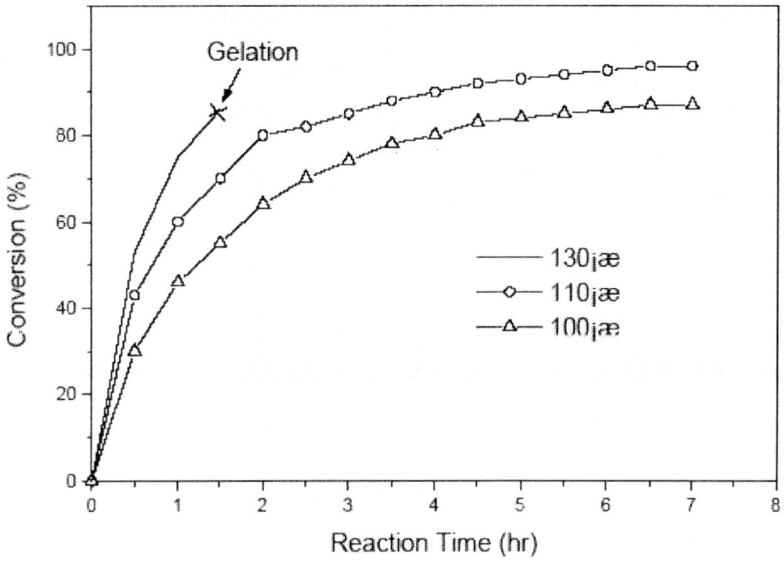

Figure 3. The effect of reaction temperatures on epoxy conversion in hydrosilylation reaction of PSE with AA

2.2.3. Effects of the Concentration of Reactants

With the reaction going on, PSE and AA's concentration will not equal to each other in the end point. It can be found that acid value conversion curve and epoxy value conversion curve do not coincide with each other. Furthermore, the difference increases with the reaction time. At last, rudimental epoxy value is higher than the acid value. This phenomenon can be explained that the acrylic groups react with the -OH groups produced in the reaction of epoxy and acrylic groups. At the beginning of those two competitive reactions, epoxy open loop reaction is chief. With reaction proceeding, hydroxyl group increases sharply and esterification is obvious. Therefor, acid value decreases more quickly than epoxy value. In order to complete epoxy open loop reaction, acrylic acid must be excessive. The optimistical ratio is 1:1.08 (PSE:AA).

2.3. Properties of PSEA

2.3.1. The Molecular Weight Distribution Of PSEA

The molecular weight distribution of PESA is inspected by GPC as shown in Figure 4. It can be found that the molecular weight distribution of PSEA is narrow and the coefficient of dispersion (M_w/M_n) is 1.3, the number-average molecular of PSEA is around 45,000 when the PSE preparation reaction occurs at 130 °C. However, the shoulder peak appears when the temperature of PSE preparation reaction is higher than 130 °C, which is due to homopolymerization of AGE in hydrosilylation process. This is very harmful to the hydrosilylation and the curing speed of the end product although the amount is very small.

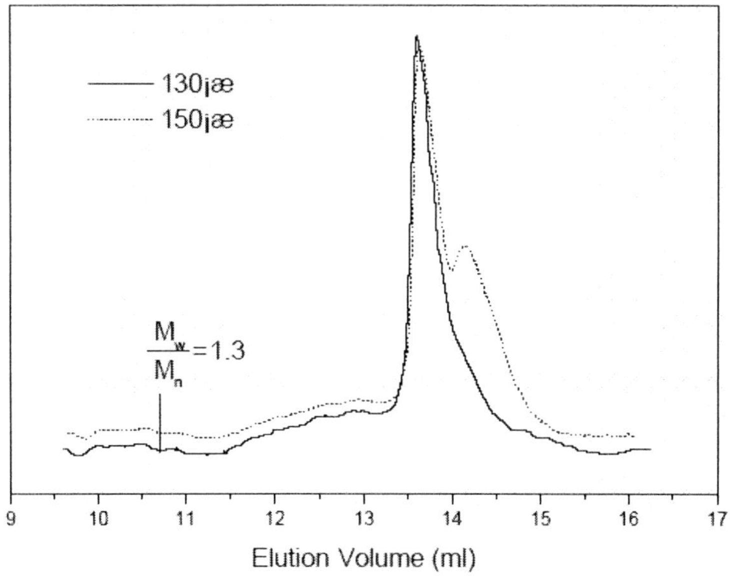

Figure 4. GPC of PSEA. (THF, 1.0 ml min^{-1}, 308K)

2.3.2. UV-spectrum of PSEA

In UV-spectrum (Figure 5), PSEA has one absorption peak at 208 nm, which attributes to the unsaturated carbon double bond. This kind of structure will not shield the process that BDK absorbs UV-light to form free radical. Therefor, PSEA has high photopolymerization rate in the presence of BDK as a photoinitiator in PSEA curing system. When blending BDK into PSEA, it disappears completely soon, which is due to its good compatibility with PSEA, and sensitive to UV-light as well.

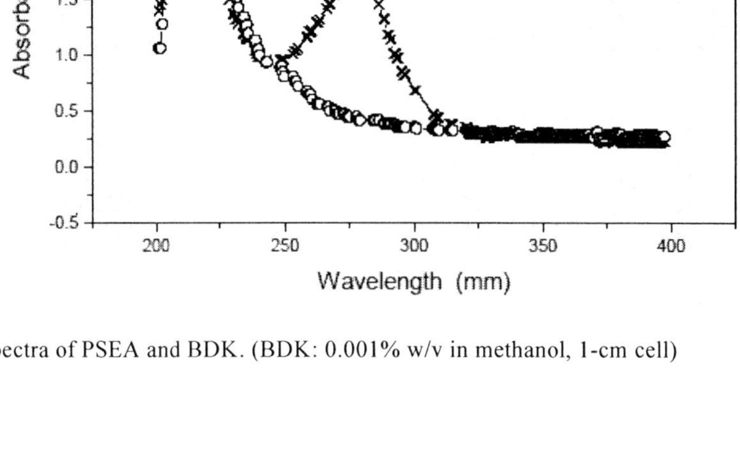

Figure 5. UV spectra of PSEA and BDK. (BDK: 0.001% w/v in methanol, 1-cm cell)

2.3.3. Photocuring Process Was Inspected by the FT-IR Spectrum

Figure 6 is FT-IR spectra of PSEA, curve *a* is cured film and curve *b* is non-cured liquid film. We see clearly from curve *b* that $SiCH_3$ (800cm^{-1}, 1260 cm^{-1}), SiO (1080 cm^{-1}) are characteristic peaks of silicone. At 2140 cm^{-1}, SiH has no peak, which shows the hydrogen of silicon is completely added by AGE. The peak of 910 cm^{-1} is not existence, which is the absorption of epoxy groups formed by AGE and SiH. In the curve *a* and *b*, the common characteristic peaks of SiH_3(800 cm^{-1}, 1260 cm^{-1}), SiO(1100 cm^{-1}) and COO(1272 cm^{-1}) are obvious. However, the differences are samely obvious. In curve *b*, 1625 cm^{-1}、 1610 cm^{-1} and 930 cm^{-1} are peaks of unsaturated double bond, which show the content of acryloyl oxygen groups (CH_2=CHCOO—) are large amounts. After curing, we can find from the curve *a* that those peaks disappear completely. It due to curing is almost absoluteness. The double bond is nonexistent. If we calculated the conversion of carbon double bond by absorption, the percent is 87%. This result is accorded to the value which is gained by chemical method. This proved the preparation and structure of PSEA is right.

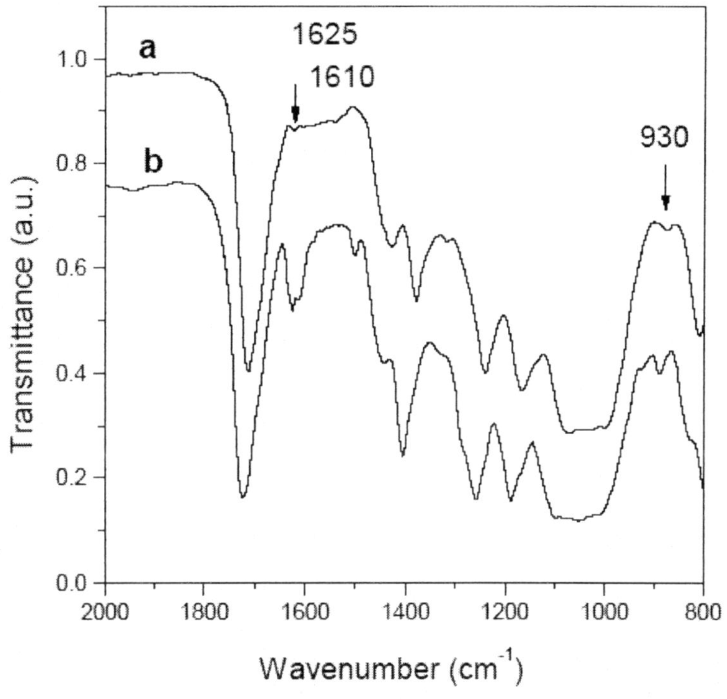

Figure 6. FT-IR spectra of PSEA, before curing (b) and cured (a)

2.3.4. Thermal Behaviors Analyzed by TGA and DMA

PSEA's thermolysis is shown in Figure 7. The curve indicates that PSEA has the similar thermal stability to that of silicone rubber, decomposition temperature of PSEA is about 300 °C. This is the result of the somewhat unusual molecular structure of the polymer chains. The (Si-O) bond is one of the most thermally stable bonds formed by silicon and has a bond dissociation energy of 462 kJ/mol compared with 359 kJ/mol for (C-O), 347 kJ/mol for (C-C) and 319 kJ/mol for (Si-C) bond. The thermal stability of siloxanes is thus partly a consequence of the high bond strength. When other organic radicals, such as hydrogen replace methyl groups on silicon atoms, the properties described above are usually affected to a certain degree. The extent of the change in the properties is a function of the level, type and nature of the substituent.

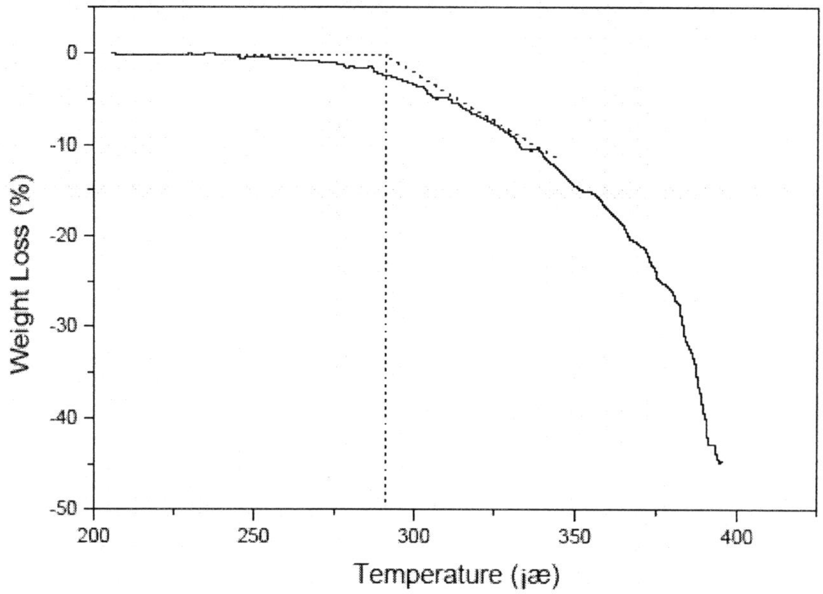

Figure 7. TGA of PSEA film

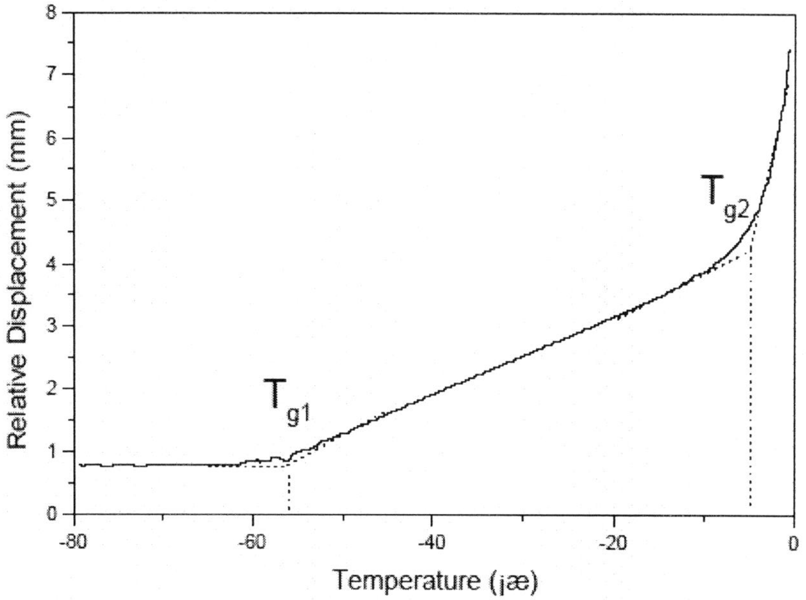

Figure 8. Relative displacement Vs temperature of PSEA cured thin film

Figure 8 is the thermal mechanical curve of the PSEA cured film, it is clearly that there are two distinct glass transition temperatures: T_{g1}= -57°C, T_{g2}= -5°C, which declare PSEA has two-phase segregate microstructures. The lower one probably is the characteristics of the siloxane component, and the higher one may be attributed to the crosslinking site which results mainly from acrylates.

2.3.5. The Property of PSEA Film

The properties of PSEA and the UV cured film have been measured as those listed in Table 1 and Table 2. In Table 1, the viscosity of PSEA is 2~3 Pa·S at 25 °C. Its appropriate viscosity makes this product suitable for all automatic and manual application without diluent and heater, comparing with the commercial polyether urethane acrylate or polybutadiene urethane acrylate optical fiber coating. Because the optical fiber base is poorly wettable to obtain an adequate thickness when the viscosity of coating is too low. On the other hand, if the viscosity is too high, disadvantages are caused by the moving of air bubbles from the coating layer and the difficulties in obtaining an adequate thickness of the coating layer.

After curing, PSEA film owns flexibility and toughness which was proved by the data of hardness, tensile strength, elongation and tensile modulus. The excellent of ageing and weather resistance are also properties of PSEA film. The thermal weight loss of PSEA is smaller than that of other materials, such as polyether urethane acrylate coatings with thermal weight loss 7~10% after 40 min at 200 °C. It is attributed to the stability of Si-O bond of polydimethylsiloxane. The low T_g of PSEA (-57 °C) decides it can be used in atrocious weather where common optical fiber coating is useless, especially in the winter of north countries. Naked optical fiber is sensitive to water, trace water can cause the defection on fiber or destroy the transmission of light. So, the resistance of water is critical in application, such as marine cables. Comparing with the water absorption (4~6%, 24hrs) of polyether urethane acrylate coating, PSEA cured film has low water absorption (2%, 24hrs). The electrical data in Table 1 show PSEA film has excellent electrical performance.

Table 1. properties of PSEA

Liquid Coating (25℃)	Typical Properties
Density (kg/m^3 × 10^3)	1.1131
Viscosity (Pa.s)	2~3
Refractive Index (25℃)	1.4698
Molecular (Mn)	4,5000
Cured Coating (25℃)	Typical Properties
Density (kg/m^3 × 10^3)	1.2129
Tensile Strength (MPa)	6.9
Elongation(%)	20
Tensile Modulus (MPa)	3.4
Refractive Index(25℃)	1.4918
Weight Loss (40min. 300 °C)	5%
Tg1 (°C)	-57
Tg2 (°C)	-5
Hardness(A)	18
Water Absorption (%, 24hr)	2
Electric Broken (KV/mm)	50.5
Dielectric Constant	6.76
Dielectric Loss (50Hz)	0.006048

In Table 2, PSEA cured film is immerged in different solvents to test the solvent resistance. The results show alkaline liquor completely decomposes the film and the coating can not protect optical fiber. In other solvents listed in Table 2, PSEA film is perfect but the weight changes are different. In organic solvents, PSEA film absorbs solvent and the weights increase the in order of acetone, THF, xylene, DMF, toluene and ethanol. However, in inorganic solvents (20% aqueous solution), the weights decrease in order of sulfuric acid, hydrochloric and sodium carbonate.

Table 2. Resistance to solvents of PSEA film

Solvent	weight change(%)	appearance
toluene	9.0	good
xylene	6.9	good
acetone	3.6	good
ethanol	11	good
DMF	7.9	good
THF	6.8	good
20%H_2SO_4	-0.02	good
20%HCl	-0.7	good
20%HNO_3	-7.0	good
20%Na_2CO_3	-10	good
20%NaOH	/	broken
20%KOH	/	broken

As discussed above, PSEA not only maintains the properties of polydimethylsiloxane, but also honors fast photocured speed, high tensile modulus, low elongation, low T_g, heat and solvents resistance, and excellent electrical performance. Furthermore, it compatible with others material very well, which does not owned by general polydimethylsiloxane. So, PSEA can be used as a basic rein for optical fiber coating.

3. Photostabilization of UV-Curable PSEA

Figure 9 shows the change of carbonyl groups of photocured PESA without and with three HALS stabilizers in PAS-FTIR spectra after photoaging for different times. The appearance of the peak at 1725 cm^{-1} after irradiation is mainly attributed to –C=O groups as oxidation products in blank sample. Its absorption intensity increases rapidly with increasing irradiation time during first 50h. Then, the absorption intensity of carbonyl groups decrease after irradiated more than 50h. After three stabilizer are added into PSEA resin, they greatly decrease the rate of formation of carbonyl groups with an initial sharp at 50h and then a gradual decrease during prolonged photoaging time.

Figure 10 shows the change of carboxyl groups of photocured PESA without and with three HALS stabilizers in PAS-FTIR spectra after photoaging for different times. The obvious change is the peak at 3224 cm^{-1}, it due to –OH stretching absorption in carboxylic groups. The carboxyl index increase quickly in first 300h and keep steadily in prolonged time. Figure 11 shows the change of hydroxyl groups of photocured PESA without and with three HALS stabilizers in PAS-FTIR spectra after photoaging for different times. The main change is at

3440 cm^{-1} due to –OH stretching in hydroperoxide and hydroxyl groups. The change trends are similar to Figure 11.

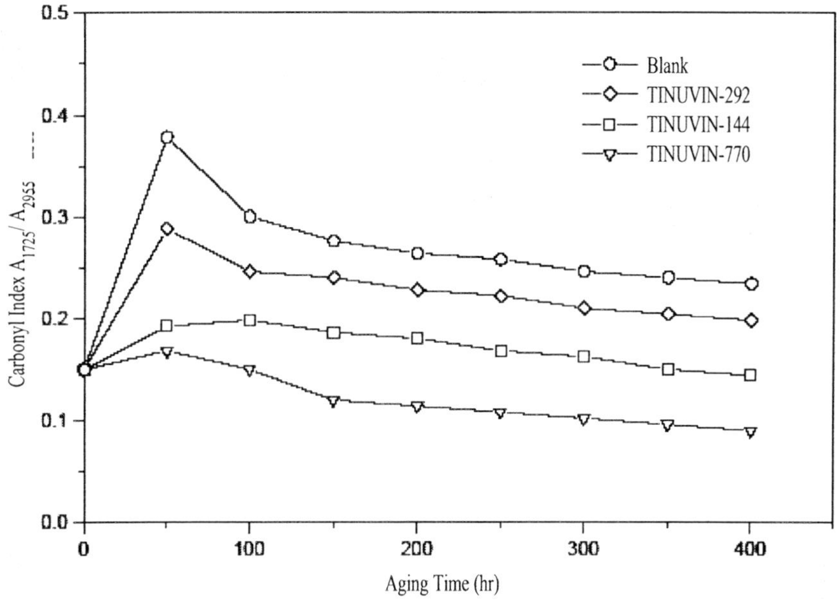

Figure 9. The change of carbonyl groups of UV-cured PESA during photo-aging in air

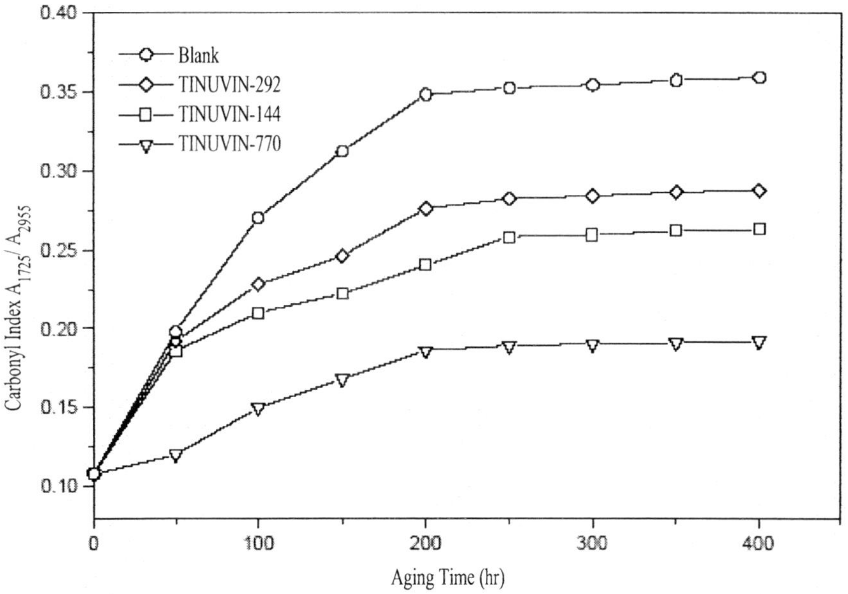

Figure 10. The change of carboxyl groups of UV-cured PESA during photo-aging in air

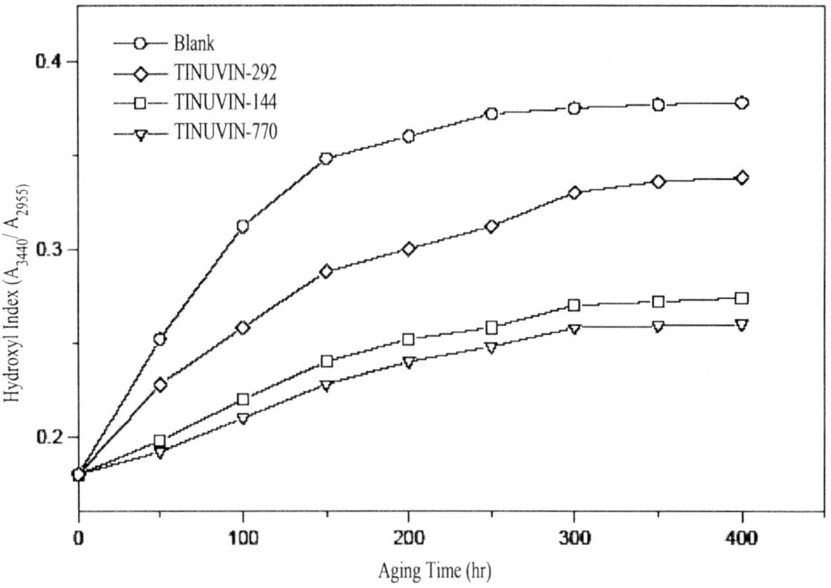

Figure 11. The change of hydroxyl groups of UV-cured PESA during photo-aging in air

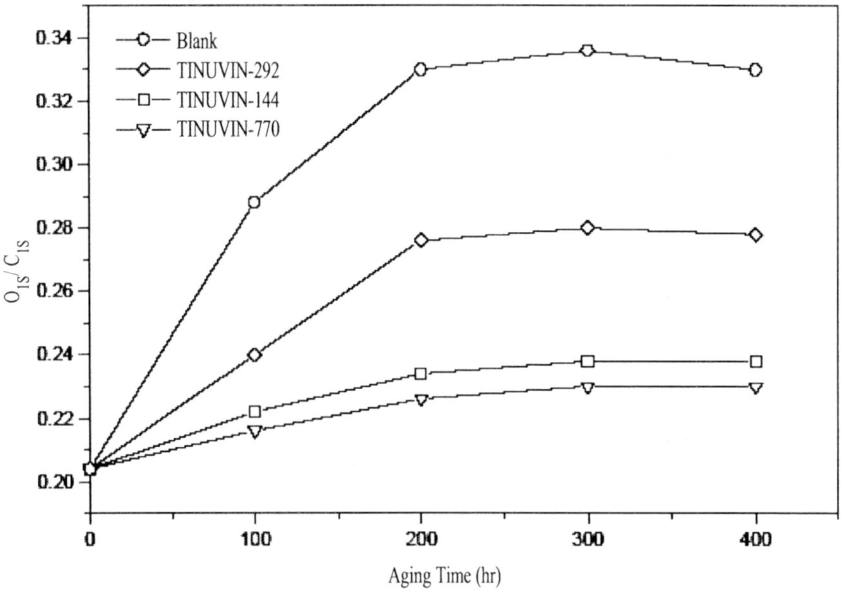

Figure 12. Change of O_{1S}/C_{1S} ratio in XPS spectra of UV-cured PESA during photo-aging in air

With TINUVIN-770 added, the carbonyl index decrease slowly as a result of photoaging. This is interpreted to mean that only a few –CO– groups are formed and some disappear as CO and CO_2 are split off. The formation of carboxyl and hydroxyl groups during photoaging is considerably decreased by the addition of the three stabilizers (Figure 10 and Figure 11).

The stabilizer TINUVIN-770 is the most efficent also in this case. Prolonged photoaging gives no further increase in carboxyl groups and only a small increase in hydroxyl groups.

The XPS measurements(Figure 12) refer to the structure of thin surface layers (5-10 nm), which is a much lower penetration depth into a polymer surface than the FTIR-PAS measurements .carbons in –CO–, –C=O and –OC=O groups contribute to the XPS peaks at 288.0, 289.2 and 291.1 eV are assigned to oxygens in –C=O, –CO– and –OC=O groups. The relative intensity of oxygen to carbon peak is a measure of surface oxidation. Considering the overall reaction process, the O_{1S}/C_{1S} peak intensity ratios were plotted in Figure 12 as a function of aging time. The effect of added stabilizers is a pronounced decrease in oxidation rate. After prolonged photoaging time, the oxidation reaches a plateau and shows no further increase after 200 h.

These results are reasonably well interpreted and understood in terms of a free radical mechanism based on spectroscopic analysis. During UV irradiation, the surface photo-oxidation of photocured PSEA aggravates with increasing irradiation time. Hydroperoxides are formed firstly, then decomposed to •OH and RO• radicals. Furthermore, the R• radicals are produced in the process, which can easily react with oxygen absorbed on PSEA surface to produce the polymeric peroxy (ROO•) radicals. These ROO• radicals react further with PSEA to form hydroperoxide and hydroxyl compounds. The chain scission of PSEA exposed to UV-irradiation will cause splitting off of CO and CO_2 molecules. The above reactions form a recycle process as shown in Scheme 1 (A). Hydroperoxides are the most important intermediates in the surface photo-oxidation process of PSEA. Under UV irradiation, hydroperoxides decompose quickly to form various oxidation products, such as ketone, carboxylic, lactone, and aldehydes, etc, as shown in Scheme 1(B).

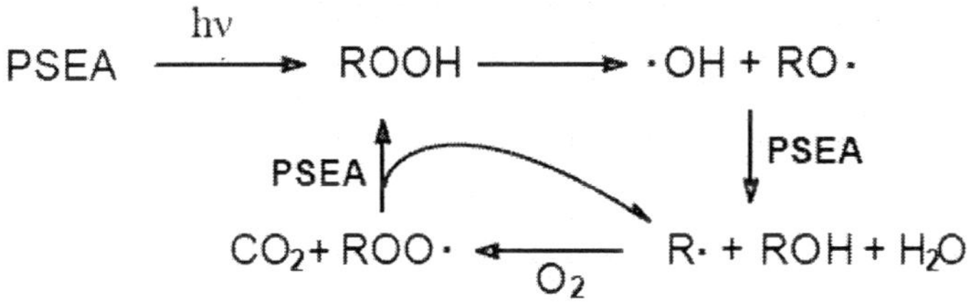

(A)

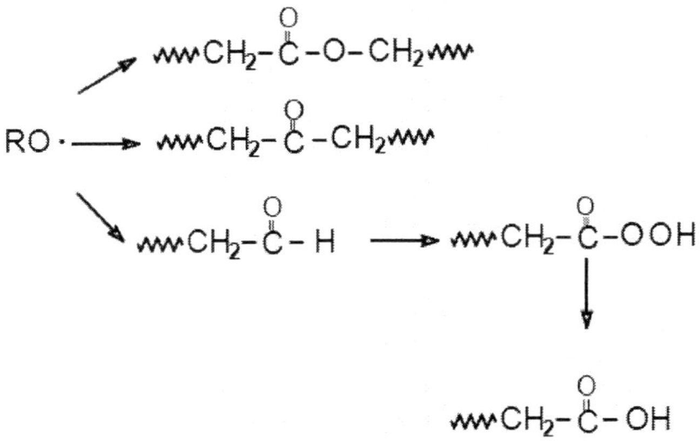

(B)

Scheme 2

The initial increase of the carbonyl absorption (Figure 9) can be interpreted as being due to the rapid formation of –C=O groups from the photo-oxidation of the PSEA. However, this increase is partly balanced by evolution of CO and CO_2 associated with chain scission, which leads to a net decrease of total carbonyl group content at a later stage of the photoreaction.

Photo-oxidation is accompanied by an increase of hydroxyl group content (Figure 11) which is largely attributed to -OH groups from hydroxyl radical substitution in the hydroperoxide group formation. Th increase of carboxyl group content (Figure 10) is interpreted as being due to the formation of olefinic and carboxyl groups via cyclic transition state.

After prolonged UV-irradiation, a leveling off of the O_{1S}/C_{1S} XPS peak intensity ratio is observed for samples without and with stabilizer. This indicates that a steady state situation is reached after 200 hours of irradiation with the evolution of oxidation products balancing the ratio of oxidation of the polymer surface. In the deeper layers of the photocured PSEA films, oxidized groups continue to increase as shown in Figure 10 and Figure 11 . XPS spectra refer to a penetration depth of only 5-10 nm while FTIR-PAS penetrates about 100 times deeper.

The photostabilizing mechanism of HALS is considered to be concerned with nitroxide radicals (⟩NO·), generated by oxidation of the parent amine (⟩NH·), acting as efficient radical traps, which may scavenge chain- or chain-end-radicals (P·) formed in the photo-oxidative process of photocured PSEA to produce substituted hydroxylamine (⟩NOP) groups. Further reactions with peroxy radicals regenerate the nitroxide groups. The high basicity of HALS molecules enhances the association of stabilizer with hydroperoxides, which leads to a strong local concentration of stabilizing species in the most vulnerable oxidized regions rich in –OOH groups. This process prevents the destructive hydrogen-abstraction from the substrate by trapping the PO· radicals formed by decomposition of hydroperoxide groups [7].

All the above results have shown that TINUVIN-770 is the most efficient stabilizer, which is due to the fast conversion rate of the amine (⟩NH) to nitroxide radicals (⟩NO·). These results are consistent with those obtained by Bauer et al [8]. For TINUVIN-292 which is the least effective stabilizer against photodegradation of UV-cured PSEA film. The TINUVIN-144 molecule is a combination of two hindered amine groups as photostabilizer and one hindered phenol group as antioxidant. The interaction between hindered phenol and hindered piperidine groups for stabilizing polyolefin films has been studied extensively. Both synergism and antagonism have been reported, depending on the polyolefin/stabilizer system under study. No mechanisms of chemical interaction have been proposed in a HALS/phenol system from a product study. In the case of the UV-cured PSEA resin film, it seems that the synergistic effect dominates, which results in high inhibiting efficiency with TINUVIN-144 than with TINUVIN-292 for stabilization against photo-oxidation of the UV-cured PSEA film. The detailed mechanism needs to be studied further [9].

4. Formulations Based on UV-Curable PSEA for Optical Fiber Coating

4.1. Optimizing Formulation of Coatings

Table 3 is a series of prepolymer mixtures that are designed and formulated according to orthogonal table to optimize the formulation for optical fiber coating. Its analysis data are list in table 4.

Table 3. Optimizing the formulation for three prepolymers [a]

Number	Component ratio (wt %) [b]			Tensile Strength	Elongation	Modulus	Gel content [c]
	PGUA	PSEA	TMPTA	(MPa)	(%)	(MPa)	(%)
1	8 (1)	3 (1)	2 (1)	8.0	132	24.3	31.6
2	9 (2)	3 (1)	3 (2)	10.9	156	27.6	41.4
3	10 (3)	3 (1)	4 (3)	9.6	147	26.2	43.1
4	8 (1)	4 (2)	4 (3)	18.1	139	53.2	41.4
5	9 (2)	4 (2)	2 (1)	6.7	110	24.3	51.3
6	10 (3)	4 (2)	3 (2)	10.9	156	23.1	55.8
7	8 (1)	5 (3)	2 (1)	14.1	100	56.5	52.8
8	9 (2)	5 (3)	4 (3)	17.4	86	80.8	42.7
9	10 (3)	5 (3)	3 (2)	6.4	84	30.9	58.9

a. Every mixture coating is prepared according to the method mentioned. Cast films are irradiated for 30 sec.

b. The weight ratios of components are outside the parentheses. The test levels are the numbers inside parentheses.

c. Films are irradiated by UV-lamp for two seconds.

Table 4. Data analysis for Table3

Properties	Levels	PGUA	PSEA	TMPTA
	1	40.3	28.5	21.1
Tensile Strength	2	32.3	35.7	35.9
(MPa)	3	26.9	35.3	26.8
	Δ	-15.4	6.8	21.4
Elongation	1	371	435	326
(%)	2	350	405	412
	3	387	268	370
	Δ	37	-167	86
	1	133.0	78.1	79.5
Modulus	2	132.7	99.6	107.2
(MPa)	3	80.2	168.2	159.2
	Δ	-52.8	90.1	79.7
	1	125.8	115.8	141.6
Gel content	2	135.4	148.5	150.0
(%)	3	157.6	154.2	127.2
	Δ	31.8	38.4	-29.2

The data show: (1) increasing the content of PGUA increases the elongation and cure rate, but decreases the tensile strength and modulus of film; (2) increasing the content of

PSEA increases the tensile strength, modulus and cure rate but decreases the elongation of film. PGUA and PSEA possess the contrary effect on the properties of cured film except for the high cure speed, which provide an opportunity to blend them together to obtain different coating with desired properties. (3) TMPTA can improve the tensile strength, elongation and modulus except for the cured speed.

After above comparison, resin mixture NO. 6 optical fiber coating (OFC-6) was selected as a candidate for a single coating for optical fiber and was studied in later section. Such coating make optical fiber manufacture easier and less expensive than processes involving a primary and a secondary coating. Moreover, mixture NO. 1 also could be selected as the candidate for a primary buffer coating.

4.2. Photoinitiators of Coating

The choice of photoinitiators is generally governed by: (1) absorb the irradiation wavelengths and exhibit the highest extinction coefficients in order to use low amounts of product; (2) the requirement for a high curing speed; (3) in the case where a long-term stability of the material is needed; moreover, one has also to consider the influence of the photolysis products of the initiators upon the photochemical time evolution of the polymerized resin.

Following with above principals, we choose BDK/BP mixture photoinitiators to initiate the polymerization of coating. The combination of photoinitiators exhibit a synergic effect, the quantum yield of polymerization is higher than the sum of the yields corresponding to the two modes of initiation. BDK generates the benzoyl and methyl free radicals through photoscission process, these free radicals are more efficient initiators in OFC-6 than BP because they are reactive enough to initiate polymerization of vinyl groups. Moreover, BDK has a broad absorption peak at 315 nm to 390 nm, which is in the same range as the UV spectrum of the HPM-15 lamp, as shown in Figure 13. However, BDK involving make the stability of coating storage deterioration or gelation through "dark reaction". This problem can be solved by mixing appropriate content BP/triethylamine, hydroxyanisol and additives into coating. Figure 13 is the UV spectra of the photoinitiators in chloroform (10^{-3} M), and CFO-6 without photoinitiators and emission spectrum from Philips HPM-15 lamp. It shows OFC-6 has absorption band at 210 nm, which can not shield BDK or BP to form free radical, and the photoinitiators absorb completely the UV light at 350~370 nm of HPM-15 for their appropriate chemical structure.

4.3. Rheological Behavior of Coating

The viscosity of oligomers is an important parameter in UV-curing technology that affects the cure process and properties of the cured films. When the viscosity of coating is too low, the quartz glass made optical fiber base is poorly wettable with the difficulty in obtaining an adequate thickness layer. On the other hand, when the viscosity is too high, disadvantages are caused in the removal of air bubbled from the coating layer as well as in respect of the difficulties in obtaining an adequate thickness of the coating layer and decreased workability. The viscosity of a polymer is related to the dynamic extension of the molecules and the segment density within the volume of a molecule and intermolecular chain entanglement. Figure 14 is the effect of temperature on the viscosity of optical fiber coating OFC-6, 950-075 and 950-042. It shows OFC-6 has low viscosity because the PSEA resin maintains the properties of polydimethylsiloxane, comparing with 950-705 and 950-042 coating.

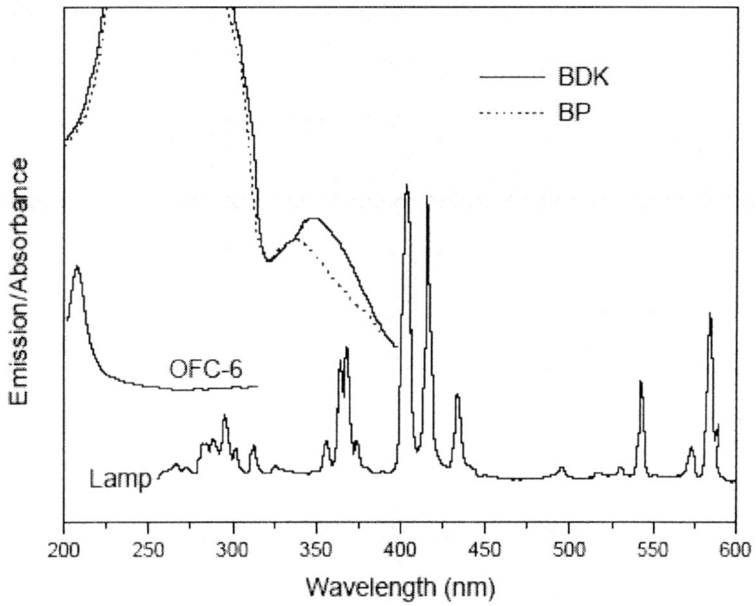

Figure 13. UV spectra of the photoinitiators in chloroform (10^{-3} M), and CFO-6 without photoinitiators and emission spectrum from Philips HPM-15 lamp

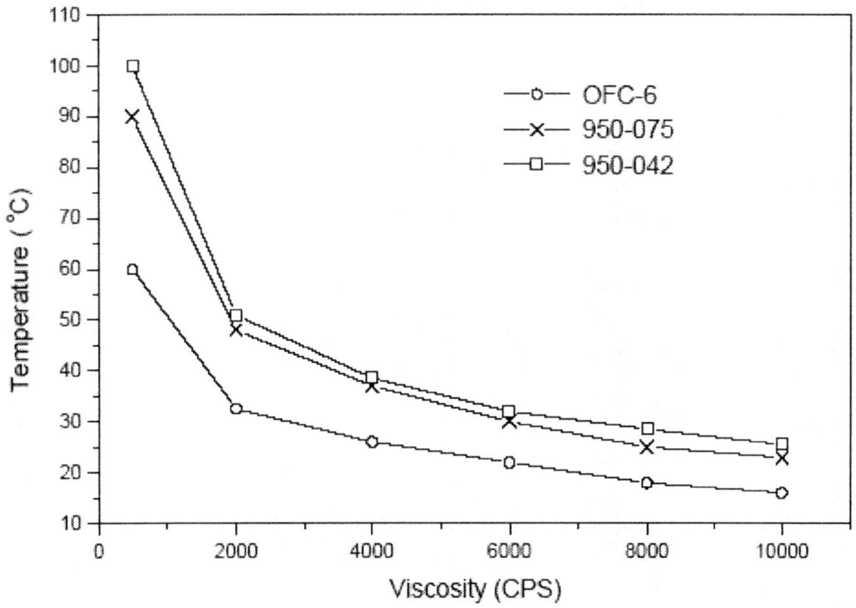

Figure 14. The effect of temperature on the viscosity of optical fiber coatings

The viscosity of OFC-6 changes sharply from 10,000 cps at 15 °C to 2,000 cps at 32 °C, the difference is five times. The viscosity changes of 950-075 and 950-042 are similar with small difference in degree as shown in Figure 14, which go down from 10, 000 cps at 28 °C to almost 2,000 cps at 50 °C. The changes are also obvious but the scope of temperature is higher than that of OFC-6, which may induce some difficulties at low temperature operation.

4.4. Photocuring Process Was Inspected by the FT-IR Spectrum

Figure 15 is FT-IR spectra of OFC-6 optical fiber coating before irradiation and irradiated for 5 s. We see clearly that $SiCH_3$ (800cm^{-1}, 1260 cm^{-1}), SiO (1090 cm^{-1}) are characteristic peaks of silicone. The peaks of NH (3310 cm^{-1}), CH_2 (2930 cm^{-1}), Ph (1600 cm^{-1}), and C=O (1720 cm^{-1}) are characteristic peaks of PGUA. Most important are the peaks at 1632 cm^{-1} and 1614 cm^{-1}, which are characteristic peaks of unsaturated double bond. This shows that the content of acryloyl oxygen groups (CH_2=CHCOO—) are large amounts. After UV irradiation for 5 s, it can be found that those peaks disappear completely. It due to curing is almost absoluteness. The double bond is nonexistent. If we calculated the conversion of carbon double bond by absorption, the percent is 92%. This result is higher than the value which is gained by solvent extraction method as show in Figure 16. This proved the results obtained from FTIR are believable and time saving.

Figure 16 is the gel content of OFC-6 cured films with different irradiation times. It shows the gel contents obtained from FTIR are higher than that from solvent extraction. OFC coating film touch free after irradiation for 2 s. Extending irradiation time more than 5 s, the gel content change is less. The results are confirmed by the application of OFC coating in spreader (HEATH WAY, UK), the coating speed are 120 m in one minute.

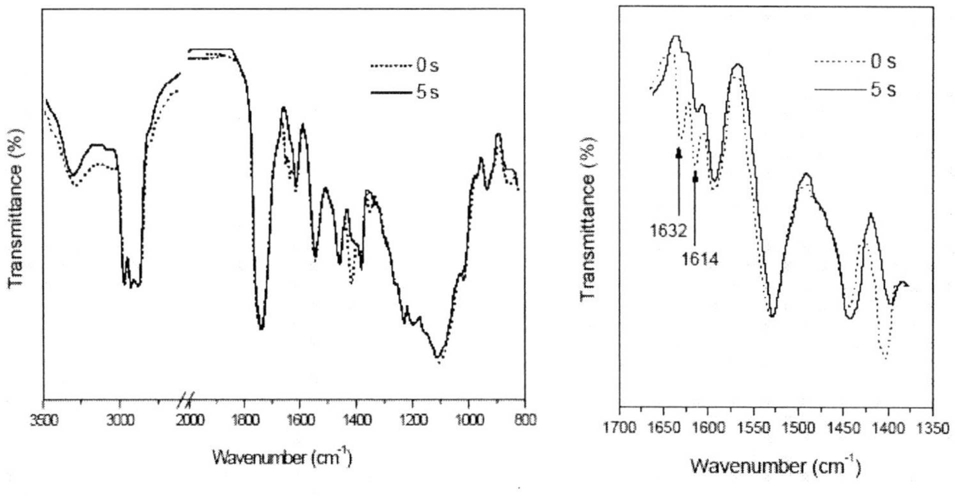

Figure 15. FTIR spectra of OFC-6 optical fiber coating before UV irradiation and irradiated for 5 s

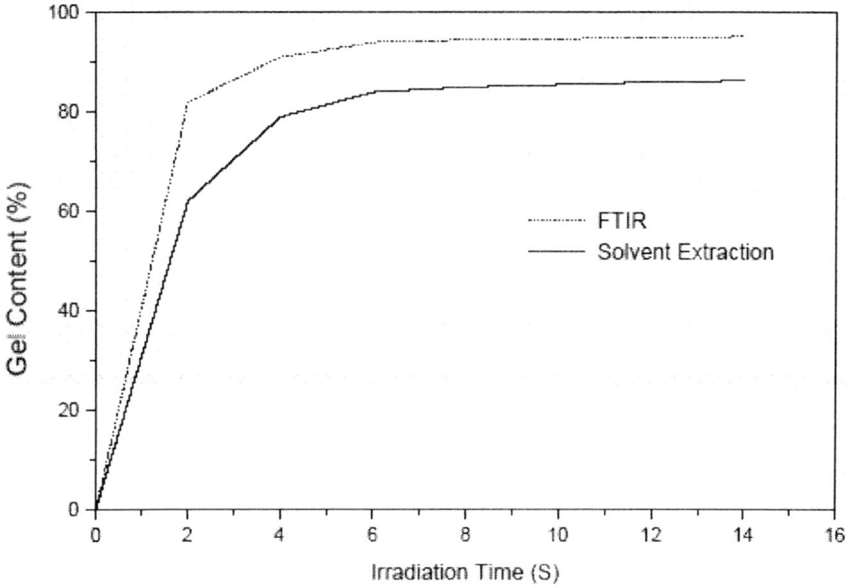

Figure 16. The gel content of OFC-6 cured films with different irradiation times

4.5. Thermal Behaviors Analyzed by TGA and DMA

OFC-6's thermolysis is shown in Figure 17. The curve shows OFC-6 has the similar thermal stability to silicone rubber, decomposition temperature is about 280 °C. This is the result of the somewhat unusual molecular structure of the polymer chains, the (Si-O) bond is one of the most thermally stable bonds formed by silicon and has a bond dissociation energy of 462 kJ/mol compared with 359 kJ/mol for (C-O), 347 kJ/mol for (C-C) and 319 kJ/mol for (Si-C) bond. The thermal stability of OFC-6 is thus partly a consequence of the high bond strength of PSEA. When the proportion of coating varying, such as increasing the content of PGUA and TMPTA, the thermal properties described above usually decreases in some degree.

Most polymeric materials can be considered to be viscoelastic solids that exhibit both viscous and elastic properties. DMA is an efficient tool to measure the viscoelastic properties and obtain information about the microstructure of crosslinked networks. The storage modulus (E′) and the loss factor (tan δ) for OFC-6 cured film as a function of temperature are shown in Figure 18. The change of loss factor shows that there are two distinct glass transition temperatures: T_{g1}= -80 °C, T_{g2}= +20 °C, which declare PSEA has two-phase segregate microstructures. The lower one probably is the characteristics of the siloxane component, and the higher one may be attributed to the crosslinking site which resulted mainly from acrylates.

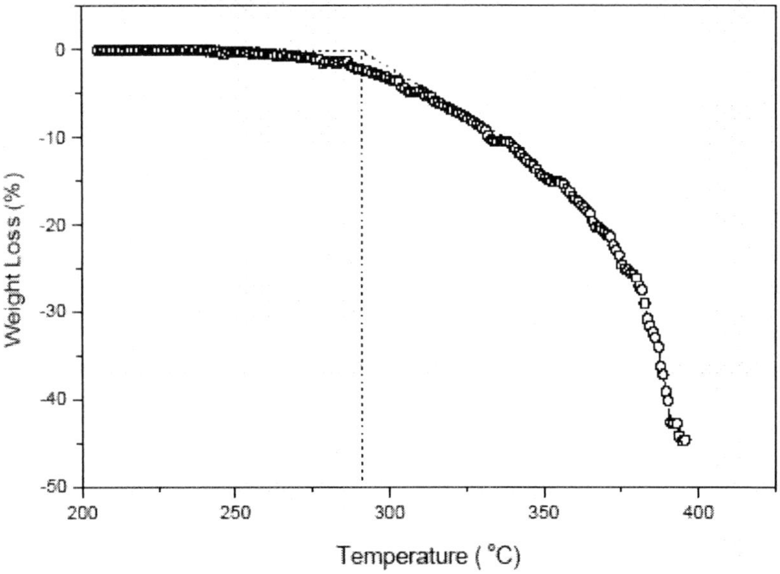

Figure 17. TGA thermogram of OFC-6 cured film at a heating rate of 10 K min^{-1}

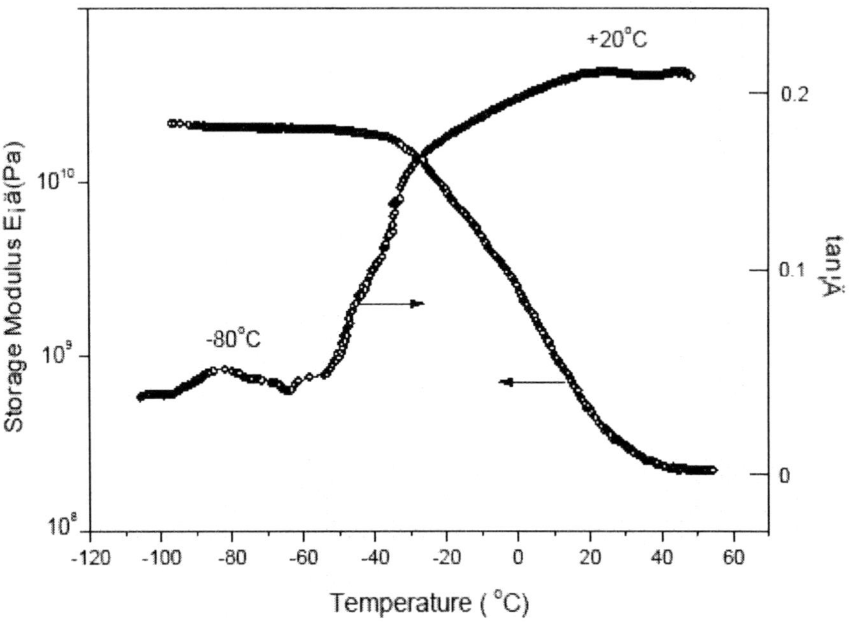

Figure 18. Dynamic viscoelastic behavior as a function of temperature for OFC-6 cured film.

5. Conclusion

In this study, photosensitive resins are successfully synthesized for the formulation of optical fiber coating. The conclusions are summarized as follows:

1. Hydrosilylation reaction of AGE with H-PDMS at 130°C gives a nearly 100% yield. The optimistical conditions of acrylic acid modified PSE to prepare PSEA are that the ratio of PSE and AA is 1:1.08, the concentration of H_2PtCL_6 catalyst is 0.8 wt% of reactants, and the reaction temperature is 110°C. The Mn of PSEA is 45,000, and the coefficient of dispersion (M_w/M_n)is 1.3. The cured film has two T_g (T_{g1} = -57°C, T_{g2} = -5°C) ; the decomposition temperature is 280°C; the conversion percentage of double bond is 87% in cured PSEA film, and the cured film has good solvent resistance.
2. PSEA are cured without and with HALSs to form thin films. UV-irradiation in a UV-CON irradiator causes a surface oxidation of the UV-cured thin film, measured by FTIR-PAS and XPS spectra up to a certain level. The reaction proceeds as a free radical mechanism as previously shown. The addition of hindered amine photostabilizers largely decrease the rate of surface oxidation of PSEA. TINUVIN-770 is the most efficient HALS in this study.
3. When the proportion was PGUA: PSEA:TMPTA = 10:4:3, it was used for a single coating. The proportion PGUA: PSEA:TMPTA = 8:3:2 was used for a primary buffer coating. BDK and BP are reactive photoinitiators for OFC-6, this coating has low viscosity and high curing speed. FTIR and gel content measurement show OFC-6 film is completely cured after irradiated 5 s. The excellent high and low temperature behaviors of the cured film (T_{g1} = -80 °C; the decomposition temperature is 280 °C) are the advantages of OFC-6.

References

[1] Iojoiu, C; Abadie, JM; Harabagiu, V; et al. Synthesis and photocrosslinking of benzyl acrylate substituted polydimethylsiloxanes. *Eur Polym J* 2000, 36, 2115-2123.

[2] Pinteala, M; Harabagiu, V; Cotzur, C; et al. Functional polysiloxanes .1. polysiloxanes containing chloromethylphenethyl groups. *Eur Polym J* 1994**,** 30 (3), 309-312.

[3] Iskender, Y. Advances In Polymer Science. Berlin Heidelberg: Spring-Verlag Press; 1988.

[4] Suzuki, T. Composition for coating optical communication glass fiber. US 4755577, 1988.

[5] Lee, C; Lutz, M. Adhesion promoter for UV curable siloxane compositions and compositions containing same. US5188864,1993.

[6] Feng, S; Cui, M. Study of polysiloxanes containing epoxy groups: I. Synthesis and characterization of polysiloxanes containing 3-(2,3-epoxypropoxy)propyl groups. *React and Funct Polym* 2000, 45, 79-83.

[7] Vyprachticky, D; Pospisil, j; Sedlar, L. Possibilities for cooperation in stabilizer systems containing a hindered piperidine and a phenolic antioxidant-a review. *Polym Degrad Stabil* 1990, 27, 227-255.
[8] Bauer, DR; Gerlock, JL; Mielewski, DF. Photodegradation and photostabilization in organic coatings containing a hindered amine light stabilizer. 6 ESR measurements of nitroxide kinetics and mchanism of stabilization. *Polym Degrad Stabil* 1990, 28,115-129.
[9] Gijsman, P; Hennekens, JT. The mechanism of action of hindered amine light stabilizers. *Polym Degrad Stabil* 1993, 39, 225-233.

In: Recent Developments in Polymer Research
Editor: Antony V. Hopper, pp. 95-133
ISBN 1-60021-346-4

Chapter 5

SORPTION PHENOMENA OF ORGANIC SOLVENTS ON POLYMER SURFACES

Fatemeh Sabzi** *and Ali Boushehri
Department of Chemistry, College of Sciences, Shiraz University,
Shiraz 71454, Iran

ABSTRACT

Sorption of ordinary fluids on polymer surfaces is of interest in various technologies including packaging and coating technology, membrane separation of gases and vapors, controlled drug release systems, biocompatible materials for biomedical applications and contact lens industry. In this work we use the vapor-sorption equilibrium data to show the degree of solvent upturn in each solvent-polymer system.

For this purpose, seventy-seven isothermal data for copolymer+solvent binaries and for some of their parent homopolymer+solvent binaries have been used in the temperature range of 23.5-80°c.

Solvents studied are acetone, acetonitrile, benzene, 1-butanol, carbon tetrachloride, 1,2-dichloroethane, chloroform, cyclohexane, hexane, methanol, octane, pentane, and toluene.

Copolymers studied are poly(acrylonitrile-co-butadiene), poly(styrene-co-acrylonitrile), poly(styrene-co-butadiene), poly(vinyl acetate-co-ethylene), poly(vinyl acetate-co-vinyl chloride), and poly(n-tetracosane- ethylene oxide -n-tetracosane).All copolymers are random copolymers except one triblock copolymer.

Some homopolymers are also studied: polyacrylonitrile, poly(cis-1,4-butadiene), polyisobutylene, poly(ε-caprolactone), poly(ethylene oxide), n-heptadecane, polystyrene, poly(vinyl acetate), poly(vinyl chloride), poly(vinyl methyl ether), and n-tetracosane.

According to these data sets, solvent weight fraction in the polymer is plotted against solvent vapor activity that is calculated assuming an ideal gas phase of pure solvent vapor neglecting the vapor pressure of the polymer.

For thermodynamic reasons, most polymer pairs are immiscible and their degree of compatibility is of underlying importance to the microphase structure and consequently, to the mechanical properties of the blend. The Flory-Huggins χ interaction parameter for

---

* Author to whom correspondence should be addressed.Fax: ++98-711-2280926; Email: fche1_sabzi@yahoo.com

the polymer pair plays a dominant role in explaining critical phase behavior of a compatible pair and in estimating interfacial tension and interfacial thickness for semicompatible or incompatible pairs. Direct measurement of this parameter is not always possible and we use the Flory-Huggins theory to obtain this dimensionless parameter.

Solvent molecules have a tendency to form clusters when absorbed on polymer surfaces. To provide a measure for clustering, the Zimm and Lundberg clustering function G_{ss}/V_s, that is the ratio of the clustering integral to the partial molecular volume, and the quantity $\varphi_s(G_{ss}/V_s)$ as the mean number of solvent molecules in the neighborhood of a given solvent molecule is used. Finally the non-ideality factor, Γ, can be calculated from the clustering function.

1. Introduction

Sorption is a generalized term to describe the penetration and dispersal of the molecules of an ambient gas, vapor, or liquid onto and throughout a polymeric solid to form a mixture. The sorption process can be described phenomenologically as the distribution of the mobile component between two (or more) phases to include adsorption (that is usually divided into chemisorption and physical adsorption), absorption, incorporation into microcavities, and other modes of mixing.

In the strictest sense, adsorption refers to cases where the sorbate molecules remain on the surface of the solid; the thermodynamic properties of the underlying bulk solid are essentially unaffected by the sorbed material on the surface. Adsorption only on external solid surfaces would be favored for exceptionally large sorbate molecules and for pure, crystalline solids containing relatively few cracks or other defects.

Absorption refers to cases of penetration of the sorbate into the bulk structure of the polymer such that a mixture similar to that of a true solution is created. The thermodynamic properties of the whole polymer are altered as represented by various theories of polymer solution. However, dividing line between absorption and adsorption onto internal or molecular surfaces is not usually clear. The total sorption process in polymeric solids can be considered in terms of different types of absorption and internal adsorption. Insufficient knowledge regarding the true state of the process is the major rationale leading to the use of the nonspecific descriptive term sorption.

The equilibrium amount of penetrant sorbed by a polymer under given conditions is governed principally by the thermodynamics of the system. In general, sorption will occur whenever the partial molar free energy (i.e. the chemical potential) of the penetrant in the external phase exceeds that of the penetrant in the polymer. The first stage of dissolution of a polymer involves sorption of solvent molecules by the dry polymer solid. If a polymer is to be used in a liquid or vapor environment, it is desirable and often essential to have knowledge of the sorption characteristics of the system, since the imbibition of penetrant can induce profound changes in the structure, physical and mechanical properties, morphology, and useful life of the polymer for a given application; for example, the sorption water by a polymer can drastically affect its usefulness as an electrical insulator.

The phenomena of sorption by polymers are manifested in a variety of ways, some of which are very important in technical senses such as packaging and coating technology, membrane separation of gases and vapors, controlled drug release systems, biocompatible materials for biomedical applications and contact lens industry [1-4].

This chapter concerns sorption of some organic solvent vapor on 14 homopolymer and 14 copolymer films in temperature range 23.5-80°c. Tables 1 and 2 give a summary of all polymers and copolymers with their solvents that are considered in this chapter.

Table 1. Summary of Polymers with their Relevant Solvents

Polymer	Mw	V_{sp} (cm^3g^{-1})	Solvent	T (°c)	P_s (KPa)	V_{sp} (cm^3g^{-1})
Poly(acrylonitrile	150,000[a]	0.845[a]	Acetonitrile	60	50.729[c]	1.374[c]
Poly(acrylonitrile	150,000	0.845	1,2-Dichloroethane	70	66.650[d]	0.849[e]
Poly(acrylonitrile)	150,000	0.845	1,2-Dichloroethane	80	93.310[d]	0.861[e]
Poly(acrylonitrile)	150,000	0.845	Hexane	60	76.360[f]	1.608[f]
Poly(acrylonitrile)	150,000	0.845	Pentane	60	214.50[f]	1.709[f]
Poly(cis-1,4-butadiene)	250,000[a]	1.099[a]	Acetonitrile	60	50.729	1.374
Poly(cis-1,4-butadiene)	250.000	1.099	Chloroform	60	106.30[h]	0.709[e]
Poly(cis-1,4-butadiene)	250,000	1.099	Cyclohexane	60	48.059[f]	1.351[f]
Poly(cis-1,4-butadiene)	250,000	1.099	Hexane	60	76.360	1.608
Poly(cis-1,4-butadiene)	250,000	1.099	Pentane	60	214.50	1.709
Polyisobutylene	2700[b]	0.949[b]	Pentane	25	68.333[b]	1.183[b]
Polyisobutylene	1350[b]	0.949[b]	Pentane	25	68.333	1.183
Poly(ε-caprolactone)	33,000[b]	0.769[b]	Carbontetrachloride	65	69.981[b]	0.487[b]
Poly(ethylene oxide)	600[b]	0.753[b]	Benzene	55	43.583[b]	0.895[b]
Poly(ethylene oxide)	100,000[b]	0.753[b]	Benzene	70	73.437[b]	1.199[f]
Poly(ethylene oxide)	5,000,000[a]	0.826[a]	Chloroform	60	106.30	0.709
n-Heptadecane	240[b]	1.063[b]	Pentane	25	68.333	1.183
Polystyrene	800[b]	0.810[b]	Chloroform	25	26.222[b]	0.512[b]
Polystyrene	800	0.810	Benzen	25	12.664[b]	0.886[b]
Polystyrene	110,000[a]	0.952[a]	1,2-Dichloroethane	70	66.650	0.849
Polystyrene	110,000	0.952	1,2-Dichloroethane	80	93.310	0.861
Poly(vinyl acetate)	194,000[a]	0.847[a]	1-Butanol	80	21.804[g]	1.307[g]
Poly(vinyl acetate)	194,000	0.847	Chloroform	60	106.30	0.709
Poly(vinyl acetate)	194,000	0.847	Methanol	80	160.69[f]	1.359[f]
Poly(vinyl chloride)	77,300[b]	0.624[b]	Carbon tetrachloride	65	69.981	0.487
Poly(vinyl methyl ether)	14,000[b]	0.828[b]	Chloroform	25	26.222	0.512
Poly(vinyl methyl ether)	14,000	0.828	Benzene	25	12.664	0.886
n-Tetracosane	338[b]	1.054[b]	Benzene	55	43.583	0.895

[a] These values are taken from [8]
[b] These values are taken from [19].
[c] These values are taken from [20].
[d] This value is taken from [21].
[e] These values are taken from [22].
[f] These values are taken from [23].
[g] These values are taken from [24].
[h] These values are taken from [25].

Table 2. Summary of Copolymers with their Relevant Solvents

Copolymer	V_{sp}[a] (cm^3g^{-1})	Solvent	T (^{o}c)	P_s (KPa)	V_{sp} (cm^3g^{-1})
Poly(21% acrylonitrile-co-butadiene)	1.0309	Acetonitrile	60	50.729[b]	1.374[b]
Poly(33% acrylonitrile-co-butadiene)	1.0204	Acetonitrile	60	50.729	1.374
Poly(51% acrylonitrile-co-butadiene)	1.0000	Acetonitrile	60	50.729	1.374
Poly(33% acrylonitrile-co-butadiene)	1.0204	Chloroform	60	106.30[g]	0.709[e]
Poly(21% acrylonitrile-co-butadiene)	1.0309	Cyclohexane	60	48.059[c]	1.351[c]
Poly(51% acrylonitrile-co-butadiene)	1.0000	Cyclohexane	60	48.059	1.351
Poly(21% acrylonitrile-co-butadiene)	1.0309	Hexane	60	76.360[c]	1.608[c]
Poly(33% acrylonitrile-co-butadiene)	1.0204	Hexane	60	76.360	1.608
Poly(51% acrylonitrile-co-butadiene)	1.0000	Hexane	60	76.360	1.608
Poly(21% acrylonitrile-co-butadiene)	1.0309	Octane	60	10.488[c]	1.494[c]
Poly(51% acrylonitrile-co-butadiene)	1.0000	Octane	60	10.488	1.494
Poly(21% acrylonitrile-co-butadiene)	1.0309	Pentane	60	214.50[c]	1.709[c]
Poly(33% acrylonitrile-co-butadiene)	1.0204	Pentane	60	214.50	1.709
Poly(51% acrylonitrile-co-butadiene)	1.0000	Pentane	60	214.50	1.709
Poly(70% styrene-co-acrylonitrile)	0.9321	1,2-Dichloroethane	70	66.650[d]	0.849[e]
Poly(70% styrene-co-acrylonitrile)	0.9321	1,2-Dichloroethane	80	93.310[d]	0.861[e]
Poly(5% styrene-co-butadiene)	1.0989	Acetone	50	69.180[c]	1.322[c]
Poly(23% styrene-co-butadiene)	1.0695	Acetone	50	69.180	1.322
Poly(23% styrene-co-butadiene)	1.0695	Acetone	60	98.880[c]	1.343[c]
Poly(45% styrene-co-butadiene)	1.0363	Acetone	50	69.180	1.322
Poly(45% styrene-co-butadiene)	1.0363	Acetone	60	98.880	1.343
Poly(5% styrene-co-butadiene)	1.0989	Chloroform	50	71.114[g]	0.699[e]
Poly(23% styrene-co-butadiene)	1.0695	Chloroform	50	71.114	0.699
Poly(45% styrene-co-butadiene)	1.0363	Chloroform	50	71.114	0.699
Poly(5% styrene-co-butadiene)	1.0989	Cyclohexane	60	48.059	1.351
Poly(23% styrene-co-butadiene)	1.0695	Cyclohexane	23.5	10.889[c]	1.289[c]
Poly(23% styrene-co-butadiene)	1.0695	Cyclohexane	60	48.059	1.351
Poly(45% styrene-co-butadiene)	1.0363	Cyclohexane	23.5	10.889	1.289
Poly(45% styrene-co-butadiene)	1.0363	Cyclohexane	60	48.059	1.351
Poly(23% styrene-co-butadiene)	1.0695	Hexane	70	105.37[c]	1.633[c]
Poly(45% styrene-co-butadiene)	1.0363	Hexane	70	105.37	1.633
Poly(23% styrene-co-butadiene)	1.0695	Pentane	60	214.50	1.709
Poly(45% styrene-co-butadiene)	1.0363	Pentane	60	214.50	1.709
Poly(5% styrene-co-butadiene)	1.0989	Toluene	35	6.2540[c]	1.171[c]
Poly(23% styrene-co-butadiene)	1.0695	Toluene	35	6.2540	1.171
Poly(9% vinyl acetate-co-ethylene)	1.0776	1-Butanol	80	21.804[f]	1.307[f]
Poly(9% vinyl acetate-co-ethylene)	1.0776	Chloroform	60	106.30	0.709
Poly(25% vinyl acetate-co-ethylene)	1.0638	Chloroform	60	106.30	0.709
Poly(50% vinyl acetate-co-ethylene)	1.0101	Chloroform	60	106.30	0.709
Poly(70% vinyl acetate-co-ethylene)	0.9434	Chloroform	60	106.30	0.709
Poly(25% vinyl acetate-co-ethylene)	1.0638	Cyclohexane	80	92.575[c]	1.389[c]
Poly(50% vinyl acetate-co-ethylene)	1.0101	Cyclohexane	80	92.575	1.389
Poly(70% vinyl acetate-co-ethylene)	0.9434	Cyclohexane	80	92.575	1.389

Poly(10% vinyl acetate-co-vinyl chloride)	0.7353	Methanol	80	160.69[c]	1.359[c]
Poly(12% vinyl acetate-co-vinyl chloride)	0.7299	1-Butanol	80	21.804	1.307

[a] These values are taken from [18].
[b] These values are taken from [20].
[c] These values are taken from [23].
[d] This value is taken from [21].
[e] These values are taken from [22].
[f] These values are taken from [24].
[g] These values are taken from [25].

2. Vapor-Sorption Equilibria

Sorption data can be presented in three general ways: as sorption isotherms, isobars, or isochores. An isotherm illustrates amount sorbed versus pressure (or external concentration) at a given temperature. Variation of amount sorbed with temperature at constant pressure produces an isobar, and for constant amount sorbed, the variation of temperature with pressure represents an isochore. Most sorption results are given as isotherms, except when the sorption has been measured over an unusually large range of pressure.

The manner in which sorbed molecules are held and distributed within a polymer is termed the mode of sorption. Any given system may involve a number of concurrent modes of sorption, each of which may predominate as conditions are changed. The modes of sorption which are operative can have profound effects on the subsequent mobility of the penetrant in the polymer.

The simplest case is that of ideal solution behavior with sorbed penetrant randomly dispersed within the polymer such that Henry's law is obeyed. The solubility is then independent of the sorbed penetrant concentration at a given temperature and the sorption isotherm is a linear relation of the concentration of penetrant, C, versus partial pressure of the ambient vapor, P.

Three other general types of sorption isotherm are illustrated schematically in Figure 1. The figures on these isotherms also allude to the BET classification (from equations derived by Brunauer, Emmett, and Teller) of isotherms in general [5]. Types of isotherm designed IV and V also exist, but their discussion is not applicable to the material in this chapter.

In terms of the molecular-pair distribution approach, type I represents a preference for the formation of polymer-penetrant pairs at relatively small pressures with a smaller amount of sorption of more nearly ideal solution behavior at high pressures. In physical terms, this represents initial sorption on some kind of specific sites in the polymer. When the sites are nearly all occupied, a small amount of penetrant dissolves in the polymer with a more-or-less random distribution. This situation is exemplified by the sorption of some dyes by ionic polymers or polymers containing polar groups. It also may occur when certain gases and vapors are sorbed in composite heterogeneous media and when gases are sorbed at high pressures by glassy polymers containing voids.

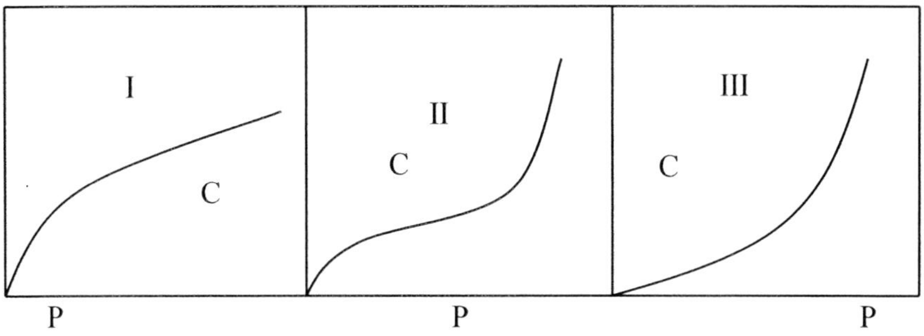

Figure 1. BET isotherm classification

Type III isotherms represent a preference for the formation of penetrant-penetrant pairs such that the solubility increases continuously with pressure. There are two principal physical interpretation of this behavior. One is that the first molecules sorbed tend to loosen the polymer structure locally and make it easier for subsequent molecules to enter in the neighborhood of the first than to go elsewhere. This interpretation implies that the sorbed penetrant effectively plasticizes the polymer, and type III isotherms are observed when a liquid or vapor penetrant is a strong solvent or swelling agent for the polymer. Although polymer-penetrant interactions are relatively strong, they are not specific in the sense of site-penetrant interactions.

Another physical interpretation of type III behavior is reserved for systems in which penetrant-penetrant interactions are inherently stronger than the corresponding polymer-penetrant interactions. It might be anticipated that stable clusters or aggregates of sorbed penetrant molecules would be relatively less mobile in comparison with isolated molecules. Therefore, the proportion of clustered molecules increases with increasing sorbed concentration C, as implied by a type III isotherm.

Type II isotherms are essentially a combination of type I at low pressures and type III at higher pressures. Initially, solvent molecules are strongly sorbed on sites corresponding to similar groups in the polymer. At higher relative vapor pressures, solution or clustering processes predominate. It is to be emphasized that the above physical interpretations are probably over-simplified. In particular, it is likely that two or more modes of sorption will occur concurrently. No information on the degree of overlap of the two modes can be deduced from the simple molecular-pair distribution treatment.

The sorption of solvent vapor by polymers often involves large vapor uptakes and consequent swelling. In order to interpret the sorption process, it is convenient to express solubility in terms of vapor activities. When the vapor pressure is sufficiently low and the temperature is sufficiently high, the activity can be approximated by the usual ratio of the partial pressures of the volatile component,

$$a_s = \frac{P}{P_s} \tag{1}$$

where P is the measured pressure and P_s is the solvent saturation pressure at the temperature of the experiment.

In this chapter we use 77 isothermal vapor-liquid equilibria (VLE) data-sets to show the degree of solvent upturn and to examine the effect of copolymer composition or weight fraction of polymer on VLE in each system. According to these data-sets, solvent weight fraction is plotted against solvent vapor activity that is calculated assuming an ideal gas phase of pure solvent vapor neglecting the tiny vapor pressure of the polymer.

3. FLORY-HUGGINS THEORY

The failure of regular solution theory to account for the large negative deviations from Raoult's law observed in polymer solutions, despite the fact that large exothermic heats of mixing do not occur, led to the suggestion [6] that the entropy of mixing in polymer solution must be larger than the ideal value given by equation:

$$\Delta S_m = -k[n_1 \ln X_1 + n_2 \ln X_2] \tag{2}$$

where n_1 and n_2 are the number of molecules of solvent and polymer, and X_1 and X_2 are their mole fractions respectively. Eq. (2) can be expressed in molar units as, i.e., letting N_1 and N_2 be the mole numbers, we have:

$$\Delta S_m = -R[N_1 \ln X_1 + N_2 \ln X_2] \tag{3}$$

In view of the derivation of Eq. (2) [6], which requires the molecules of the two liquids to have (at least approximately) the same size, it is hardly surprising that Eq. (2) fails, since the condition of equal molecular sizes fails conspicuously for polymer solutions, where the molar volume ratio V_2/V_1 may exceed values of 10^3. The basis for the abnormally large increase in entropy that occurs when polymer chains mix with solvent molecules is to be found in the requirement that all of the units of a flexible chain must necessarily be connected by chain bonds. Every chain configuration requires a contiguous series of spatial locations that are occupied neither by segments of other chains nor by solvent molecules. For a large collection of chains in an undiluted sample this is a restriction that seriously lowers the total number of configurations that would obtain in the absence of this requirement. For a given collection of chains there is a larger number of configurations accessible if some solvent molecules are present. The presence of the latter, which are not constrained to occupy adjacent spatial locations, allow the collection of chains to assume all of those configurations that were accessible in the undiluted sample, plus those new configurations that can occur in the mixed state. This effect, which has to do entirely with the configurational entropy, becomes more important as the chain length increases.

The lattice treatment of this effect was originally developed independently by Flory [7] and Huggins [8], following a suggestion of Meyer [9,10] that a brief version of the theory [11], due to Flory will be outlined here.

It is assumed that a lattice exists in which a site accommodates either a solvent molecule or a segment of a polymer molecule, where the latter is not necessarily to be identified with a monomer unit but is a portion of the chain whose volume is approximately equal to the volume of a solvent molecule. Formally, r, the number of segments in a chain, is defined by the ratio of the molar volumes of the polymer V_2, and of the solvent V_1:

$$r = V_2/V_1 \tag{4}$$

with Flory's theory [11], the entropy of mixing is thus given by:

$$\Delta S_m = -k[n_1 \ln \varphi_1 + n_2 \ln \varphi_2] \tag{5}$$

where φ_1 and $\varphi_2 = 1 - \varphi_1$ are the volume fractions of the solvent and of the polymer:

$$\varphi_1 = n_1/(n_1 + rn_2) \;\; \varphi_2 = rn_2/(n_1 + rn_2) \tag{6}$$

Alternatively, this can be written in molar units, i.e., letting N_1 and N_2 be the mole numbers of the solvent and of the polymer, we have

$$\Delta S_m = -R[N_1 \ln \varphi_1 + N_2 \ln \varphi_2] \tag{7}$$

Equation (5) has the same form as Equation (2) for an ideal solution, the important difference being the replacement of the mole fractions X_i with volume fractions φ_i. If the molar volume of polymer and solvent are equal, then r =1, $\varphi_i = X_i$ and Equations (2) and (5) are identical. If, however, $r > 1$, as is the case for high polymers dissolved in low molecular weight liquids, then for a given composition of the solution the configurational entropy of mixing will be larger than the ideal value $\Delta S_m > \Delta S_m^{ideal}$.

By differentiating Equation (7), we obtain an equation for the entropy of dilution

$$\Delta S_{1m} = \left(\frac{\partial \Delta S_m}{\partial N_1} \right)_{N_2}$$
$$\Delta S_{1m} = -R[\ln(1-\varphi_2) + (1-(1/r)\varphi_2] \tag{8}$$

If the polymer solution mixes athermally, ΔH_1 is zero, so the configurational contribution to the free energy of dilution is $(\mu_1 - \mu_1^0)_m = -T(\Delta S_{1m})$, and accordingly the following equation holds:

$$(\mu_1 - \mu_1^0)_m = RT\;[\ln(1-\varphi_2) + (1-(1/r)\varphi_2] \tag{9}$$

This equation predicts negative deviations from Raoult's law which increase in magnitude with an increase in the ratio of molar volumes, $r \equiv V_2/V_1$. This type of deviation from solution ideality was foreseen by Guggenheim [12] as early as 1936. It is impossible to test Eq. (9)

since there can be entropic contributions to the free energy of the dilution in addition to the configurational entropy. Even within the confines of the lattice model, for example, it is assumed that the contributions of the entropy and enthalpy to the contact free energy of dilution have the same composition dependence, then in analogy with regular solution theory, the total contact free energy of dilution, can be written as:

$$(\mu_1 - \mu_1^0)_{contact} = RT\chi_{12}\varphi_2^2 \tag{10}$$

Thus, $RT\chi_{12}$ is that part of the change in free energy attributable to segment-solvent contacts which occurs upon transferring a mole solvent from pure solvent to a large amount of polymer $(\varphi_2 = 1)$. Combining Eqs. (9) and (10), the total free energy of dilution for mixtures of molecules of unequal size becomes:

$$(\mu_1 - \mu_1^0) = (\mu_1 - \mu_1^0)_m + (\mu_1 - \mu_1^0)_{contact}$$

$$(\mu_1 - \mu_1^0) = RT\left[\ln(1-\varphi_2) + (1-(1/r)\varphi_2 + \chi_{12}\varphi_2^2\right] \tag{11}$$

that is sometimes called the Flory-Huggins equation. The parameter χ_{12} is a measure of the solvent power for the system. Since $(\mu_1 - \mu_1^0)$ increases with increasing χ_{12}, large values of χ_{12} correspond to poor solvent media, and low values of χ_{12} to good solvent media. The historical importance of the theory just described relates not only to its ability to account for negative deviations from Raoult's law, but to its reasonably successful applications to such topics as liquid-liquid phase separation (coacervation equilibrium), fractionation by precipitation, and crystallization from solution [11]. In general, wherever equations for activities of polymer solutions are required we have:

$$(\mu_1 - \mu_1^0) = RT\ln a_1 \tag{12}$$

$$\ln a_1 = \ln\varphi_1 + \left(1 - \frac{1}{r}\right)\varphi_2 + \chi_{12}\varphi_2^2 \tag{13}$$

For the polymer r is so large that we set the ratio of 1/r equal to zero. If absorbant is a mixture of polymers, in the general case, it is deemed necessary to use an analogous term $\chi_{1,23}$. This term considerably simplifies the presentation of data for ternary systems and provides with a general formalism for the treatment of polymer solutions. For a ternary system (solvent(1)-polymer(2)-polymer(3)) the solvent activity is given by:

$$\ln a_1 = \ln\varphi_1 + \left(1 - \frac{1}{r_{23}}\right)\varphi_{23} + \chi_{1,23}\varphi_{23}^2$$

$$\varphi_{23} = \varphi_2 + \varphi_3 = 1 - \varphi_1 \tag{14}$$

$r_{23} = X_{2p}r_2 + X_{3p}r_3$ where φ_2 and φ_3 indicate volume fractions of component 2 and 3 in the polymer mixture, X_{2p} and X_{3p} show mole fractions of these components and again the ratio of $1/r_{23}$ can be set zero because of large size of polymers. In this chapter we show Flory-Huggins parameter χ_{12} or $\chi_{1,23}$ with χ for all systems.

The χ value provides a good (inverse) indication of the solvent power of the penetrant for the polymers. Solvent is a good penetrant for ($\chi \cong 0$), a non-solvent for (χ>1) and when χ is very close to 1, there is a borderline for solvent character.

4. The Clustering Function

As mentioned, the simplest mode of sorption is that of ideal solution behavior with sorbed penetrant randomly dispersed within the polymer such that Henry's law is obeyed. The solubility coefficient σ is then a constant independent of the sorbed penetrant concentration at a given temperature and the sorption isotherm is a linear relation of concentration versus pressure. The term randomly dispersed refers to the molecular scale such that, if the molecular pair distribution function F_2 is considered, neither polymer-penetrant nor penetrant-penetrant pairs are preferred. In a mixture of molecules of type 1 and 2 (where 1 refers to the penetrant) the probability that molecules i and j are at the positions specified by the coordinates (i,j) in the range of these coordinates d(i) and d(j) is

$$(1/V^2)F_2(i,j)d(i)d(j)$$

where V is the total volume of the mixture. The cluster integral for penetrant-penetrant pairs G_{11} is defined by equation:

$$G_{11} = (1/V)\int\int[F_2(i,j) - 1]d(i)d(j) \qquad (15)$$

where i and j now both refer to penetrant molecules. G_{11} is related to equilibrium sorption isotherm data for an incompressible system [13-14] by equation:

$$G_{11}/v_1 = -(1-\varphi_1)\left[\partial(a_1/\varphi_1)/\partial a_1\right] - 1 \qquad (16)$$

where v_1 and $\gamma_1=(a_1/\varphi_1)$ are respectively partial molecular volume and activity coefficient. For an ideal solution, a_1 varies linearly with φ_1, and it follows from Eq. (16) that $G_{11}/v_1=-1$.This means that a penetrant molecule excludes its own volume to other penetrant molecules but does not otherwise affect their distribution. The quantity G_{11}/v_1 is a measure of the clustering tendency, and $G_{11}/v_1=0$ means that clustering is just sufficient to overcome the excluding effect of the central solvent molecule, such that when $G_{11}/v_1>0$ [13] (sorption isotherm convex towards the pressure axis) there is an overall tendency for molecular pair association or clustering of penetrant molecules to occur. This means that the first penetrant molecules enter the polymer structure and make it easier for subsequent molecules to enter in the neighborhood of the first than to go elsewhere. Conversely, when $G_{11}/v_1<-1$ (isotherm concave towards the pressure axis) there is an overall tendency for polymer-penetrant pairs to be preferred and tendency of like molecules to remain isolated and increasingly so as G_{11}/v_1 becomes more negative. To calculate G_{11}/v_1 or G_{ss}/v_s [15], we first fit our experimental equilibrium data (φ_1 versus a_1) with Eq. (13), where we assume $\chi = C_1\varphi_1 + C_2$ with C_1 and C_2 constants. We then differentiate the resulting smooth function. Although the calculated values may vary somewhat with differing smoothing procedures [15],the following conclusions remain valid.

Also, in this chapter, we calculate the quantity $\varphi_s(G_{ss}/v_s)$ that is the mean number of solvent molecules in excess of the mean solvent concentration in the specific neighborhood of a given solvent molecule [14,17] and a thermodynamic non-ideality factor, Γ, [16] defined by:

$$\Gamma = \frac{\partial \ln a_s}{\partial \ln \varphi_s} \tag{17}$$

that can be calculated from the clustering function:

$$\Gamma = \frac{1-\varphi_s}{1+\varphi_s G_{ss}/v_s} \tag{18}$$

For an ideal system, the clustering function is equal to -1; in that event, Γ is equal to 1. Because Gss/vs is greater than -1 when solvent tends to cluster, the non-ideality factor decreases with rising solvent content in system with a tendency to cluster. Conversely, Γ is an increasing function of composition when Gss/vs is less than -1, i.e. for systems where solvent molecules tend to isolate.

5. Application to Experimental Data

5.1. System: Polystyrene (PS)-Poly (Vinyl Methyl Ether) (PVME) At 25°c in Chloroform (Chcl$_3$)

Fig. 2 gives solvent weight fraction in the polymer against solvent vapor activity calculated from Eq. (1). Solvent sorption is largest in PVME, is lowest in PS and for their mixtures lie between those for pure polymers. Solvent sorption increases with increasing its activity and decreases with increasing weight fraction of PS in these systems. Isotherms concave towards the activity axis, as the percentage of PS decreases. Then there is a tendency of solvent molecules to remain isolated.

Fig. 3 shows Flory-Huggins parameter, χ, calculated from Eq. (13) and indicates that chloroform is a good solvent for all systems that have PVME in their structure. Also concentration dependency of χ is clear for all systems except for PS. As it is shown by the value of χ, the nature of the intermolecular interactions between solvent and two polymers is considerably different. Our results obtained for χ also indicate that when the polymer-solvent interaction is strong, (as for PVME), chloroform is a good solvent for the polymer, and chloroform molecules do not cluster in the polymer matrix.

In Figs. 4 and 5 clustering function G_{ss}/V_s and mean cluster size $\varphi_s(G_{ss}/V_s)$ are calculated from Eq. (16). In accordance with these figures, for all systems G_{ss}/V_s is less than zero and chloroform does not cluster.

Fig. 6 displays Γ, calculated from Eq. (18), is a decreasing function of solvent vapor activity. This behavior can be explained with plasticization. In fact, large degree of solvent uptake that was mentioned in Fig. 2 is in general due to a rearrangement of the polymer chain following plasticization induced by a solvent molecule.

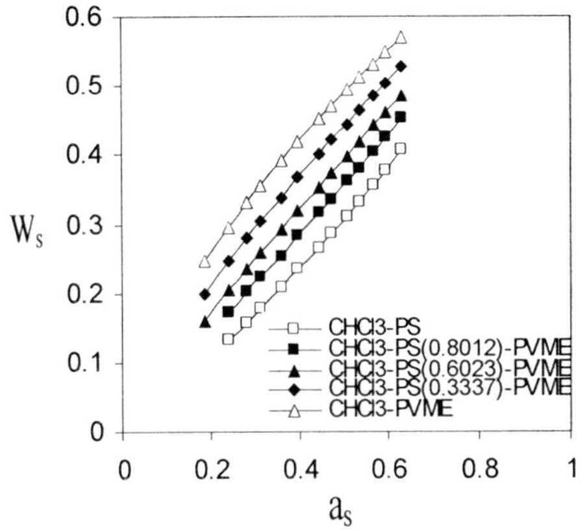

Figure 2. Solvent vapor sorption equilibria for PS-PVME at different weight fractions at 25°c in $CHCl_3$

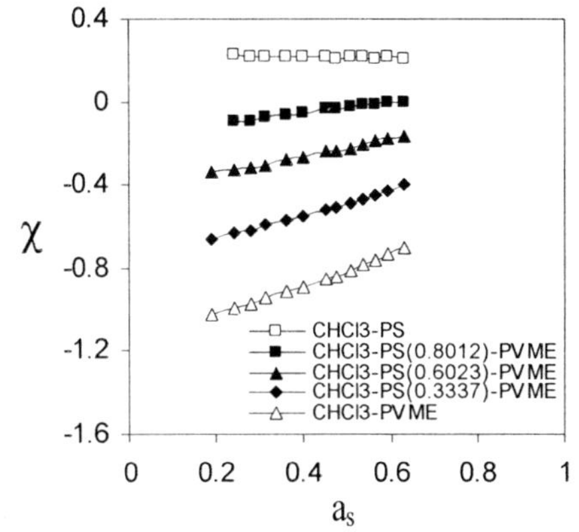

Figure 3. Flory interaction parameter χ versus solvent vapor activity for PS-PVME at different weight fractions at 25°c in $CHCl_3$

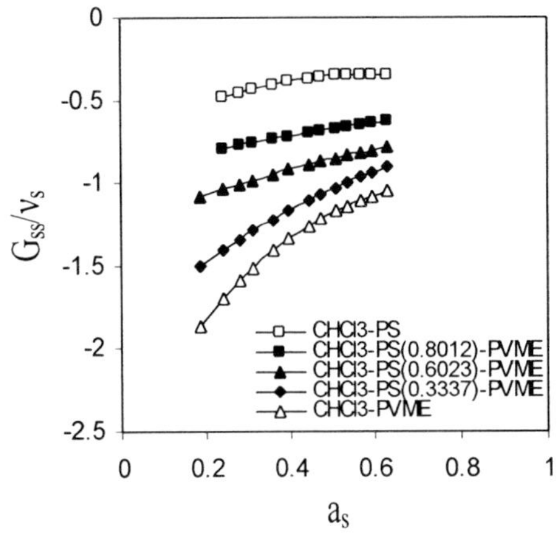

Figure 4. Clustering function G_{ss}/v_s at various solvent-vapor activities for PS-PVME at different weight fractions at 25°c in $CHCl_3$

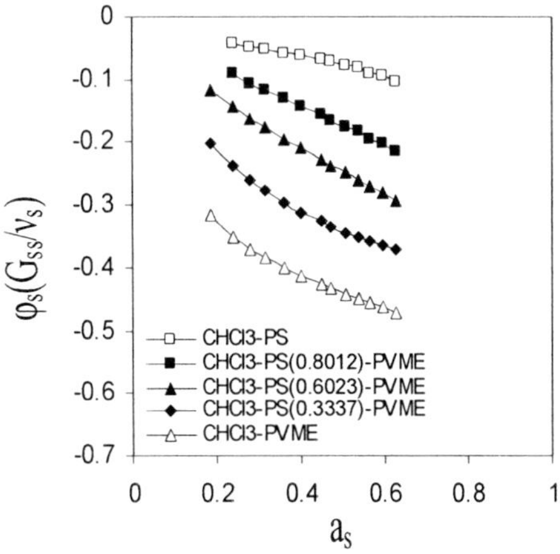

Figure 5. Mean cluster size $\varphi_s(G_{ss}/v_s)$ at various solvent-vapor activities for PS-PVME at different weight fractions at 25°c in $CHCl_3$

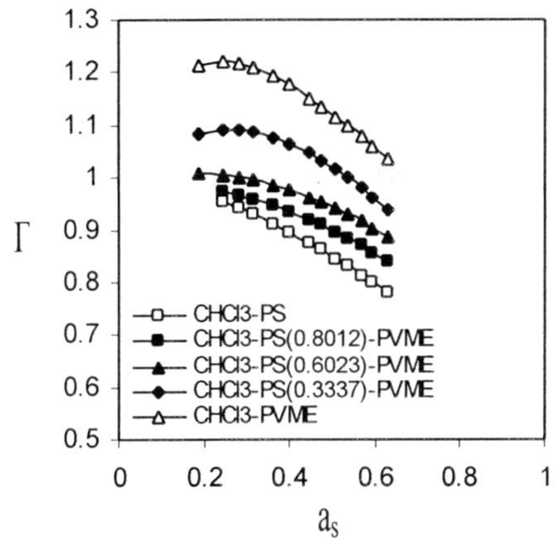

Figure 6. Thermodynamic factor Γ at various solvent-vapor activities for PS-PVME at different weight fractions at 25°c in $CHCl_3$.

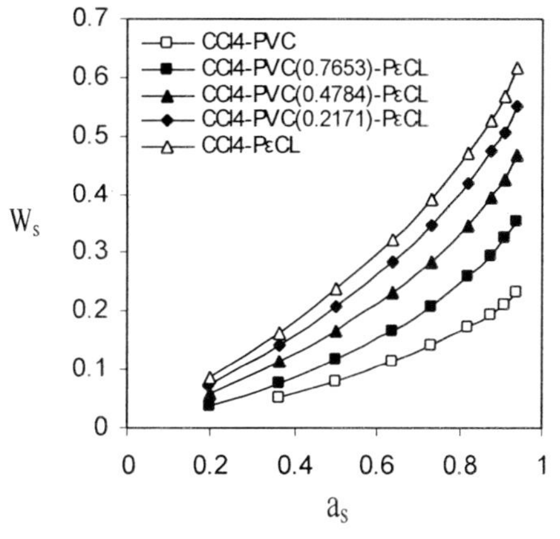

Figure 7. Solvent vapor sorption equilibria for PVC-PεCL at different weight fractions at 65°c in CCl_4

5.2. System: Poly (Vinyl Chloride) (PVC)-Poly (E-Caprolactone) (Pεcl) at 65°c in Carbon Tetrachloride (Ccl_4)

In Fig. 7 solvent weight fraction in the polymer is plotted against solvent vapor activity. According to the BET classification, all the systems exhibit a type III isotherm. The measured solvent uptake is lowest in PVC and increases with increasing percentage of PεCl and solvent concentration.

Fig. 8 displays χ as a function of solvent vapor activity. Carbon tetrachloride is a non-solvent for PVC (χ>1) and for systems with large amount of PVC. When PεCl concentration increases, a moderate interaction between solvent and polymer can be seen. Also it is obvious that χ is concentration-dependent for all systems, increasing with solvent activity for pure PεCl and for systems with PVC weight fraction less than 0.5.

Fig. 9 compares the values of the clustering function G_{ss}/V_s. This value is less than zero for all systems with large amount of PεCl and suggesting that solvent molecules do not cluster in these systems. As PVC concentration increases, clustering function goes towards positive values, and has a maximum about a_s=0.85. Fig. 10 shows mean cluster size, $\varphi_s(G_{ss}/V_s)$, as a function of solvent vapor activity. For PVC and PVC(0.7653)-PεCl, $\varphi_s(G_{ss}/V_s)$, is small and negative, except at very high activity, where it becomes slightly positive in agreement with data of Fig. 9. Both figures indicate that systems with PVC weight fraction almost greater than 0.5, have a higher tendency to form clusters when absorbed in a polymer at high solvent concentration.

Finally in Accordance with Fig. 11 non-ideality factor, Γ, decreases with rising solvent content in systems , because G_{ss}/V_s is greater than -1 for all systems. Combined with the large upturn in the equilibrium isotherm, this trend suggests that the polymer matrix is plasticized by the solvent.

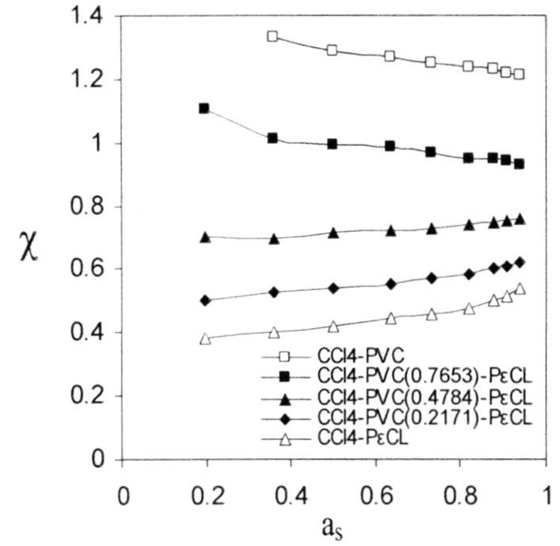

Figure 8. Flory interaction parameter χ versus solvent vapor activity for PVC-PεCL at different weight fractions at 65°c in CCl_4

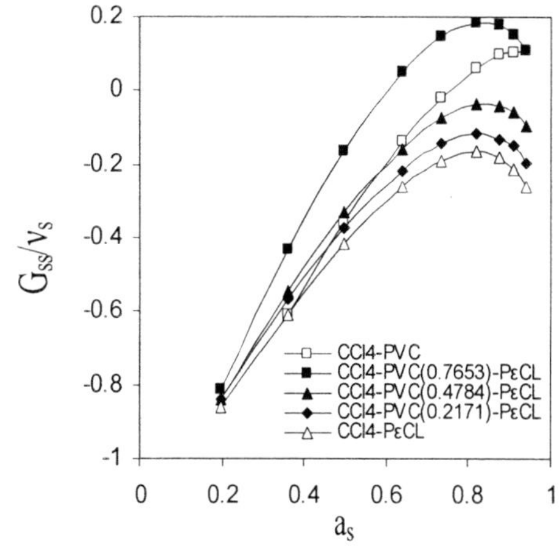

Figure 9. Clustering function G_{ss}/v_s at various solvent-vapor activities for PVC-PεCL at different weight fractions at 65°c in CCl_4

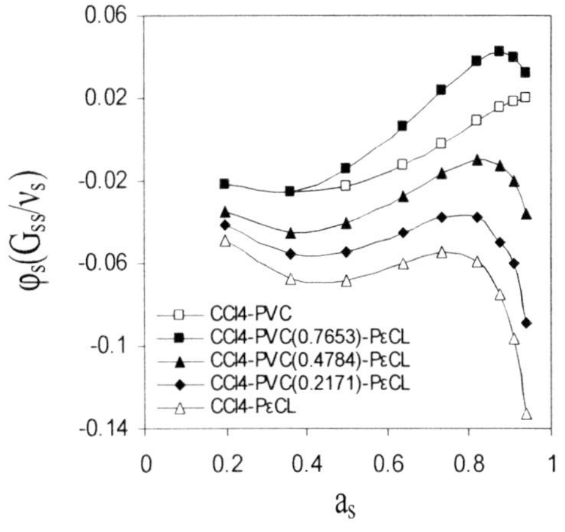

Figure 10. Mean cluster size $\varphi_s(G_{ss}/v_s)$ at various solvent-vapor activities for PVC-PεCL at different weight fractions at 65°c in CCl_4

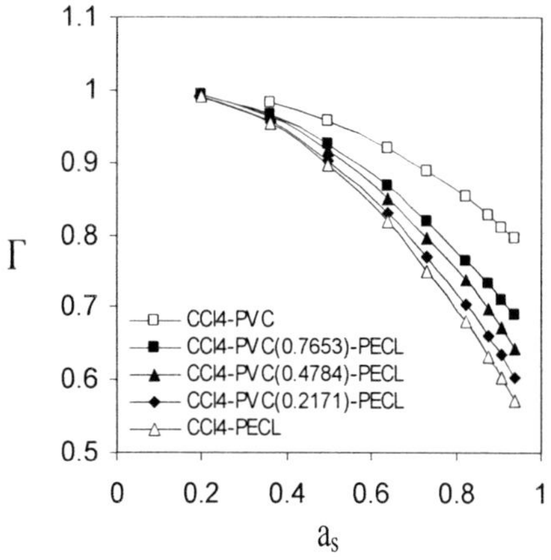

Figure 11. Thermodynamic factor Γ at various solvent-vapor activities for PVC-PεCL at different weight fractions at 65°c in CCl_4

5.3. System: Polyisobutylene (PIB)-N-Heptadecane (Nhd) At 25°c in Pentane (C_5H_{12})

The equilibrium isotherms in Fig. 12 show a large degree of upturn at high activities and all systems obey a type III isotherm. Pentane has a strong interaction with nHD because the hydrocarbon segments of pentane better liked by the similar hydrocarbon segments of nHD. Fig. 13 indicates Flory-Huggins parameter, χ, is smaller for pure nHD than other systems, in agreement with largest sorption of solvent by this material that is evident in Fig. 12. Pentane is approximately a moderate solvent for all systems because of χ<1. For systems studied here, Flory interaction parameter is dependent on molecular weight of polymers considered.

Clustering function G_{ss}/V_s and mean cluster size $\varphi_s(G_{ss}/V_s)$ are illustrated in Figs. 14 and 15 suggesting that pentane does not cluster in these polymers. As shown in Fig. 16 non-ideality factor, Γ, is a decreasing function of solvent vapor activity and is a confirmation on the tendency of solvent molecules to plasticize when absorbed in a polymer largely.

5.4. System: Polystyrene (PS)-Poly (vinyl methyl ether) (PVME) at 25°c in Benzene (C_6H_6)

Fig. 17 displays the vapor-sorption equilibrium data. Solvent weight fraction in the polymer is plotted against solvent vapor activity. There is no difference between two polymers in sorption of benzene because both of them are soluble in benzene and their sorption strength goes high with increasing content of benzene in the system.

Also it is evident from Fig. 18 that benzene suited to these polymers as a solvent. Results obtained using chloroform as a solvent suggest that polystyrene and poly(vinyl methyl ether) are more compatible than results obtained when benzene is used as a solvent. On the other hand, in both cases the compatibility is shown enhanced with increasing PVME concentration.

Again, negative amount of clustering function, G_{ss}/V_s and mean cluster size $\varphi_s(G_{ss}/V_s)$ represent the tendency of benzene molecules to remain isolated and decreasing pace of thermodynamic factor, Γ, can be explained by plasticization. As a result, PS has a weak interaction with benzene and solvent molecules are more liable to do cluster in this polymer.

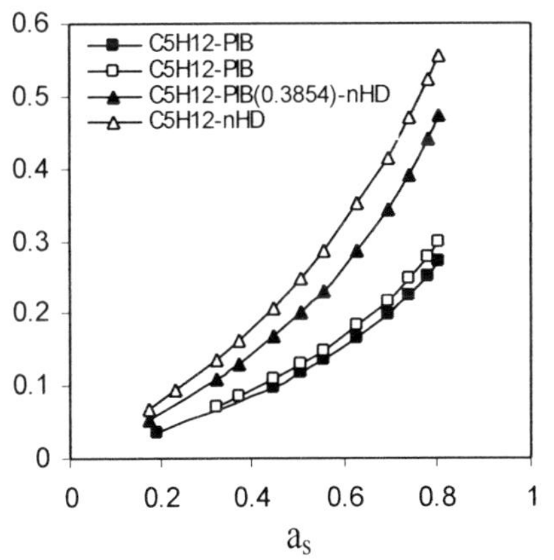

Figure 12. Solvent vapor sorption equilibria for PIB-nHD with PIB(Mw=2700(■)and Mw=1350 (□,▲)) at different weight fractions at 25°c in C_5H_{12}

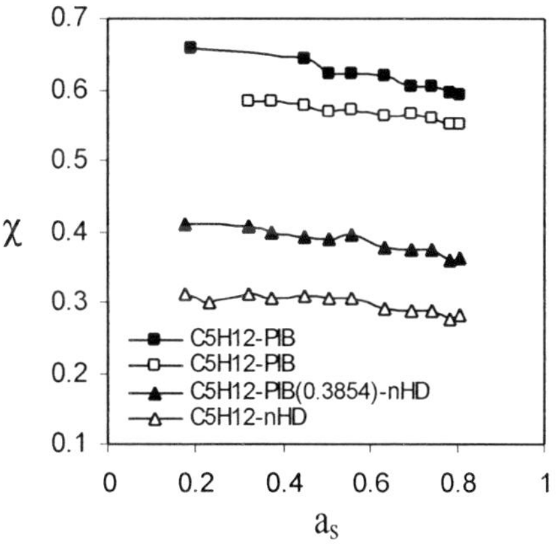

Figure 13. Flory interaction parameter χ versus solvent vapor activity for PIB-nHD with PIB(Mw=2700(■) and Mw=1350 (□,▲))at different weight fractions at 25°c in C_5H_{12}.

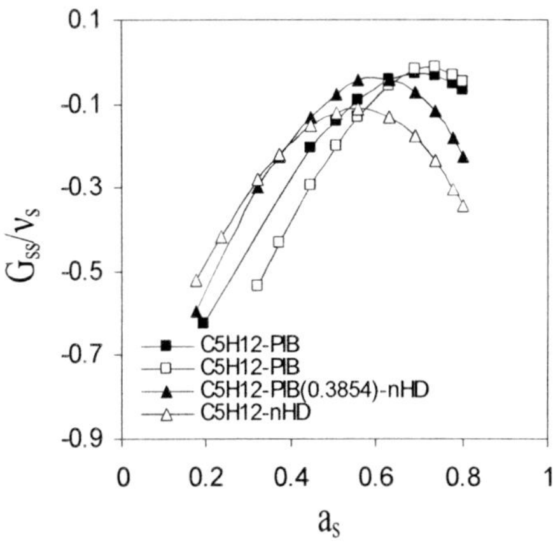

Figure 14. Clustering function G_{ss}/v_s at various solvent-vapor activities for PIB-nHD with PIB(Mw=2700(■) and Mw=1350(□,▲)) at different weight fractions at 25°c in C_5H_{12}

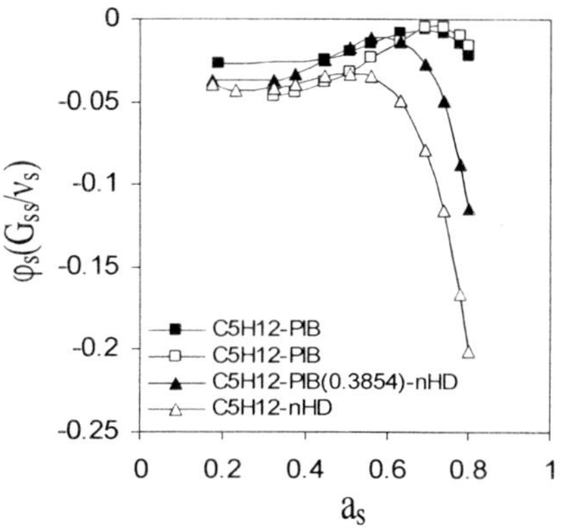

Figure 15. Mean cluster size $\varphi_s(G_{ss}/v_s)$ at various solvent-vapor activities for PIB-nHD with PIB(Mw=2700(■)and Mw=1350(□,▲)) at different weight fractions at 25°c in C_5H_{12}

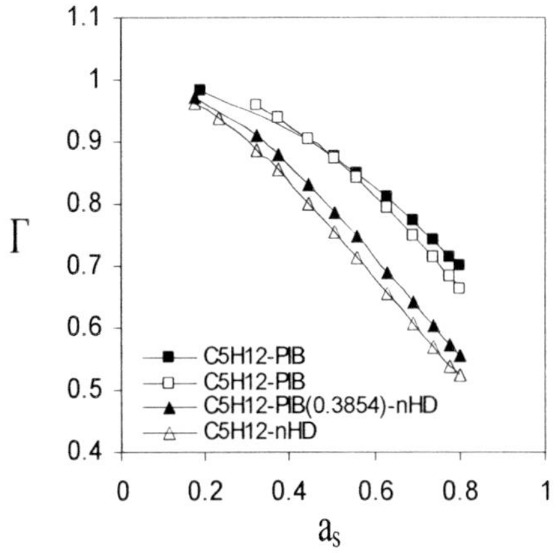

Figure 16. Thermodynamic factor Γ at various solvent-vapor activities for PIB-nHD with PIB(Mw=2700(■) and Mw=1350(□,▲)) at different weight fractions at 25°c in C_5H_{12}

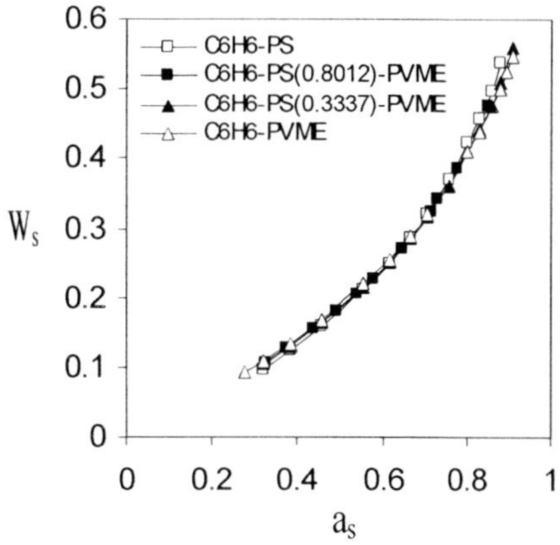

Figure 17. Solvent Vapor Sorption Equilibria for PS-PVME at different weight fractions at 25°c in C_6H_6.

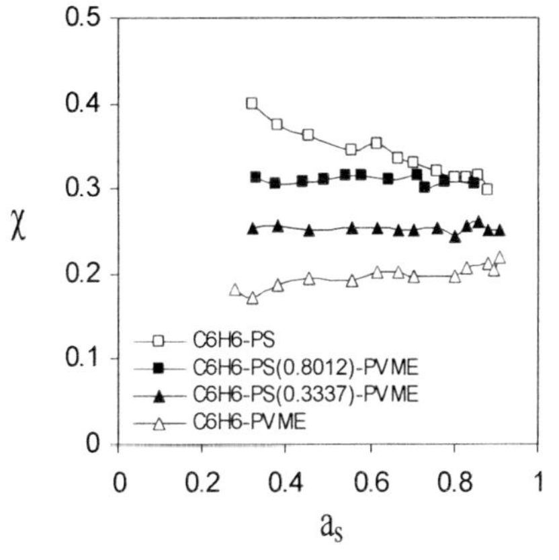

Figure 18. Flory interaction parameter χ versus solvent vapor activity for PS-PVME at different weight fractions at 25°c in C_6H_6

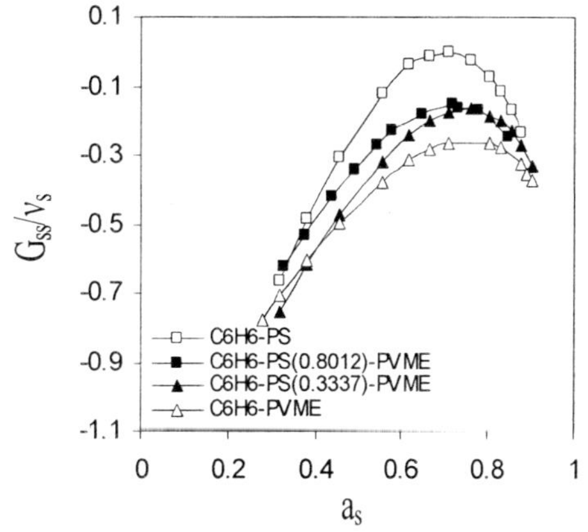

Figure 19. Clustering function G_{ss}/v_s at various solvent-vapor activities for PS- PVME at different weight fractions at 25°c in C_6H_6

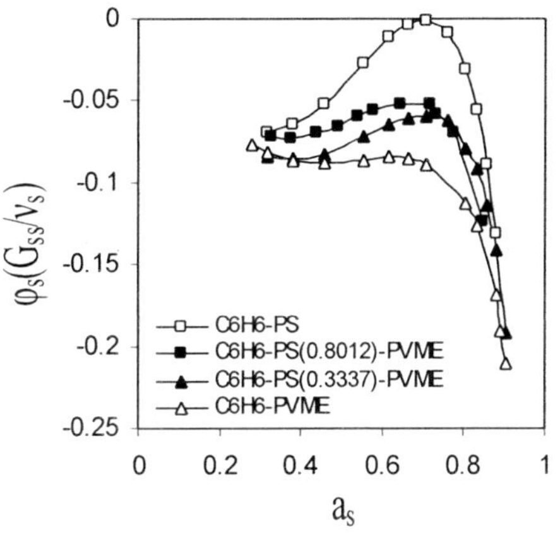

Figure 20. Mean cluster size $\varphi_s(G_{ss}/v_s)$ at various solvent vapor activities for PS- PVME at different weight fractions at 25°c in C_6H_6.

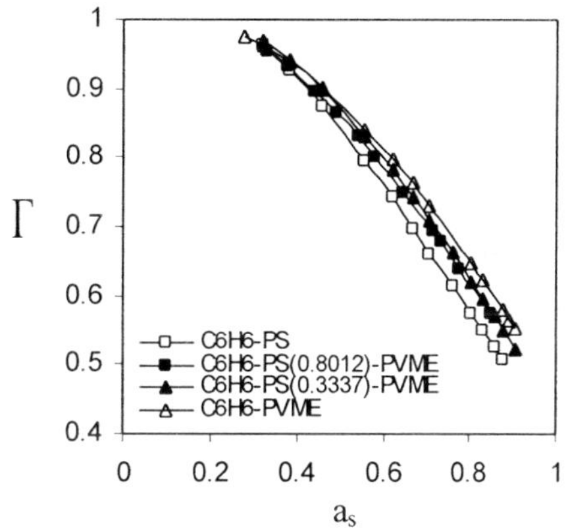

Figure 21. Thermodynamic factor Γ at various solvent vapor activities for PS- PVME at different weight fractions at 25°c in C_6H_6.

5.5. System: N-Tetracosane (Ntc)-Poly (Ethylene Oxide) (PEO)- N-Tetracosane (Ntc) At 55°c And PEO at 70°c in Benzene (C_6H_6)

A different way for determining polymer-solvent interaction is by studying copolymer-solvent systems. For such informations to be extracted, one has also to study the constituent homopolymer-solvent systems. For this purpose, a triblock copolymer has been chosen. Its end blocks are n-tetracosane while the central block is poly (ethylene oxide) with an average molecular weight 600 at 55°c. Also, a high molecular weight (100,000) poly (ethylene oxide) has been used at 70°c.

In Fig. 22 type III isotherms represent the formation of polymer-solvent pairs such that solubility increases continuously with solvent activity.

Flory parameter, χ, in Fig. 23 indicates that benzene is a good solvent for PEO or any system consisting of PEO but has a weak interaction with nTC because of difference in nature of two aliphatic and aromatic materials.

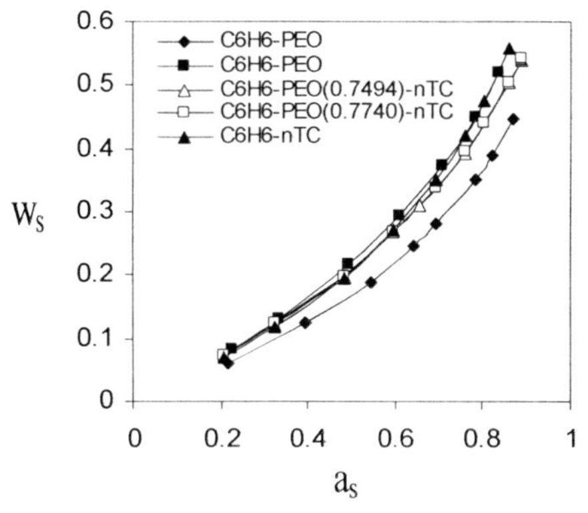

Figure 22. Solvent vapor sorption equilibria for nTC-PEO-nTC with PEO(MW=100,000(♦)at 70°c and MW=600(■,Δ,□) at 55°c) at different weight fractions in C_6H_6

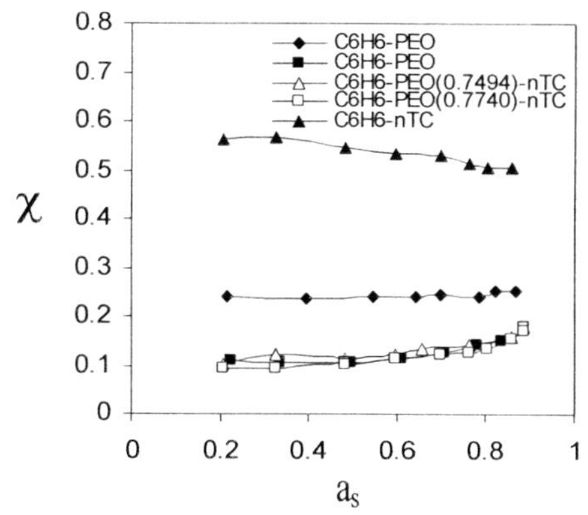

Figure 23. Flort interaction parameter χ versus solvent vapor activity for nTC-PEO- nTC with PEO(MW=100,000(♦)at 70°c and MW=600(■,Δ,□) at 55°c) at different weight fractions in C_6H_6.

Negative values of clustering function G_{ss}/V_s and mean cluster size $\varphi_s(G_{ss}/V_s)$ in Figs. 24 and 25 show that benzene molecules do not cluster in systems that contain PEO. Results for nTC give negative G_{ss}/V_s and $\varphi_s(G_{ss}/V_s)$ for benzene activities up to 0.5, whereas at high activity these values become positive, suggesting that benzene tends to cluster in the polymer. Finally plasticization of the polymer matrix by solvent sorption is likely responsible for decreasing pace of, Γ, non-ideality factor in Fig. 26.

5.6. System: Poly (styrene(St)-co-Butadiene(Bu)) at 50 and 60°c in Acetone ($(CH_3)_2CO$)

Fig. 27 gives the solvent weight fraction in the copolymer versus solvent vapor activity. Due to insolubility of this copolymer in acetone, only small amounts of solvent can be absorbed. However, for poly(St(0.23)-Bu) and poly(St(0.45)-Bu)at 60°c, the measured solvent uptake is larger, as a result of temperature increasing. For these systems solubility is not a simple function of copolymer composition.

Fig. 28 displays χ as a function of solvent vapor activity. In agreement with above discussion, acetone is a non-solvent for this copolymer (χ>1) and have the most interaction with it at temperature of 60°c.

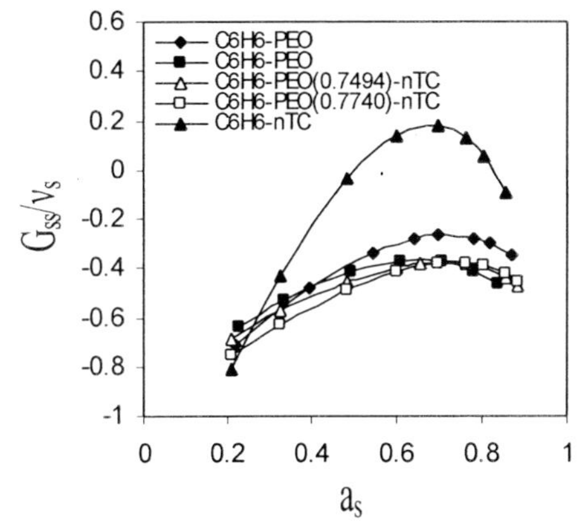

Figure 24. Clustering function G_{ss}/v_s at various solvent-vapor activities for nTC-PEO-nTC with PEO(MW=100,000(♦)at 70°c and MW=600(■,Δ,□) at 55°c) at different weight fractions in C_6H_6

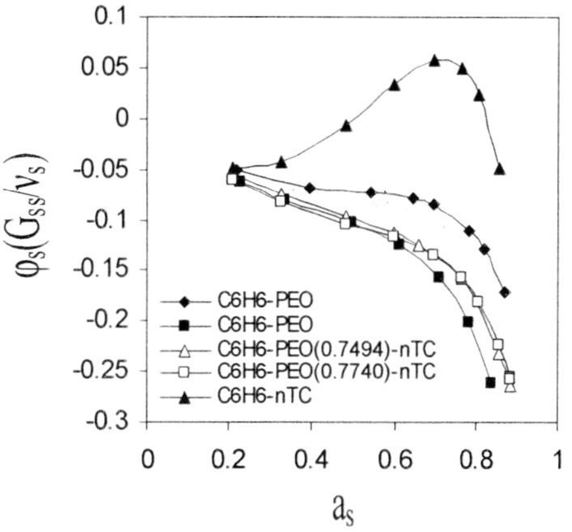

Figure 25. Mean cluster size $\varphi_s(G_{ss}/v_s)$ at various solvent-vapor activities for nTC-PEO-nTC with PEO(MW=100,000(♦)at 70°c and MW=600(■,Δ,□) at 55°c) at different weight fractions in C_6H_6.

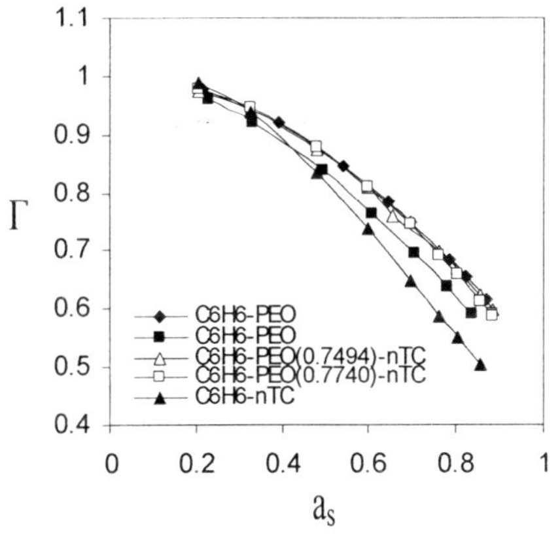

Figure 26. Thermodynamic factor Γ at various Solvent-vapor activities for nTC-PEO-nTC with PEO(MW=100,000(♦)at 70°c and MW=600(■,Δ,□) at 55°c) at different weight fractions in C_6H_6

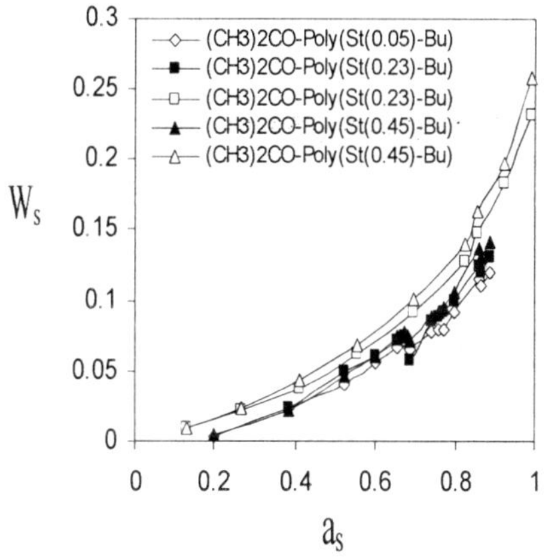

Figure 27. Solvent vapor sorption equilibria for poly(St-co-Bu) at different weight fractions at (◊,■,▲) 50°c and (□,Δ) 60°c in $(CH_3)_2CO$

Figs. 29 and 30 compare the values of the clustering function, G_{ss}/V_s, and the mean cluster size $\varphi_s(G_{ss}/V_s)$ for different weight fractions of this copolymer. Results for poly(St(0.23)-Bu) at 50 and 60°c and for poly(St(0.45)-Bu) at 60°c give negative G_{ss}/V_s and $\varphi_s(G_{ss}/V_s)$ for solvent activities up to 0.5, whereas at higher activities, these values become positive, suggesting that acetone cluster in these compositions only at high solvent activities. Conversely, poly(St(0.05)-Bu) and poly(St(0.45)-Bu) at 50°c exhibit positive values for

G_{ss}/V_s and $\varphi_s(G_{ss}/V_s)$, indicating that solvent molecules tend to cluster in these compositions, as expected for poor solvent that was mentioned in Fig.28 with the least interaction between these copolymers and acetone.

Fig. 31 shows the thermodynamic factor, Γ, calculated from the equilibrium solvent uptake. The non-ideality factor decreases with rising solvent content in system with a tendency to cluster that confirms results of Figs. 29 and 30.

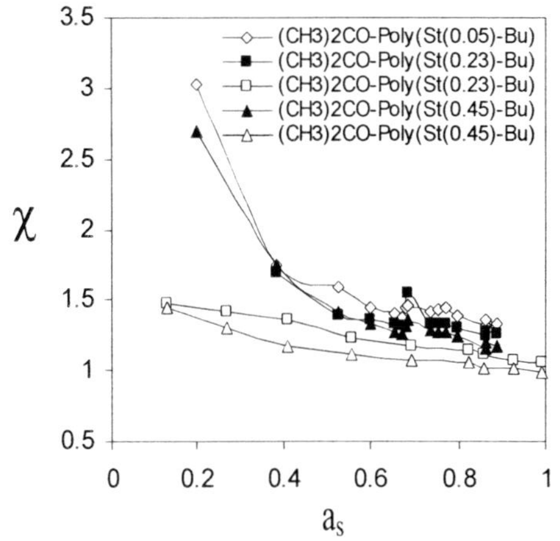

Figure 28. Flory interaction parameter χ versus solvent-vapor activity for poly(St-co-Bu) at different weight fractions at (◊,■,▲) 50°c and (□,Δ) 60°c in $(CH_3)_2CO$

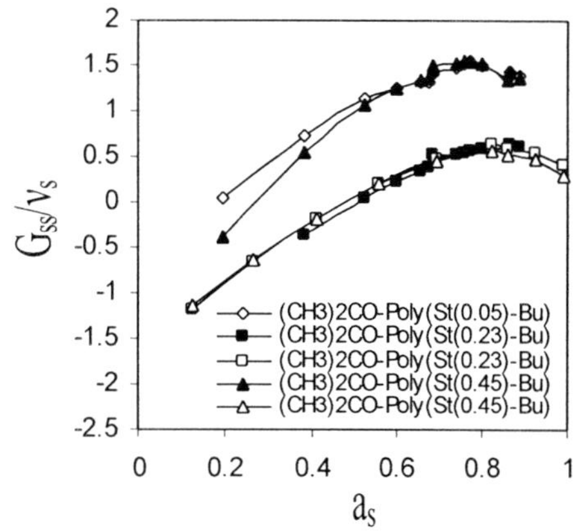

Figure 29. Clustering function G_{ss}/v_s at various solvent-vapor activities for poly(St-co-Bu) at different weight fractions at (◊,■,▲) 50°c and(□,Δ) 60°c in $(CH_3)_2CO$

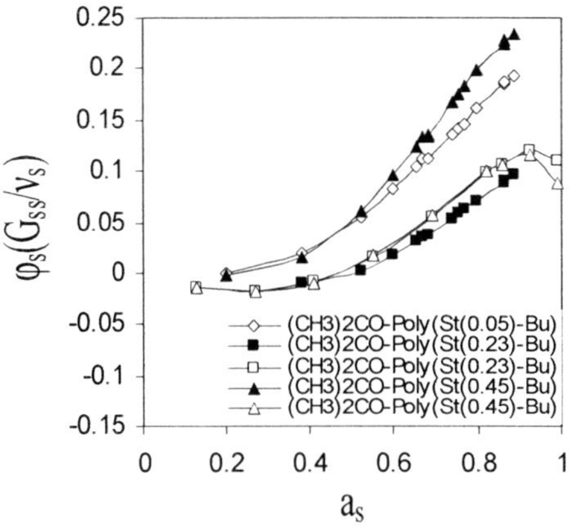

Figure 30. Mean cluster size $\varphi_s(G_{ss}/v_s)$ at various solvent-vapor activities for poly(St-co-Bu) at different weight fractions at (◊,■,▲) 50°c and(□,Δ) 60°c in $(CH_3)_2CO$

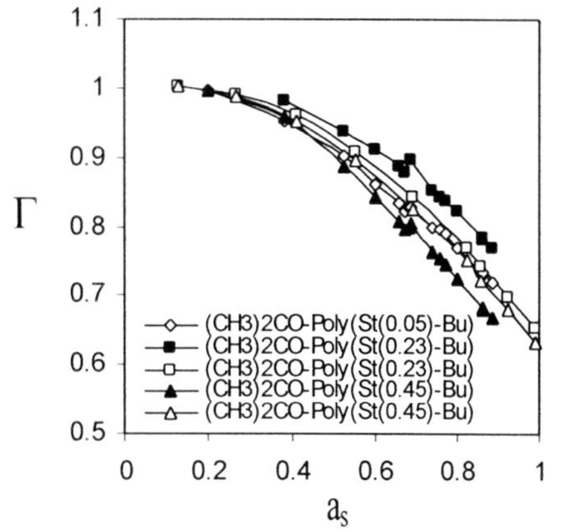

Figure 31. Thermodynamic factor Γ at various solvent-vapor activities for poly(St-co-Bu) at different weight fractions at (◊,■,▲) 50°c and(□,Δ) 60°c in $(CH_3)_2CO$

5.7. System: Poly(Acrylonitrile)(PAN), Poly(Cis-Butadiene)(Cis-Bu), and Poly (Acrylonitrile (AN)-Co-Butadiene (Bu)) At 60°c in Acetonitrile (CH_3CN)

Fig. 32 shows VLE data for poly(AN-co-Bu) and parent homopolymers. At a given activity, solvent absorption in polyacrylonitrile is higher than in polybutadiene because polar acetonitrile molecules prefer polar segments of polyacrylonitrile to hydrocarbon segments of polybutadiene. Intuitively, one might expect that the copolymer curve should lie between the two homopolymer curves, but Fig. 32 shows that the copolymer curve lie beyond the bound of the two homopolymer curves.

At a given pressure, acetonitrile solubility in a copolymer is much higher than that in the corresponding homopolymers and is highest in poly(AN(0.51)-Bu). This non-intuitive behavior is attributed to intramolecular repulsion between unlike segments of the copolymer. This repulsive interaction is weakend when acetonitrile molecules are in the vicinity of unlike copolymer segments, favoring copolymer+solvent miscibility.

Again, in Fig.33, we can see that in spite of insolubility of parent homopolymers in acetonitrile (χ>1), but copolymers are soluble in this solvent and their solubility increases with enhancement of acrylonitrile weight fraction because of interaction between polar groups.

$G_{ss}/V_s \approx -1$, negative amount of $\varphi_s(G_{ss}/V_s)$ and $\Gamma=1$ for parent homopolymers in Figs. 34, 35 and 36 indicate that we encounter an ideal solution where solvent molecules prevent other molecules from entering their own volume. Thus solvent molecules tend to remain isolated and do not cluster.

For poly(AN(0.21)-Bu) and poly(AN(0.51)-Bu), due to attractive interaction between polar groups, the first solvent molecules enter the polymer structure and make it easier for subsequent molecules to enter in the adjacent position than to go elsewhere, and we see clustering in these copolymers. There is an exception in the case of poly(AN(0.33)-Bu) that does not cluster. Considerable amount of solvent uptake and decreasing function Γ, in this copolymer can be a sign of plasticization.

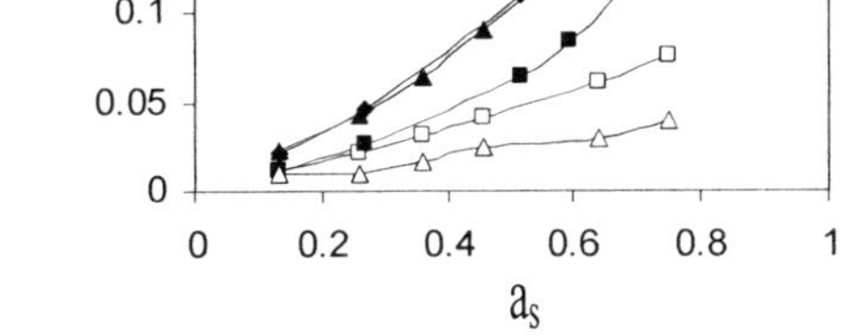

Figure 32. Solvent vapor sorption equilibria for poly(AN-co-Bu) at different weight fractions at 60°c in CH_3CN

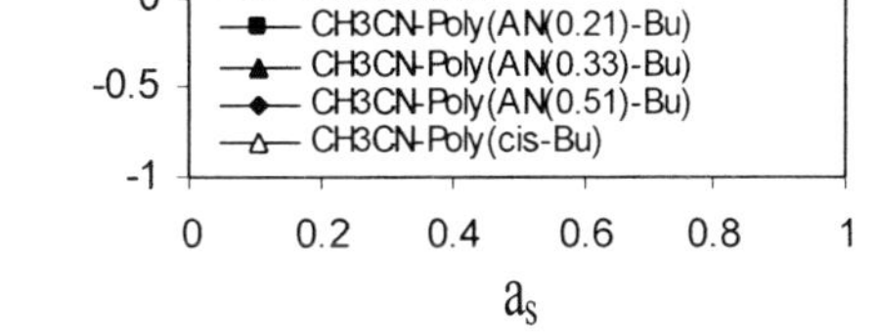

Figure 33. Flory interaction parameter χ varsus solvent-vapor activity for poly(AN-co-Bu) at different weight fractions at 60°c in CH_3CN

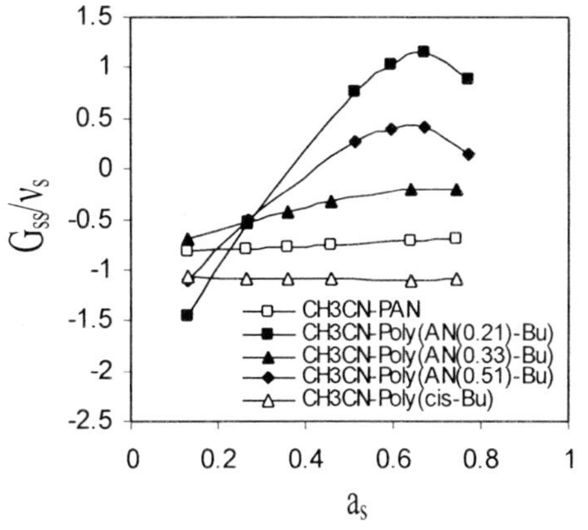

Figure 34. Clustering function G_{ss}/v_s at various solvent-vapor activities for poly(AN-co-Bu) at different weight fractions at 60°c in CH_3CN

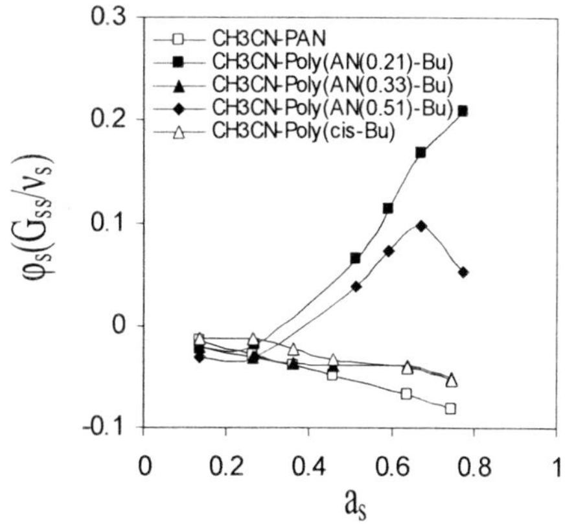

Figure 35. Mean cluster size $\varphi_s(G_{ss}/v_s)$ at various solvent-vapor activities for poly(AN-co-Bu) at different weight fractions at 60°c in CH_3CN

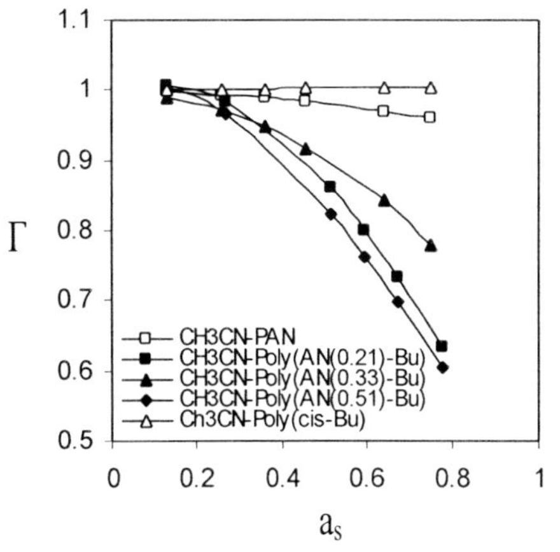

Figure 36. Thermodynamic factor Γ at various solvent-vapor activities for poly(AN-co-Bu) at different weight fractions at 60°c in CH_3CN

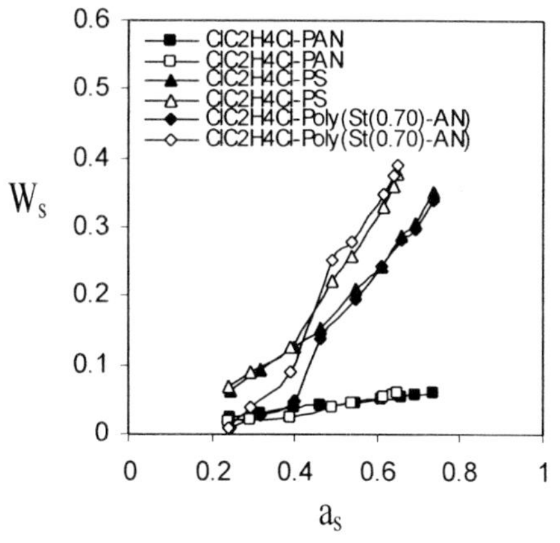

Figure 37. Solvent vapor sorption equilibria for poly(St-co-AN) at different weight fractions at (■,▲,♦) 70°c and at (□,Δ,◊) 80°c in ClC_2H_4Cl.

5.8. System: Polystyrene (PS), Poly(Acrylonitrile) (PAN) and Poly (Styrene(St)-Co-Acrylonitrile(AN)) at 70 and 80oc in 1,2-Dichloroethane (Clc2h4cl)

Fig. 37 displays solvent weight fraction in the polymer versus solvent vapor activity. At a given activity, the solubility of 1,2-dichloroethane decreases dramatically with replacement of polystyrene(PS) with polyacrylonitrile(PAN). 1,2-dichloroethane has negligible solubility in PAN due to strong repulsion between nonpolar 1,2-dichloroethane molecules and highly polar acrylonitrile segments. 1,2-dichloroethane also has some repulsive interaction with styrene segments, but these are not as strong as those with acrylonitrile segments. The solubility of 1,2-dichloroethane rises in presence of PS or poly(St-co-AN).

Flory interaction parameter, χ, in Fig. 38 indicates that 1,2-dichloroethane is a non-solvent for PAN (χ>1), when styrene segments add to PAN to build a copolymer, χ decreases such that 1,2-dichloroethane can slightly dissolve PS and poly(St(0.70)-AN) especially at high activities.

Figs. 39 and 40 show values of the clustering function G_{ss}/V_s and the mean cluster size $\varphi_s(G_{ss}/V_s)$. In spite of repulsive interaction of PAN with 1,2-dichloroethane, solvent molecules prefer to remain isolated (G_{ss}/V_s<0) in this polymer, because solvent uptake is low. In respect of PS and poly(St(0.70)-AN), a weak polymer/solvent interaction accompany with large solvent upturn causes solvent molecules associate with each other and form cluster. Larger amounts of clustering at low or moderate activities for poly(St(0.70)-AN) can be explained by weaker interaction of this copolymer with solvent in this region.

Fig. 41 says that PAN with $G_{ss}/V_s \approx -1$ and $\Gamma \approx 1$ behaves similar to an ideal solution. For PS and poly(St(0.70)-AN), non-ideality factor decreases with rising solvent content in the system with a tendency of solvent molecules to cluster.

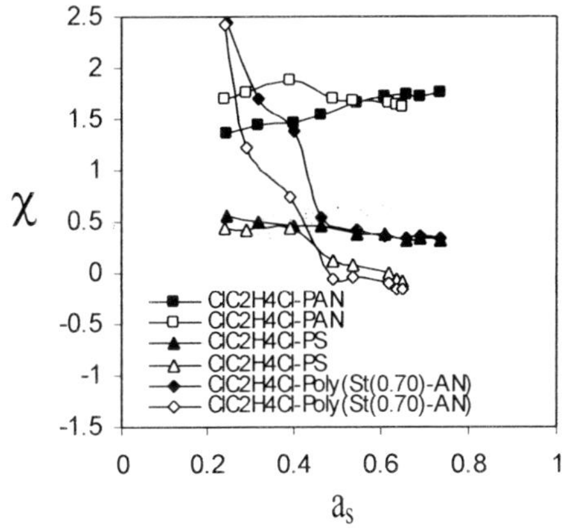

Figure 38. Flory interaction parameter χ versus solvent-vapor activity for poly(St-co-AN) at different weight fractions at (■,▲,♦) 70°c and at (□,Δ,◊) 80°c in ClC_2H_4Cl

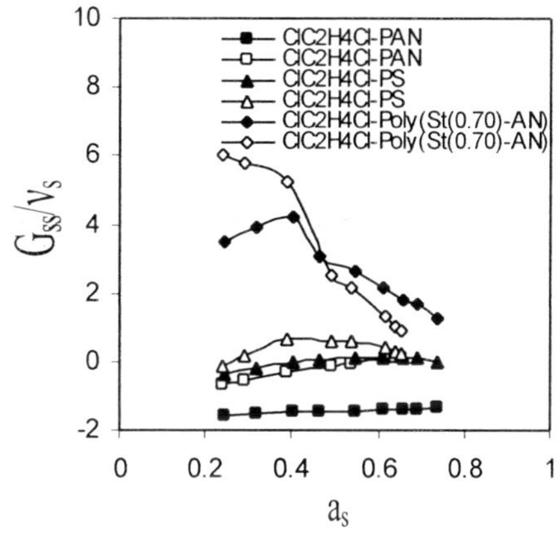

Figure 39. Clustering function G_{ss}/v_s at various solvent-vapor activities for poly(St-co-AN) at different weight fractions at (■,▲,♦) 70°c and at (□,Δ,◊) 80°c in ClC_2H_4Cl

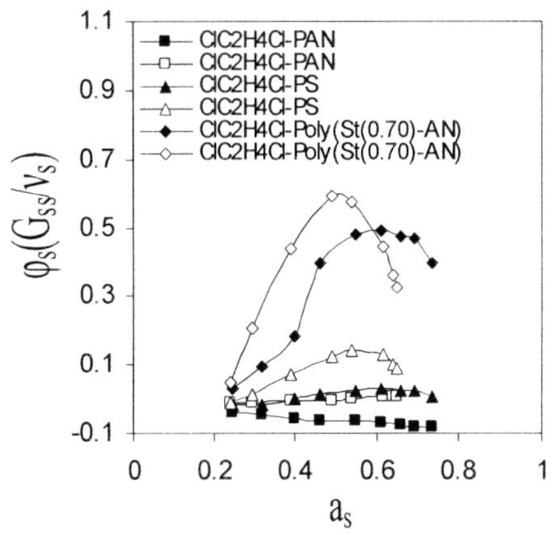

Figure 40. Mean cluster size $\varphi_s(G_{ss}/v_s)$ at various solvent-vapor activities for poly(St-co-AN) at different weight fractions at (■,▲,♦) 70°c and at (□,Δ,◊) 80°c in ClC_2H_4Cl.

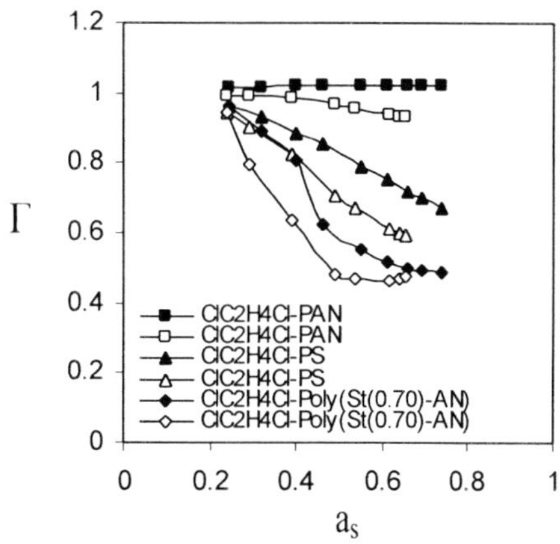

Figure 41. Thermodynamic factor Γ at various solvent-vapor activities for poly(St-co-AN) at different weight fractions at (■,▲,♦) 70°c and at (□,Δ,◊) 80°c in ClC_2H_4Cl

5.9. System: Poly (Acrylonitrile(AN)-Co-Butadiene(Bu)) and Poly (Cis-Butadiene) At 60°c and Poly(Styrene(St)-Co-Butadiene(Bu)) At 50°c in Chloroform

Fig. 42 gives the solvent weight fraction in the copolymer versus solvent vapor activity, Due to solubility of all parent homopolymers and copolymers in chloroform; the measured solvent uptake is high. However, this value for poly(AN(0.33)-Bu)and for poly(cis-Bu) at 60°c is larger. Fot these systems effect of copolymer composition on solubility is not clear.

Fig. 43 displays Flory interaction parameter, χ, as a function of solvent vapor activity. Because of $\chi<0$, chloroform is a good solvent for poly(AN(0.33)-Bu) and poly(cis-Bu). If we refer to the discussion stated in section 5.1, we will see that chloroform has also some polar interaction with styrene segments, but these are not as strong as those with the butadiene segments. Therefore, as a rule, copolymers consisting of poly(cis-butadiene) and polystyrene have a χ value between that of two homopolymers and have a fair interaction with chloroform. Effect of polystyrene percentage on degree of interaction is negligible.

In Figs. 44 and 45 negative amounts of clustering function G_{ss}/V_s and the mean cluster size $\varphi_s(G_{ss}/V_s)$for poly(cis-butadiene) and poly(AN(0.33)-Bu) indicate that clustering does not occur in these systems. But for poly(St-co-Bu) in solvent activities below 0.6, clustering is dominant that confirms lower interaction of this copolymer with chloroform at low solvent concentration. With rising solvent content in the system, polymer/solvent interaction goes high and solvent molecules separate from each other.

Fig. 46 shows non-ideality factor, Γ, calculated from equilibrium solvent upturn and determines the deviation of these systems from ideal solution. It is evident that copolymers deviate from ideality more than their parent homopolymers.

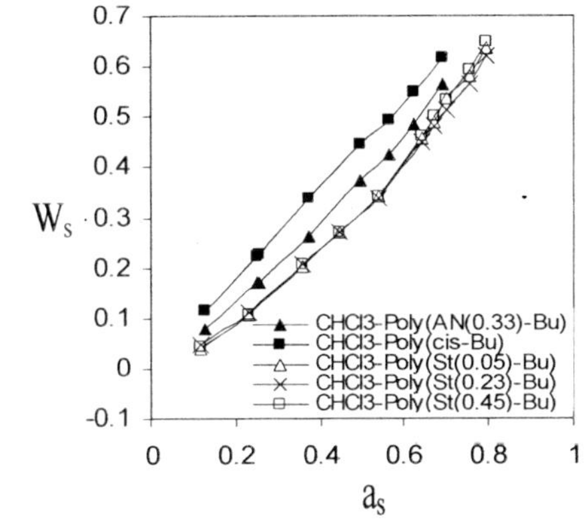

Figure 42. Solvent vapor sorption equilibria for poly(AN(0.33)-Bu) and poly(cis-Bu) at 60°c and for poly(St-co-Bu) at different weight fractions at 50°c in $CHCl_3$

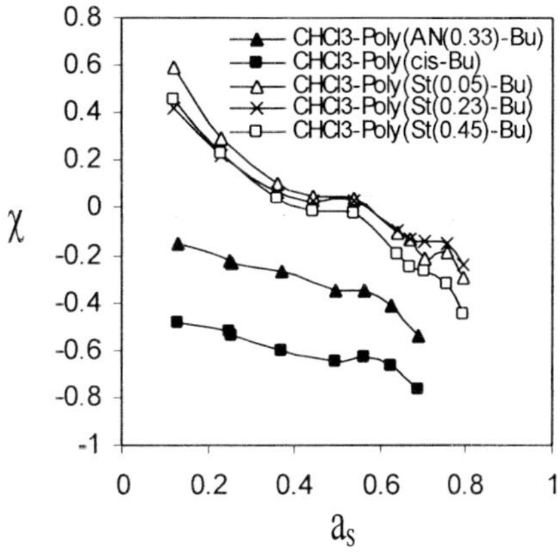

Figure 43. Flory interaction parameter χ varsus solvent-vapor activity for poly(AN(0.33)-Bu) and poly(cis-Bu) at 60°c and for poly(St-co-Bu) at different weight fractions at 50°c in $CHCl_3$

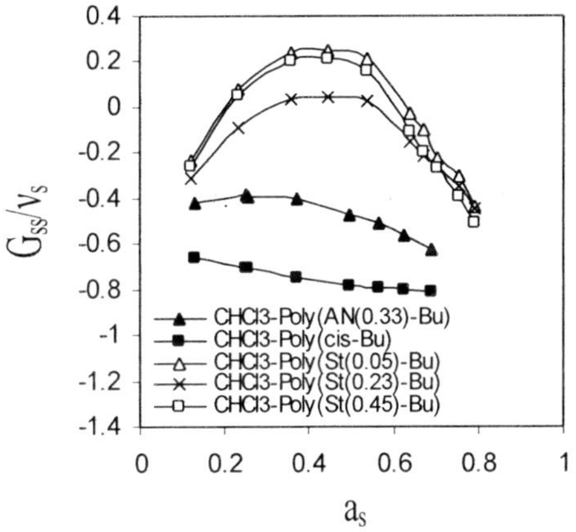

Figure 44. Clustering function G_{ss}/v_s at various solvent-vapor activities for poly(AN(0.33)-Bu) and poly(cis-Bu) at 60°c and for poly(St-co-Bu) at different weight fractions at 50°c in $CHCl_3$

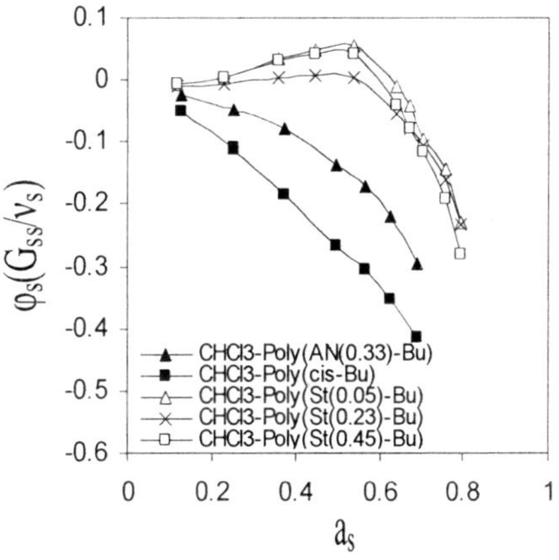

Figure 45. Mean cluster size $\varphi_s(G_{ss}/v_s)$ at various solvent-vapor activities for poly(AN(0.33)-Bu) and poly(cis-Bu) at 60°c and for poly(St-co-Bu) at different weight fractions at 50°c in $CHCl_3$

5.10. System: Poly (Ethylene Oxide)(PEO), Poly(Vinyl Acetate)(PVA) and Poly (Vinyl Acetate(VA)-Co-Ethylene(Et)) At 60°c in Chloroform

In Fig. 47 solvent weight fraction in the polymer is plotted against solvent vapor sorption. Almost all isotherms concave towards the activity axis and show a large degree of solvent uptake at relatively small activities. According to the BET classification, all systems exhibit a type I isotherm with a tendency for polymer/solvent pairs to be preferred.

Fig. 48 displays Flory interaction parameter, χ, as a function of solvent vapor activity. Chloroform is a good solvent for homopolymers, i.e. PEO and PVA, and for systems with vinyl acetate more than 50% in them. Going from poly(VA(0.09)-Et) to poly(VA(0.70)-Et) decreases value of χ, increases interaction between copolymer and solvent and causes solubility of chloroform rises in the proportion to the vinyl acetate content in the copolymer composition.

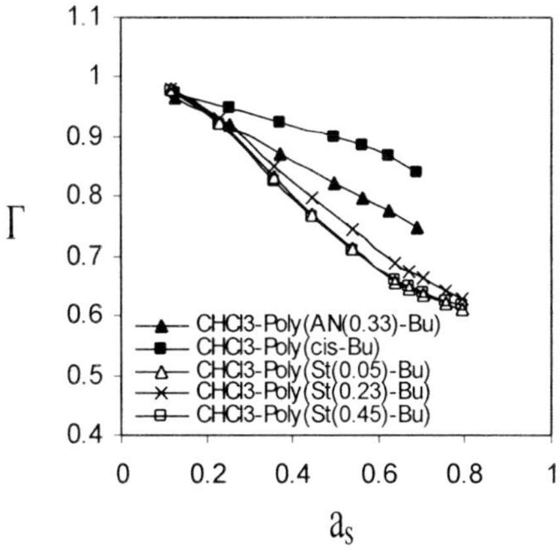

Figure 46.Thermodynamic factor Γ at various solvent-vapor activities for poly(AN(0.33)-Bu) and poly(cis-Bu) at 60°c and for poly(St-co-Bu) at different weight fractions at 50°c in $CHCl_3$

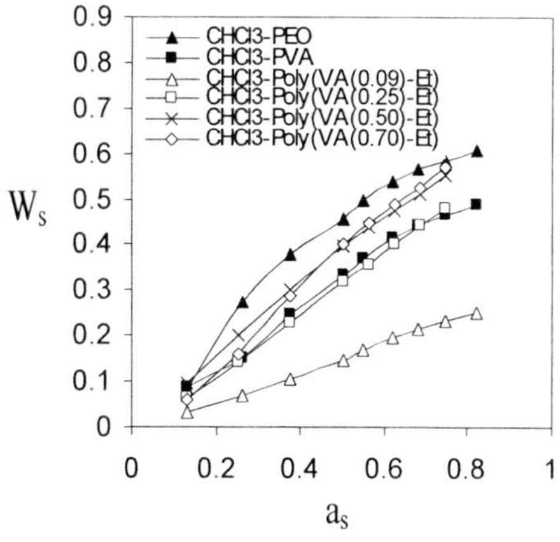

Figure 47. Solvent vapor sorption equilibria for PEO, PVA and poly(VA-co-Et) at different weight fractions at 60°c in $CHCl_3$.

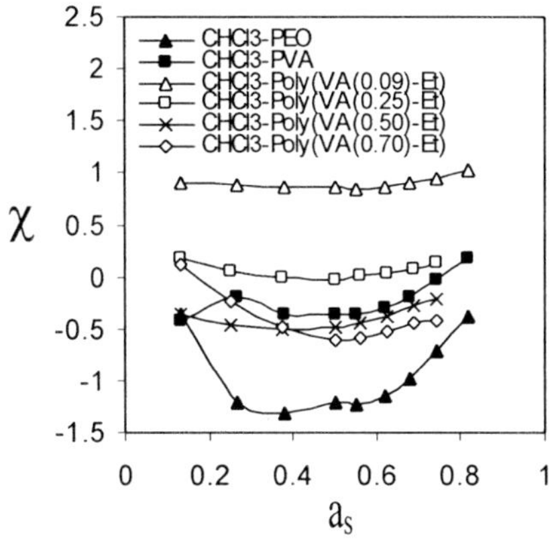

Figure 48. Flory interaction parameter χ versus solvent-vapor activity for PEO, PVA and poly(VA-co-Et) at different weight fractions at 60°c in $CHCl_3$

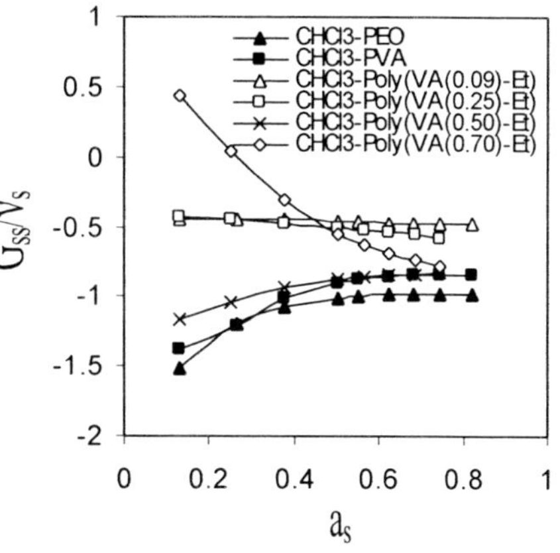

Figure 49. Clustering function G_{ss}/v_s at various solvent-vapor activities for PEO, PVA and poly(VA-co-Et) at different weight fractions at 60°c in $CHCl_3$.

Negative amounts of clustering function G_{ss}/V_s and mean cluster size $\varphi_s(G_{ss}/V_s)$, in fig. 49 and 50 indicates that chloroform does not cluster in these systems, except in the case of poly(VA(0.70)-Et) that solvent molecules slightly tend to cluster in small solvent activities. For above mentioned systems, clustering is not a simple function of copolymer composition.

When value of G_{ss}/V_s<-1, as for PEO at low activities, thermodynamic factor, Γ, in Fig. 51 will be an increasing function of solvent activity, and polymer has the least tendency to form cluster and the most interaction with solvent molecules. Deviation of copolymers from ideal behavior does not have a definite order, but the most deviation from ideality is seen in poly(VA(0.70)-Et).

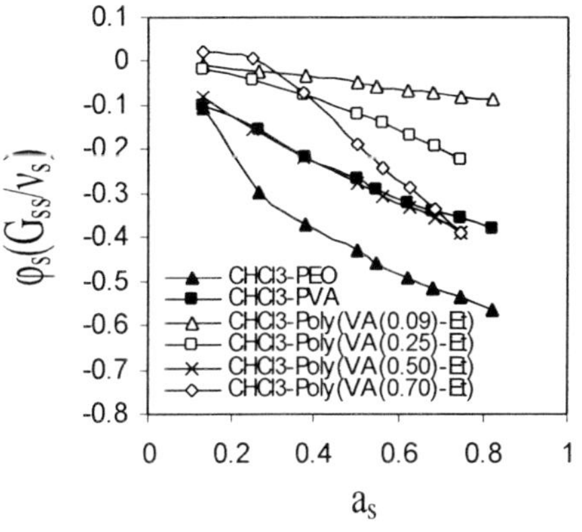

Figure 50. Mean cluster size $\varphi_s(G_{ss}/v_s)$ at various solvent-vapor activities for PEO, PVA and poly(VA-co-Et) at different weight fractions at 60°c in $CHCl_3$

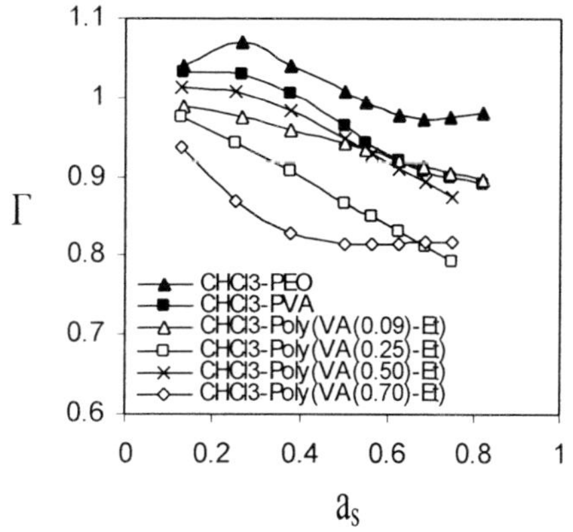

Figure 51. Thermodynamic factor Γ at various solvent-vapor activities for PEO, PVA and poly(VA-co-Et) at different weight fractions at 60°c in $CHCl_3$.

5.11. System: Poly (Cis-Butadiene)(Cis-Bu) and Poly (Acrylonitrile(AN)-Co-Butadiene(Bu)) at 60°c and Poly(Styrene(St)-Co-Butadiene(Bu)) At 23.5°c And 60°c in Cyclohexane(C_6H_{12})

Fig. 52 dispalys solvent weight fraction in the polymer versus solvent vapor activity. At a given activity, progressing from poly(cis-Bu) to poly(AN(0.21)-Bu) to poly(AN(0.51)-Bu), the solubility of cyclohexane decreases dramatically with the acrylonitrile content in the copolymer composition due to strong repulsion between nonpolar cyclohexane molecules and highly polar acrylonitrile segments. In the case of poly(St-co-Bu), solubility of cyclohexane slightly decreases in comparison with poly(cis-Bu) and decreases with styrene content in the copolymer composition, because repulsion between cyclohexane and styrene segments is not much high.

Flory interaction parameter, χ, in Fig. 53 indicates that cyclohexane is a good solvent for poly(cis-Bu) with $\chi \approx 0$, but when acrylonitrile segments add to poly(cis-Bu) to build a copolymer, χ, goes high and increasingly so as percentage of acrylonitrile in the copolymer composition reach to 51%. Then, cyclohexane is a non-solvent for this copolymer. With adding styrene segments to poly(cis-Bu), in order to build a copolymer, we will see that when styrene content is low (≈5%) or when percentage of styrene goes high up to 45% but temperature is high (60°c), interaction parameter is small and cyclohexane is a fairly good solvent for these systems, but enhancement of styrene content accompany with low temperature (23.5°c) causes copolymer/solvent interaction and solubility power of cyclohexane declines.

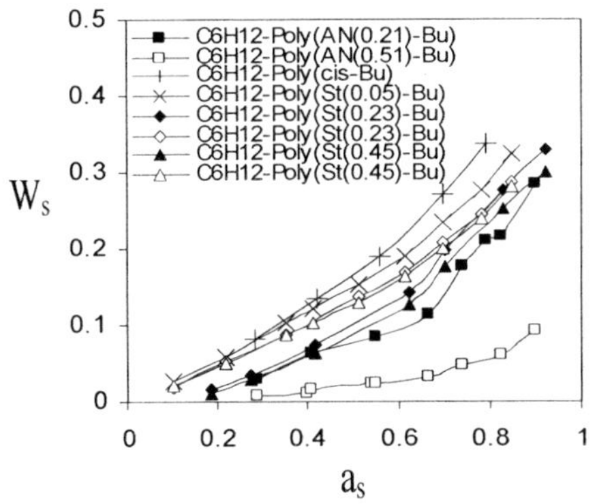

Figure 52. Solvent vapor sorption for poly(cis-Bu) and poly(AN-co-Bu) at60°c and for poly(St-co-Bu) at (♦,▲) 23.5°c and at (×,◊,Δ) 60°c at different weight fractions in C_6H_{12}.

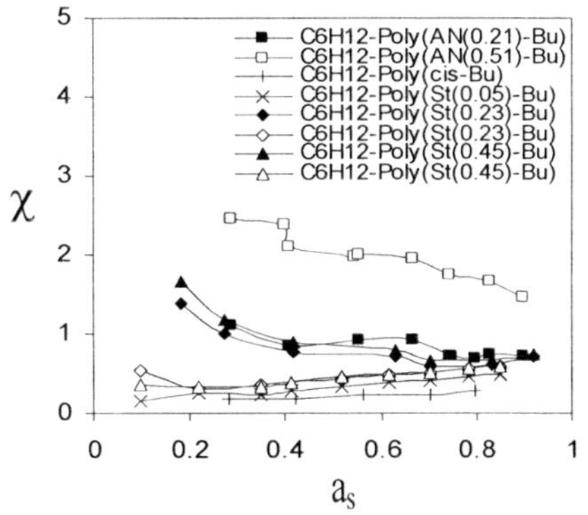

Figure 53. Flory interaction parameter χ versus solvent-vapor activity for poly(cis-Bu) and poly(AN-co-Bu) at 60°c and for poly(St-co-Bu) at (♦,▲) 23.5°c and at (×,◊,Δ) 60°c at different weight fractions in C_6H_{12}

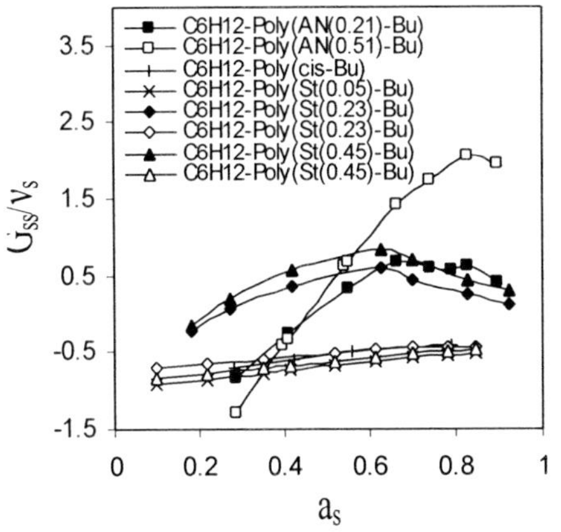

Figure 54. Clustering function G_{ss}/v_s at various solvent-vapor activities for poly(cis-Bu) and poly(AN-co-Bu) at 60°c and for poly(St-co-Bu) at (♦,▲) 23.5°c and at (×,◊,Δ) 60°c at different weight fractions in C_6H_{12}.

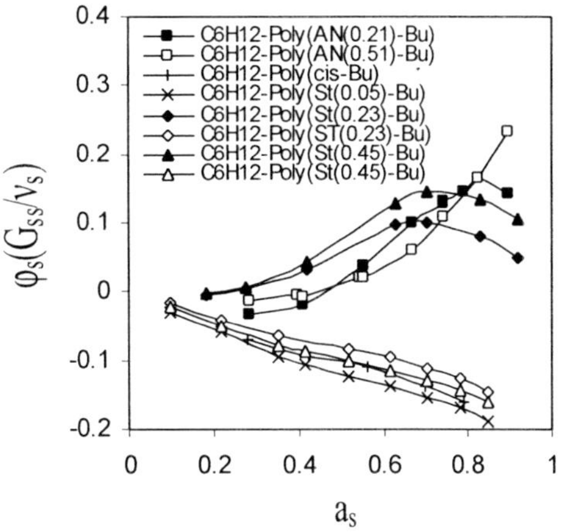

Figure 55. Mean cluster size $\varphi_s(G_{ss}/v_s)$ at various solvent-vapor activities for poly(cis-Bu) and poly(AN-co-Bu) at 60°c and for poly(St-co-Bu) at (♦,▲) 23.5°c and at (×,◊,Δ) 60°c at different weight fractions in C_6H_{12}

Figs. 54 and 55 show the values of the clustering function G_{ss}/V_s and mean cluster size $\varphi_s(G_{ss}/V_s)$ for above mentioned systems. Results for poly(AN-co-Bu) give negative values of G_{ss}/V_s and $\varphi_s(G_{ss}/V_s)$ for solvent activities up to 0.45, whereas at higher activities these values become positive, suggesting that cyclohexane molecules cluster in this region and clustering enhances with acrylonitrile content in the copolymer composition. Solvent molecules do not cluster in poly(cis-Bu) with the most interaction with cyclohexane molecules and with the most solvent uptake. Again poly(St-co-Bu)with low percentage of styrene or at relatively high temperatures has the most interaction with cyclohexane molecules and solvent molecules prefer to remain isolated. With rising styrene content in the copolymer composition and falling the temperature down to 23.5°c, there is a tendency for molecular pair association or clustering of solvent molecules to occur.

Deviation from ideality is evident in fig. 56 with decreasing function, Γ. Systems with greater amounts of styrene in them show the most deviation from ideal behavior at low temperature, but in the case of poly(AN-co-Bu) there is no definite order.

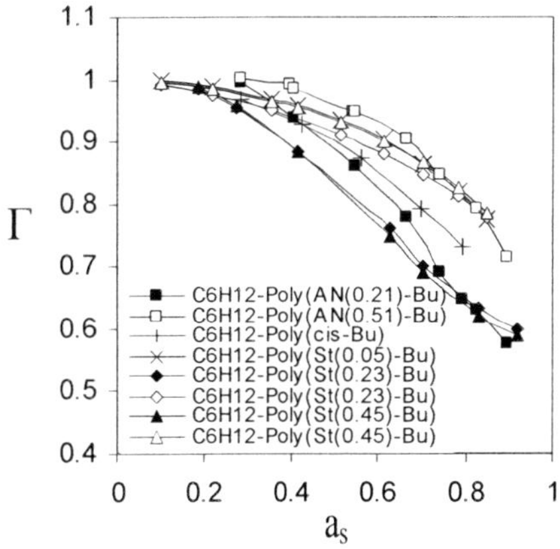

Figure 56. Thermodynamic factor Γ at various solvent-vapor activities for poly(cis-Bu) and poly(AN-co-Bu) at 60°c and for poly(St-co-Bu) at (♦,▲) 23.5°c and at (×,◊,Δ) 60°c at different weight fractions in C_6H_{12}

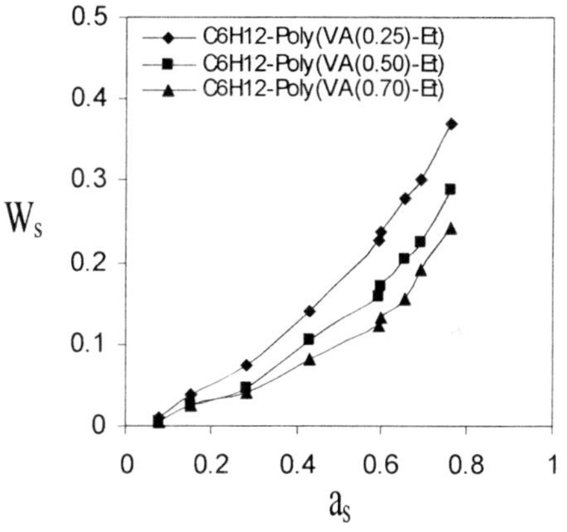

Figure 57. Solvent vapor sorption equilibria for poly(VA-co-Et) at different weight fractions at 80°c in C_6H_{12}.

5.12. System: Poly (Vinyl Acetate(VA)-Co-Ethylene(Et)) At 80°c in Cyclohexane (C_6H_{12})

Fig. 57 shows solvent weight fraction in the polymer versus solvent vapor activity. At a given activity, Solvent absorption in the copolymer increases as its ethylene content rises. This rise is expected because the the hydrocarbon segments of cyclohexane are better liked by the similar hydrocarbon segments of ethylene, whereas polar segments of vinyl acetate are disliked by solvent molecules. The solubility of cyclohexane falls in proportion to the vinyl acetate concentration in the copolymer.

Once again, influence of ethylene percentage in the copolymer on the copolymer/solvent interaction has been studied in Fig. 58. Flory interaction parameter, χ, decreases with increasing amount of ethylene and shows that cyclohexane as a solvent, plays its role well for poly(VA(0.25)-Et) especially at high activities where $\chi \approx 0$. With adding vinyl acetate to copolymer composition, interaction between cyclohexane and ethylene segments declines and cyclohexane is not a good solvent anymore.

In Fig. 59 and 60 values of the clustering function G_{ss}/V_s and the mean cluster size $\varphi_s(G_{ss}/V_s)$ have been determined. Results for poly(VA(0.25)-Et) give positive values of G_{ss}/V_s

and $\varphi_s(G_{ss}/V_s)$ for solvent activities up to 0.6 and indicate that cyclohexane cluster in this region but because of stronger interaction between solvent and copolymer at activities higher than 0.6, these values become positive, polymer/solvent pairs are preferred and clustering does not occur. In respect of copolymers with vinyl acetate more than 50%, G_{ss}/V_s and $\varphi_s(G_{ss}/V_s)$ are negative and due to weak polymer/solvent interaction, there is a tendency for clustering of solvent molecules to occur. Decreasing function, Γ, in Fig. 61 shows that deviation from ideal solution almost is the same for all systems.

5.13. System: Poly (Cis-Butadiene)(Cis-Bu), Poly(Acrylonitrile)(PAN) and Poly (Acrylonitrile(AN)-Co-Butadiene(Bu)) At 60°c and Poly (Styrene(St)-Co-Butadiene(Bu)) At 70°c in Hexane(C_6H_{14})

In Fig. 62 solvent weight fraction in the polymer is plotted against solvent vapor activity. At a given activity, progressing form PAN to poly(AN-co-Bu) to poly(cis-Bu), the solubility of hexane increases with the butadiene content in the copolymer. Hexane has negligible solubility in PAN due to strong repulsion between nonpolar cyclohexane segments and highly polar acrylonitrile segments. In the case of poly(St-co-Bu), solubility of hexane slightly decreases in comparison with poly (cis-Bu), and decreases with styrene content in the copolymer, because repulsion between hexane and styrene segments is not much high.

Flory interaction parameter, χ, in Fig. 63, indicates that hexane is a fairly good solvent for poly(cis-Bu) with χ close to zero, but when acrylonitrile segments add to poly(cis-Bu) to build a copolymer, χ increases with rising percentage of acrylonitrile in copolymer composition. Therefore, hexane is a moderate solvent for poly(AN(0.21)-Bu) ($0<\chi<1$), is a non-solvent for poly(AN(0.33)-Bu), poly(AN(0.51)-Bu) and PAN. Adding styrene segments to poly(cis-Bu) in order to build a copolymer causes hexane does not dissolve poly(St-co-Bu) as strong as poly(cis-Bu).

Figs. 64 and 65 show the values of clustering function G_{ss}/V_s and the mean cluster size $\varphi_s(G_{ss}/V_s)$ for above mentioned systems. Hexane does not cluster in poly(cis-Bu) due to $G_{ss}/V_s<0$. Results for poly(AN(0.21)-Bu) give negative values of G_{ss}/V_s and $\varphi_s(G_{ss}/V_s)$ for solvent activities up to 0.6, whereas at higher activities these values become slightly positive, suggesting that hexane plasticizes in this region and tend to form cluster at activities higher than 0.6. Clustering enhances with acrylonitrile content in the copolymer such that hexane cluster completely in poly(AN(0.33)-Bu), but poly(AN(0.51)-Bu) has a strange behavior and act in the manner of poly(cis-Bu) and both of them behave such as ideal solutions ($G_{ss}/V_s \approx -1$). Thermodynamic factor, Γ close to 1 in Fig. 66 is a confirmation on this behavior.

Also solvent molecules tend to cluster in PAN as a result of weak polymer/solvent interaction. In respect of poly(St-co-Bu), solvent-solvent pairs are preferred, solvent molecules associate with each other and clustering of hexane molecules occur. In Fig. 66, decreasing function, Γ, confirms that clustering is a dominant phenomenon in explanation of vapor-liquid equilibria of stated systems.

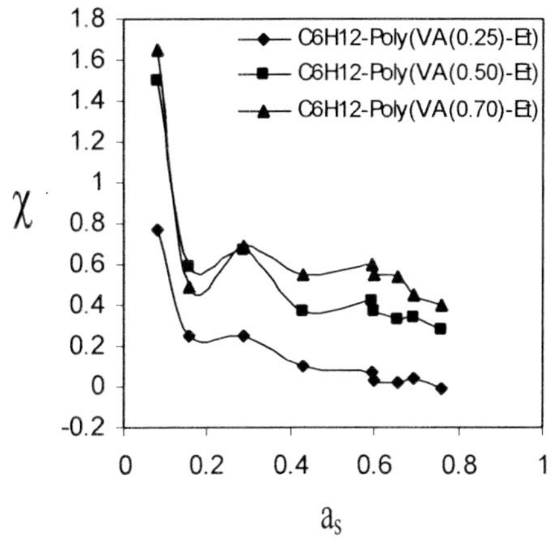

Figure 58. Flory interaction parameter χ versus solvent-vapor activity for poly(VA-co-Et) at different weight fractions at 80°c in C_6H_{12}

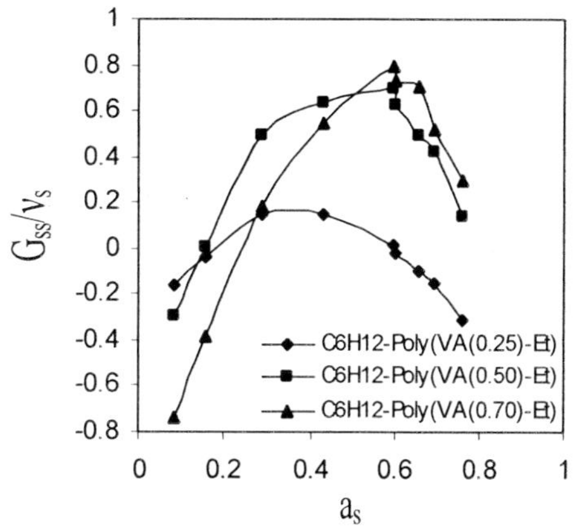

Figure 59. Clustering function G_{ss}/v_s at various solvent-vapor activities for poly(VA-co-Et) at different weight fractions at 80°c in C_6H_{12}.

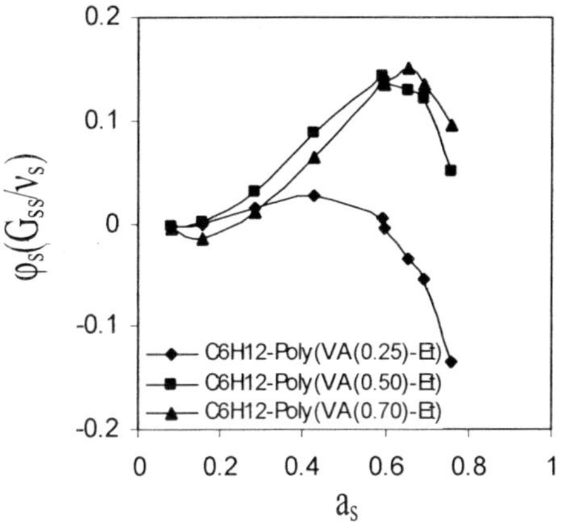

Figure 60. Mean cluster size $\varphi_s(G_{ss}/v_s)$ at various solvent-vapor activities for poly(VA-co-Et) at different weight fractions at 80°c in C_6H_{12}

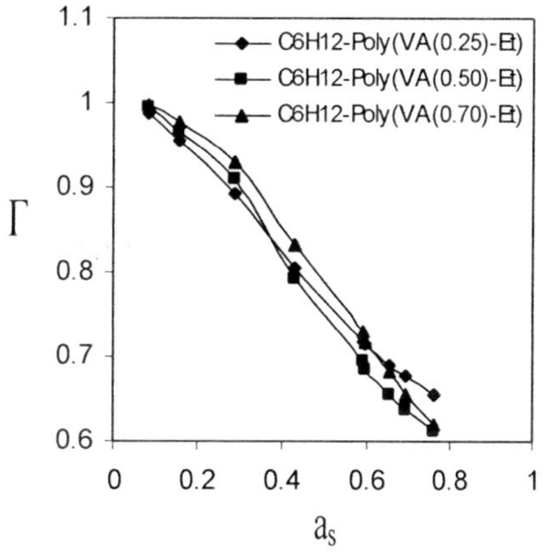

Figure 61. Thermodynamic factor Γ at various solvent-vapor activities for poly(VA-co-Et) at different weight fractions at 80°c in C_6H_{12}

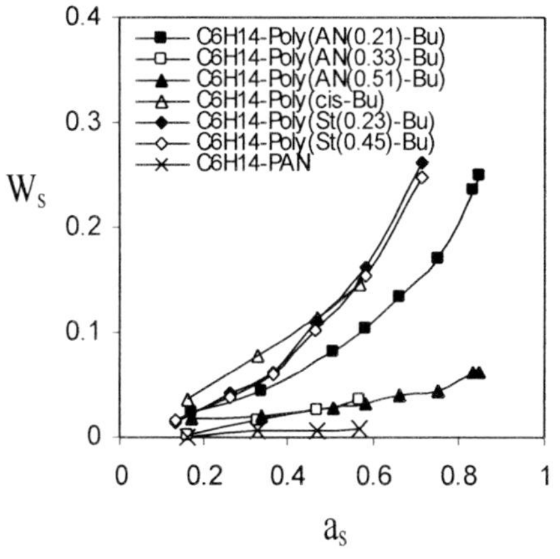

Figure 62. Solvent vapor sorption equilibria for poly(cis-Bu), PAN and poly(AN-co-Bu) at 60°c and for poly(St-co-Bu) at 70°c at different weight fractions in C_6H_{14}

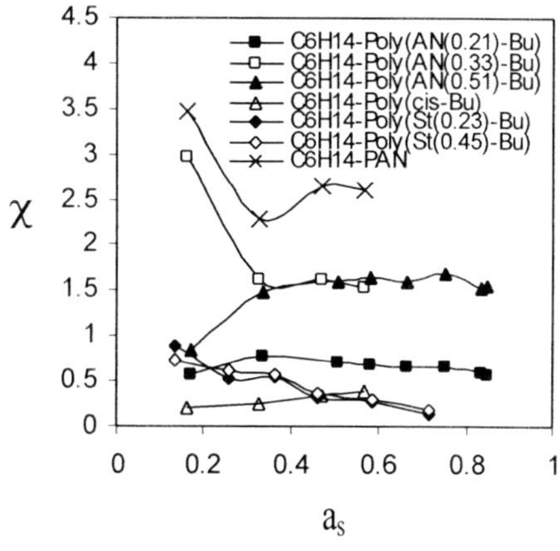

Figure 63. Flory interaction parameter χ versus solvent-vapor activity for poly(cis-Bu), PAN and poly(AN-co-Bu) at 60°c and for poly(St-co-Bu) at 70°c at different weight fractions in C_6H_{14}

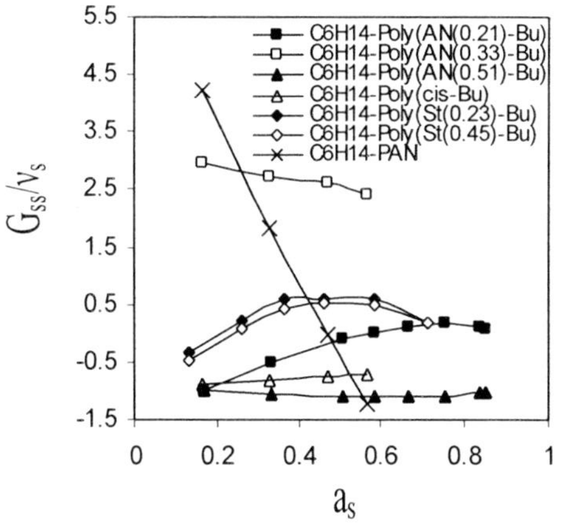

Figure 64. Clustering function G_{ss}/v_s at various solvent-vapor activities for poly(cis-Bu), PAN and poly(AN-co-Bu) at 60°c and for poly(St-co-Bu) at 70°c at different eight fractions in C_6H_{14}

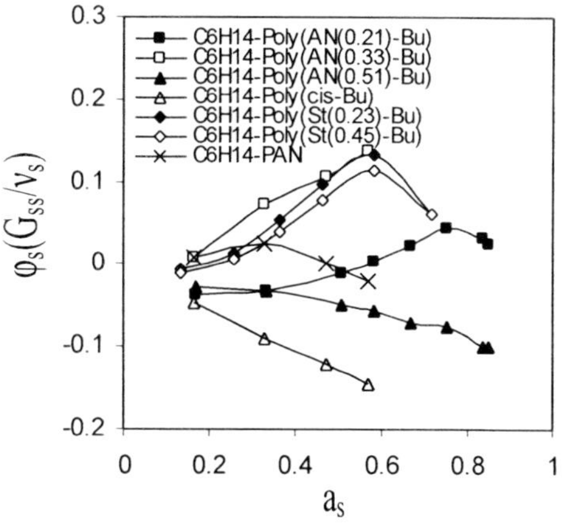

Figure 65. Mean cluster size $\varphi_s(G_{ss}/v_s)$ at various solvent-vapor activities for poly(cis-Bu), PAN and poly(AN-co-Bu) at 60°c and for poly(St-co-Bu) at 70°c at different weight fractions in C_6H_{14}.

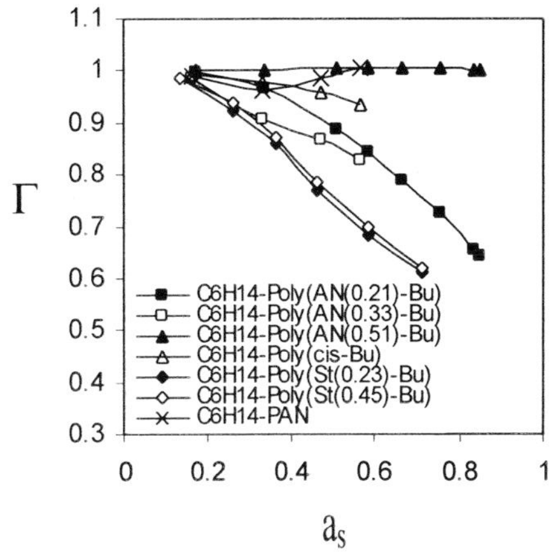

Figure 66. Thermodynamic factor Γ at various solvent-vapor activities for poly(cis-Bu), PAN and poly(AN-co-Bu) at 60°c and for poly(St-co-Bu) at 70°c at different weight fractions in C_6H_{14}

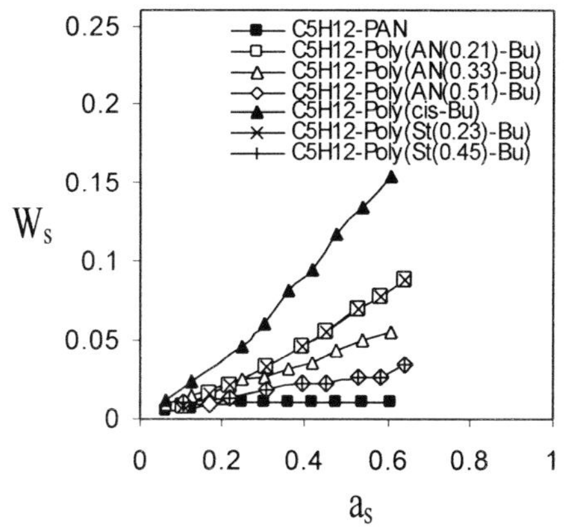

Figure 67. Solvent vapor sorption equilibria for PAN, poly(cis-Bu), poly(AN-co-Bu) and poly(St-co-Bu) at different weight fractions at 60°c in C_5H_{12}.

5.14. System: Poly(Acrylonitrile) (PAN), Poly(Cis-Butadiene) (Cis-Bu), Poly (Acrylonitrile(AN)-Co-Butadiene(Bu)) and Poly(Styrene(St)-Co-Butadiene(Bu)) At 60°c in Pentane

Fig. 67 shows solvent weight fraction in the polymer versus solvent vapor activity. Solvent absorption in the copolymer increases as its butadiene content rises. This rise is expected because the hydrocarbon segments of pentane are better liked by hydrocarbon segments of butadiene, whereas polar segments of acrylonitrile repulse nonpolar pentane molecules. Also, with presence of styrene segments to form poly(St-co-Bu) construction, because of small polarity of styrene segments, solubility of pentane decreases in comparison with poly(cis-Bu). This decline is in proportion to the styrene content in the copolymer.

Once again, Flory interaction parameter, χ, in Fig. 68 implies that with rising acrylonitrile concentration in copolymer composition, polymer/solvent interaction weakens such that pentane can highly dissolve poly(cis-Bu) and be a border solvent for poly(AN(0.21)-Bu) and dissolves poly(AN(0.33)-Bu) at low activities. In accordance with above discussion, pentane is a border solvent for poly(St(0.23)-Bu) and a non-solvent for poly(St(0.45)-Bu)

Clustering function G_{ss}/V_s and the mean cluster size $\varphi_s(G_{ss}/V_s)$ have been expressed in Figs. 69 and 70. However, there is a weak interaction between pentane and all stated systems except poly(cis-Bu), but none of them are liable to persuade clustering in them. Among these systems, those that have high percentage of butadiene, i.e. poly(cis-Bu), poly(AN(0.21)-Bu) and poly(St(0.23)-Bu), represent larger solvent uptake and their thermodynamic factor, Γ, in Fig. 71 is a decreasing function, can be a good medium for plasticizing of solvent molecules. In respect of poly(AN(0.33)-Bu), poly(AN(0.51)-Bu) and poly(St(0.45)-Bu) with $G_{ss}/V_s<-1$, non-ideality factor will be an increasing function of solvent activity and their deviation from ideal solution is greater than other systems. Finally, PAN with the lowest solvent uptake and the highest values of χ, show the most deviation from ideality. Of course, solvent molecules can not form cluster in this polymer, because solvent sorption is negligible in this system.

5.15. System: Poly (Vinyl Acetate) (PVA), Poly (Vinyl Acetate(VA)-Co-Vinyl Chloride(VC)) and Poly(Vinyl Acetate(VA)-Co-Ethylene(Et)) At 80°c in Methanol (CH_3OH) and Butanol (C_4H_9OH)

Fig. 72 gives the vapor sorption equilibrium data. Solvent weight fraction in the polymer is plotted against solvent vapor activity. Poly(VA(0.10)-VC) absorb only small amounts of methanol, but PVA has a slightly higher sorption for methanol due to attraction between polar vinyl acetate segments and methanol, and because of hydrogen bonding between OH groups. Also, small butanol sorption by poly(VA(0.09)-Et) and poly(VA(0.12)-VC), and high butanol sorption by PVA can be explained with the same reason stated for methanol.

Flory interaction parameter, χ, in Fig. 73 indicates that butanol can not dissolve poly(VA(0.09)-Et) and poly(VA(0.12)-VC), but PVA can be solved in butanol at activities higher than 0.4. Methanol is approximately a moderate solvent for poly(VA(0.10)-VC) at low activities but in the same region is a better solvent for PVA.

Because of negative values of clustering function G_{ss}/V_s and the mean cluster size $\varphi_s(G_{ss}/V_s)$ in Figs. 74 and 75, the above mentioned systems are not liable to permit solvent molecules to form cluster in them except PVA with G_{ss}/V_s and $\varphi_s(G_{ss}/V_s)$ greater than zero that is a good medium for butanol molecules to associate with each other and form cluster.

Non-ideality factor, Γ, in Fig. 76 is a decreasing function for PVA and confirms clustering of solvent molecules in this polymer. Poly(VA(0.09)-Et) and poly(VA(0.12)-VC) in butanol with G_{ss}/V_s=-1 and Γ=1 behave like ideal solutions. Poly(VA(0.10)-VC) and PVA in methanol have G_{ss}/V_s<-1 and therefore their deviation from ideality is high (Γ>1).

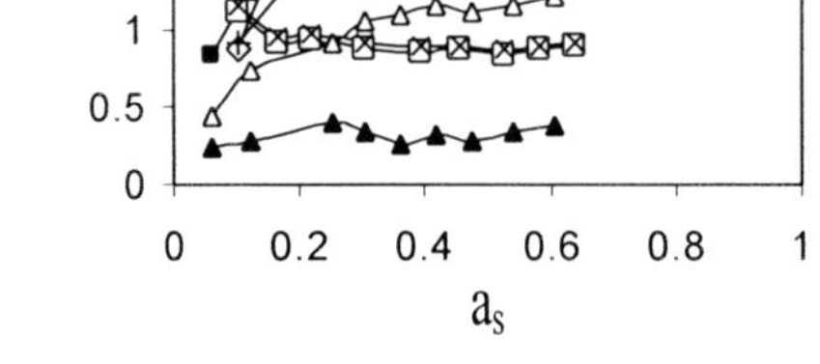

Figure 68. Flory interaction parameter χ versus solvent-vapor activity for PAN, poly(cis-Bu), poly(AN-co-Bu) and poly(St-co-Bu) at different weight fractions at 60°c in C_5H_{12}

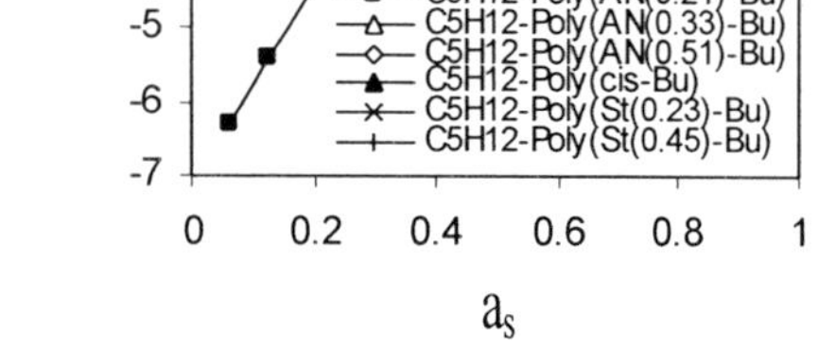

Figure 69. Clustering function G_{ss}/v_s at various solvent-vapor activities for PAN, poly(cis-Bu), poly(AN-co-Bu) and poly(St-co-Bu) at different weight fractions at 60°c in C_5H_{12}.

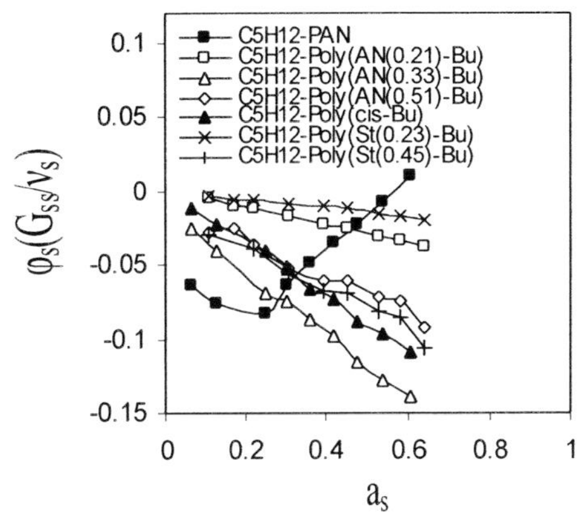

Figure 70. Mean cluster size $\varphi_s(G_{ss}/v_s)$ at various solvent-vapor activities for PAN, poly(cis-Bu), poly(AN-co-Bu) and poly(St-co-Bu) at different weight fractions at 60°c in C_5H_{12}.

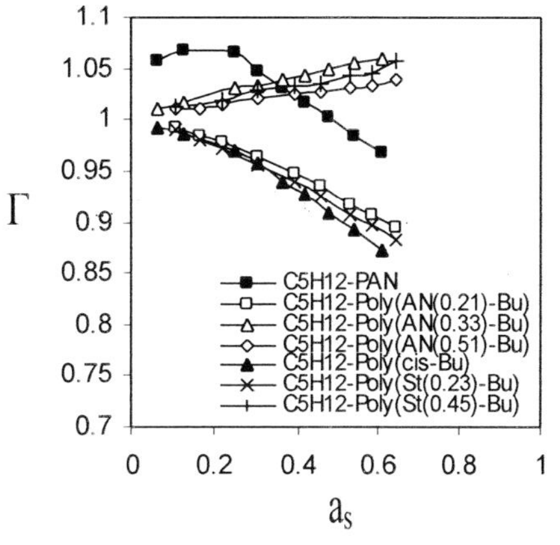

Figure 71. Thermodynamic factor Γ at various solvent-vapor activities for PAN, poly(cis-Bu) poly(AN-co-Bu) and poly(St-co-Bu) at different weight fractions at 60°c in C_5H_{12}.

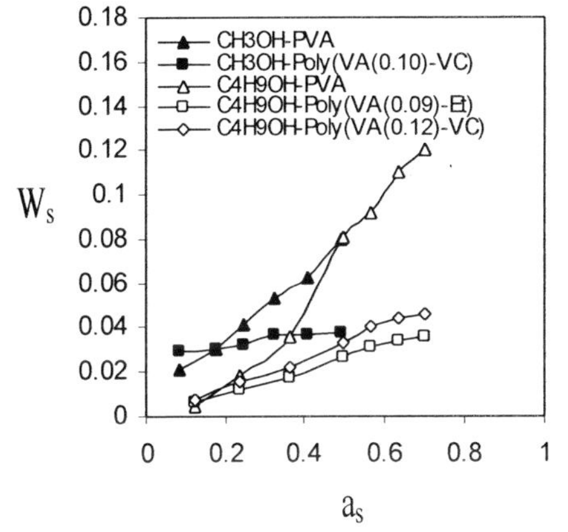

Figure 72. Solvent vapor sorption equilibria for PVA and poly(VA(0.10)-VC) in CH_3OH and for PVA, poly(VA(0.09)-Et) and poly(VA(0.12)-VC) in C_4H_9OH at 80°c

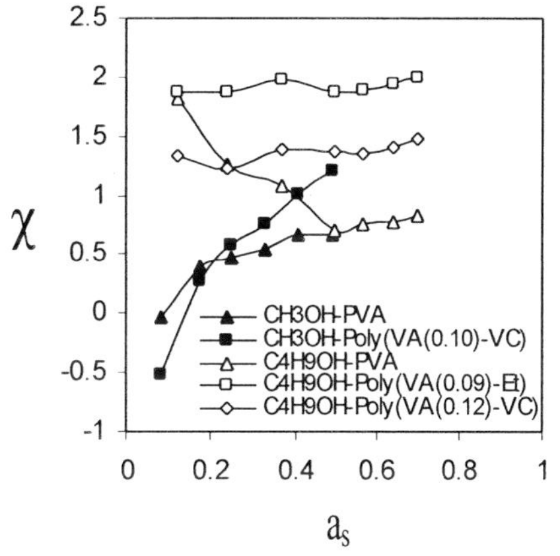

Figure 73. Flory interaction parameter χ versus solvent-vapor activity for PVA and poly(VA(0.10)-VC) in CH_3OH and for PVA, poly(VA(0.09)-Et) and poly(VA(0.12)-VC) in C_4H_9OH at 80°c.

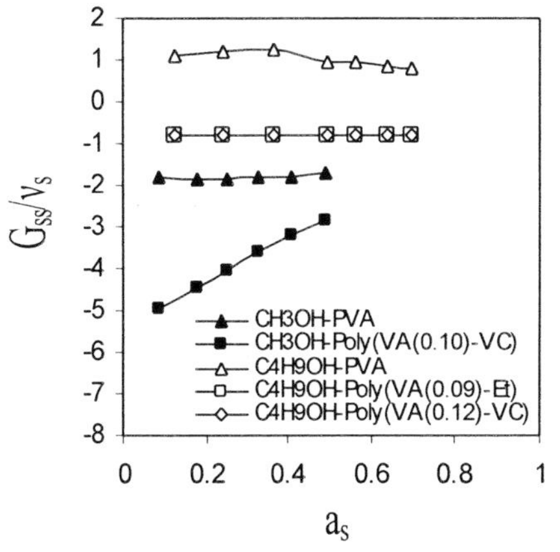

Figure 74. Clustering function G_{ss}/v_s at various solvent-vapor activities for PVA and poly(VA(0.10)-VC) in CH_3OH and for PVA, poly(VA(0.09)-Et) and poly(VA(0.12)-VC) in C_4H_9OH at 80°c.

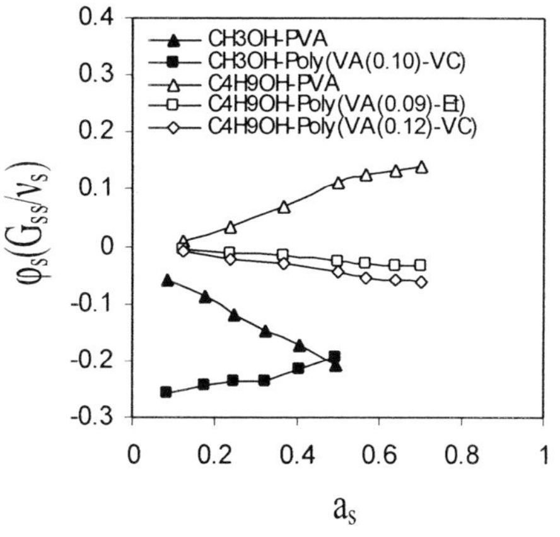

Figure 75. Mean cluster size $\varphi_s(G_{ss}/v_s)$ at various solvent-vapor activities for PVA and poly(VA(0.10)-VC) in CH_3OH and for PVA, poly(VA(0.09)-Et) and poly(VA(0.12)-VC) in C_4H_9OH at 80°c.

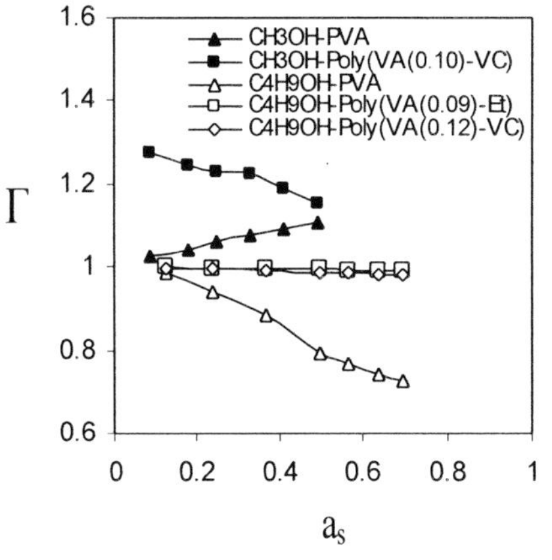

Figure 76. Thermodynamic factor Γ at various solvent-vapor activities for PVA and poly(VA(0.10)-VC) in CH_3OH and for PVA, poly(VA(0.09)-Et) and poly(VA(0.12)-VC) in C_4H_9OH at 80°c.

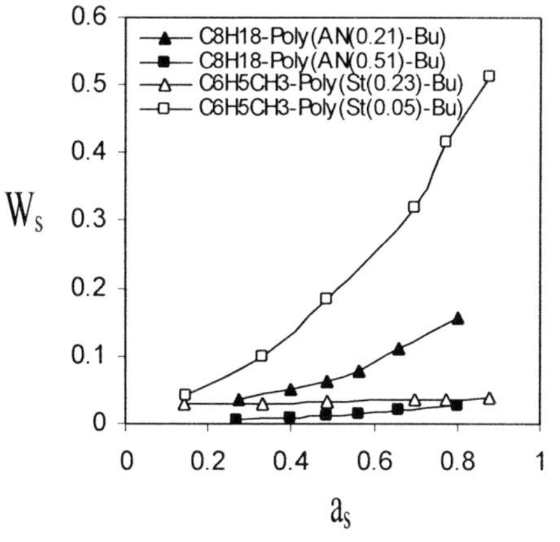

Figure 77. Solvent vapor sorption equilibria for poly(AN-co-Bu) at 60°c in C_8H_{18} and for poly(St-co-Bu) at 35°c in $C_6H_5CH_3$ at different weight fractions

5.16. System: Poly(Acrylonitrile(AN)-Co-Butadiene(Bu)) At 60°c in Octane (C_8H_{18}) and Poly(Styrene(St)-Co-Butadiene(Bu)) At 35°c in Toluene ($C_6H_5CH_3$)

Fig. 77 shows solvent weight fraction in the copolymer versus solvent vapor activity. Going from poly(AN(0.51)-Bu) to poly(AN(0.21)-Bu), solubility of octane increases as butadiene content in copolymer composition rises. This rise expected because hydrocarbon segments of octane are better liked by hydrocarbon segments of butadiene, whereas polar segments of acrylonitrile repulse nonpolar octane molecules. Also with presence of styrene segments to form poly(St-co-Bu) construction, solubility of toluene decreases as percentage of styrene increases.

Flory interaction parameter, χ, in Fig. 78 implies that octane and toluene are fairly good solvents for poly(AN(0.21)-Bu) and poly(St(0.05)0Bu) respectively. Solvent power of octane and toluene decreases with butadiene content in copolymer composition such that they are non-solvent for poly(AN(0.51)-Bu) and poly(St(0.23)-Bu) respectively.

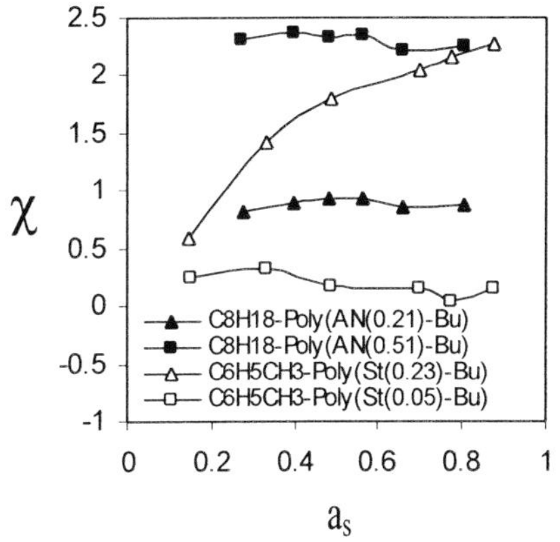

Figure 78. Flory interaction parameter χ versus solvent-vapor activity for poly(AN-co-Bu) at 60°c in C_8H_{18} and for poly(St-co-Bu) at 35°c in $C_6H_5CH_3$ at different weight fractions

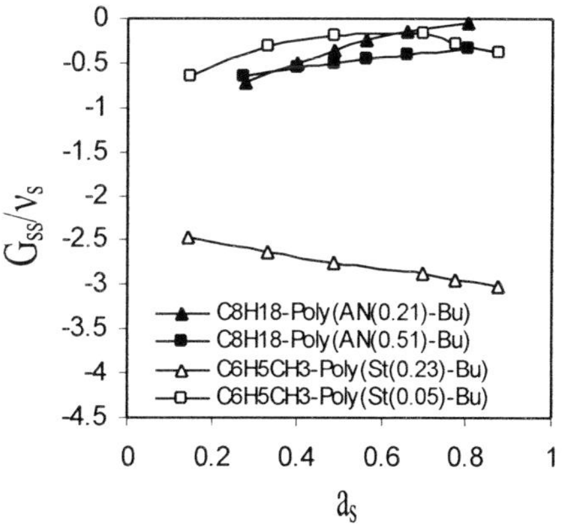

Figure 79. Clustering function G_{ss}/v_s at various solvent-vapor activities for poly(AN-co-Bu) at60°c in C_8H_{18} and for poly(St-co-Bu) at 35°c in $C_6H_5CH_3$ at different weight fractions

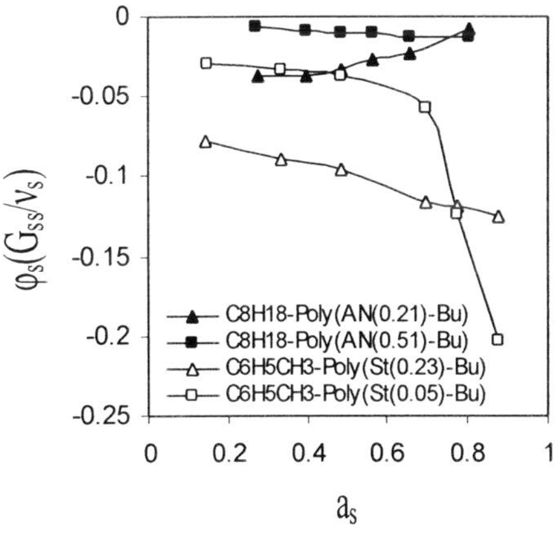

Figure 80. Mean cluster size $\varphi_s(G_{ss}/v_s)$ at various solvent-vapor activities for poly(AN-co-Bu) at60°c in C_8H_{18} and for poly(St-co-Bu) at 35°c in $C_6H_5CH_3$ at different weight fractions

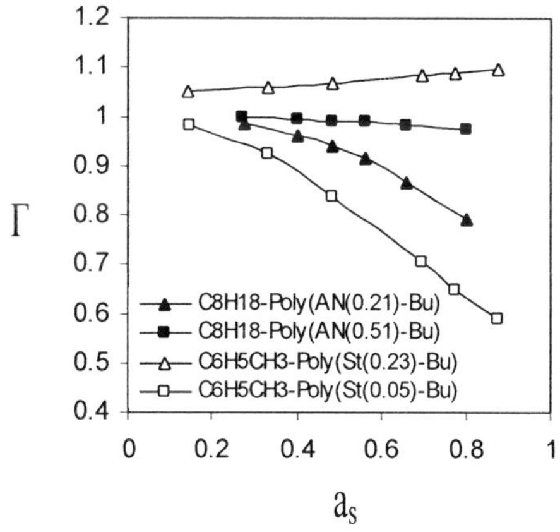

Figure 81. Thermodynamic factor Γ at various solvent-vapor activities for poly(AN-co-Bu) at60°c in C_8H_{18} and for poly(St-co-Bu) at 35°c in $C_6H_5CH_3$ at different weight fractions

Negative values of clustering function G_{ss}/V_s and the mean cluster size $\varphi_s(G_{ss}/V_s)$ in Figs. 79 and 80 indicate that solvent molecules prefer to remain isolated in both of the stated systems and clustering does not occur in these systems. For poly(AN(0.21)-Bu) and poly(St(0.05)-Bu) thermodynamic factor, Γ, is a decreasing function of activity and predicts possibility of plasticization of copolymer with sorbed solvent molecules. Poly(AN(0.51)-Bu) with Γ≈1 behaves like an ideal solution and poly(St(0.23)-Bu) with $G_{ss}/V_s<-1$ and Γ>1 deviates highly from ideal behavior.

6. Conclusion

Seventy-seven isothermal data sets for vapor-liquid equilibria (VLE) have been used for 35 polymer+solvent binaries and for 42 copolymer+solvent binaries to study sorption of solvent vapor in the polymer. The equilibrium solubility of a penetrant in a polymer depends on their mutual compatibility and, as a general rule, the well-known principle of "like dissolves like" is applicable. Equations based on theories of polymer solution tend to be more successful when there is some kind of similarity between the penetrant and the monomer repeat unit in the polymer, e.g., for nonpolar penetrants in polymers which do not contain appreciable polar groups. Expected nonideal behavior has been observed for systems such as cyclohexane+poly(ethylene-co-vinyl acetate) and especially for systems containing hydrocarbons and poly(butadiene-co-acrylonitrile).

The role of intramolecular interaction in vapor-liquid equilibria of copolymer+solvent systems is well documented for poly(acrylonitrile-co-butadiene) that have higher affinity for acetonitrile than do polyacrylonitrile or polybutadiene. As a basic assumption, when there exist a weak polymer/solvent interaction and clustering does not occur, large amounts of solvent uptake and decreasing function Γ can be interpreted by plasticization and if solvent uptake is low, neither clustering nor plasticization can not be seen.

In the discussion of sorption, we learned that when isotherms concave towards activity axis, formation of polymer-solvent pairs is preferred and solvent molecules remain isolated. For the systems studied in this chapter, it was also shown that the Flory-Huggins interaction parameter is dependent on the concentration and molecular weight of polymers considered.

Finally, it should be mentioned that the present physical interpretations are probably simplified but yield useful information concerning the physical process of the sorption phenomena.

7. References

[1] Koros, W. J.; Fleming, G. K. *J Membrane Sci* 1993, 83, 1-80.
[2] Kelkar, A. J.; Paul, D. R. *J Membrane Sci* 2001, 181, 199-212.
[3] Hoch, G.; Chauhan, A.; Radke, C. J. *J Membrane Sci* 2003, 214, 199-209.
[4] Fonn, D.; Situ, P.; Simpson, T. *Optom Visio Sci* 1999, 76, 700-704.
[5] Bikales, N. M. *Encyclopedia of Polymer Science and Technology*; John Wiley and Sons, Inc.: New York, NY, 1970; Vol. 12, PP 618-700.
[6] Atkins, P. W. *Physical Chemistry*; Oxford University Press: New York, NY, 2001, PP 163-190.
[7] Flory, P. J. *J Chem Phys* 1942, 10, 51-61.

[8] Huggins, M. L. *J Phys Chem* 1942, 46, 151-158.
[9] Meyer, K. H. Z *Phys Chem* 1939, B44, 383-391.
[10] Meyer, K. H. *Helv Chim Acta*, 1940, 23, 1063-1070.
[11] Flory, P. J. *Principles of Polymer Chemistry*; Cornell University Press: Ithaca, NY, 1953; Chap. 12.
[12] Guggenheim, E. A. *Trans Faraday Soc* 1937, 33, 151-159.
[13] Zimm, B. H. *J Chem Phys* 1953, 21, 934-935.
[14] Zimm, B. H.; Lundberg J. L. *J Phys Chem* 1956, 60, 425-428.
[15] Prausnitz, J. M.; Rodriguez, O.; Fornasiero, F.; Arce A.; Radke, C. *J. Polymer* 2003, 44, 6323-6333.
[16] Prausnitz, J. M.; Rodriguez, O.; Fornasiero, F.; Arce A.; Radke, C. *J. Phys Chem Chem Phys* 2004, 6, 103-108.
[17] Lundberg, J. L. *Pure Appl Chem* 1972, 31, 261-281.
[18] Vanzo, B. (2003). SP2. *Scientific Polymer Products*, Ontario, NY, Website: www.scientificpolymer.com.
[19] Panayiotou, C.; Vera, H. *Polymer J* 1984, 16, 89-102.
[20] Dojcansky, J.; Heinrich, *J. Chem Zvesti* 1974, 28, 157-159.
[21] Gerhartz W. *Ullmann's Encyclopedia of Industrial Chemistry*, VCH Publishers: New York, NY, 1986; Vol. A6, PP 263-271.
[22] Yaws, C. L. (2003) Yaws, *Handbook of Thermodynamic and Physical Properties of Chemical Compounds*, Electronic ISBN: 1-59-124-444-7.
[23] Vargaftik, N. B. *Handbook of Physical Properties of Liquids and Gases*; Begell House, Inc.: New York, NY, 1996; Chap. 4.
[24] Kemme, H. R.; Kreps, S. I. *J Chem Eng Data* 1969, 14, 98-102.
[25] Stull, D. R. *Ind Eng Chem* 1947, 39, 517-540.

In: Recent Developments in Polymer Research
Editor: Antony V. Hopper, pp. 135-162
ISBN 1-60021-346-4

Chapter 6

Development of Polyester Type Shape Memory Polymer and its Application to Composite Material

Yong-Chan Chung[a], Byoung Chul Chun[b], Mi-Hwa Chung[b], Yong-Sik Shim[a] and Jae Whan Cho[c]

[a]Department of Chemistry and [b]Department of Polymer Engineering,
The University of Suwon, Suwon, Korea
[c]Department of Textile Engineering, Konkuk University, Seoul, Korea

Abstract

Polyester type shape memory polymers were synthesized to improve their mechanical and shape memory properties and used for the preparation of sandwich type composite materials. Especially, poly(ethylene terephthalate) (PET) and poly(ethylene glycol) (PEG) copolymers with shape memory ability were prepared. After selecting the best composition of PET-PEG copolymer in mechanical properties, cross-linking agent such as glycerine, sorbitol, or maleic anhydride (MAH) was included for cross-linked copolymer, followed by analysis of its effect on mechanical, shape memory, and damping properties. The highest shape recovery was observed for copolymer with 2.5 mol% of glycerine, and the best damping effect indicating vibration control ability was from copolymer with 2.5 mol% of sorbitol. With the optimum copolymers in hand, sandwich-structured epoxy beam composites fabricated from epoxy beam laminate and cross-linked PET-PEG copolymer showed that impact strength increased from 1.9 to 3.7 times depending on the type of copolymer, and damping effect also increased as much as 23 times for the best case as compared to epoxy laminate beam alone. PET-PEG copolymers cross-linked with glycerol and sulfoisophthalate (SP) were also prepared to investigate the feasibility of vibration-control of composite laminate by additional ionic interaction. Composition of glycerol and SP was varied in order to get a copolymer with the best mechanical and shape memory properties. The highest shape recovery was observed for the copolymer with 2.5 mol% of glycerol and 2.5 mol% of SP. The sandwich-type copolymer composite showed improved impact strength (3.5 times) and damping effect (2.6 times) as compared to epoxy laminate beam alone. The resultant sandwich-structured epoxy beam composite can be utilized as structural composite material with vibration control ability and its glass transition temperature can be controlled by adjustment of hard segment content and cross linking agent composition.

1. INTRODUCTION

Since its introduction as a polynorborane type by Nippon Zeon company in 1984, various forms of shape memory polymers including poly(isoprene-butadiene-styrene), polyurethane, and polystyrene series were followed [1-6]. Specifically, shape memory polyurethane, developed by Nagoya Research and Development Center of Mitsubishi Company in 1988 [7], was composed of hard and soft segment and could recover the original shape in the temperature range of –30-65 °C. Hard segment and soft segment were alternatively polymerized, and various interactions among hard segments leaded to domain formation; hard segment worked as pivoting point for shape recovery and soft segment could mainly absorb external stress applied to the polymer. In order to utilize the shape memory effect, it is very important to control the structure and physical properties of shape memory polymers. Generally, deformation of polymer is much easier at the temperature above its T_g since it is in a rubbery state and has low modulus. Moreover, since there is a more than 100 times difference in modulus with respect to T_g, after the polymer deforms at above T_g and cooled back to below T_g with the load removed, it tends to retain its deformed shape. Then the deformation becomes frozen, and it is difficult to recover its original shape even though temperature is raised. Thus, normal polymers are difficult to recover its original shape. However, shape memory polymer has fixed points that can prevent plastic flow of chains and two phase structures that can reversibly change from soft to hard state depending on the temperature, thus recovering most of its original shape.

Meanwhile, active research is undergoing on smart materials such as shape memory alloy, semiconductor, polymer, and medical supply. Shape memory material, one of smart materials, has a few characteristics such as shape memory, shape retention, and impact absorption effect. In addition, it can detect thermal, mechanical, electrical, and magnetic stimulus, and then respond via property changes in shape, location, modulus, damping, and abrasion, which can be applied to various fields [8-10]. Shape memory polymer, being superior to other materials in the points of lightness (1.0-1.3 g/cm^3), high shape recovery (maximum shape recovery ratio more than 400 %), easy processing, and high damping near its T_g, has drawn our attention and is applied to manufacture of composite laminate with vibration control ability [11-14]. In our investigation, poly(ethyleneterephthalate) (PET)/polyethyleneglycol (PEG) copolymer is used as matrix material for the ultimate purpose of developing laminate composite with vibration control ability. Cross-linking agent such as glycerine, sorbitol, or maleic anhydride (MAH) was tried and its effect on the mechanical, shape memory, and damping property was compared. Comprehensive mechanical properties of various sandwich-structured laminate composite were examined, and reasons for such high vibration control were briefly discussed. Additionally, poly(ethyleneterephthalate) (PET)/polyethyleneglycol (PEG) copolymer is used as matrix material, and gylcerol and sulfoisophthalate group were adopted as covalent and ionic cross-linking agents, respectively. The effect of two different kinds of cross-linker on the mechanical, shape memory, and damping property of the copolymers was compared. Comprehensive mechanical properties of various sandwich-type composite laminate from the above copolymers were examined, and reasons for such high vibration control were briefly discussed.

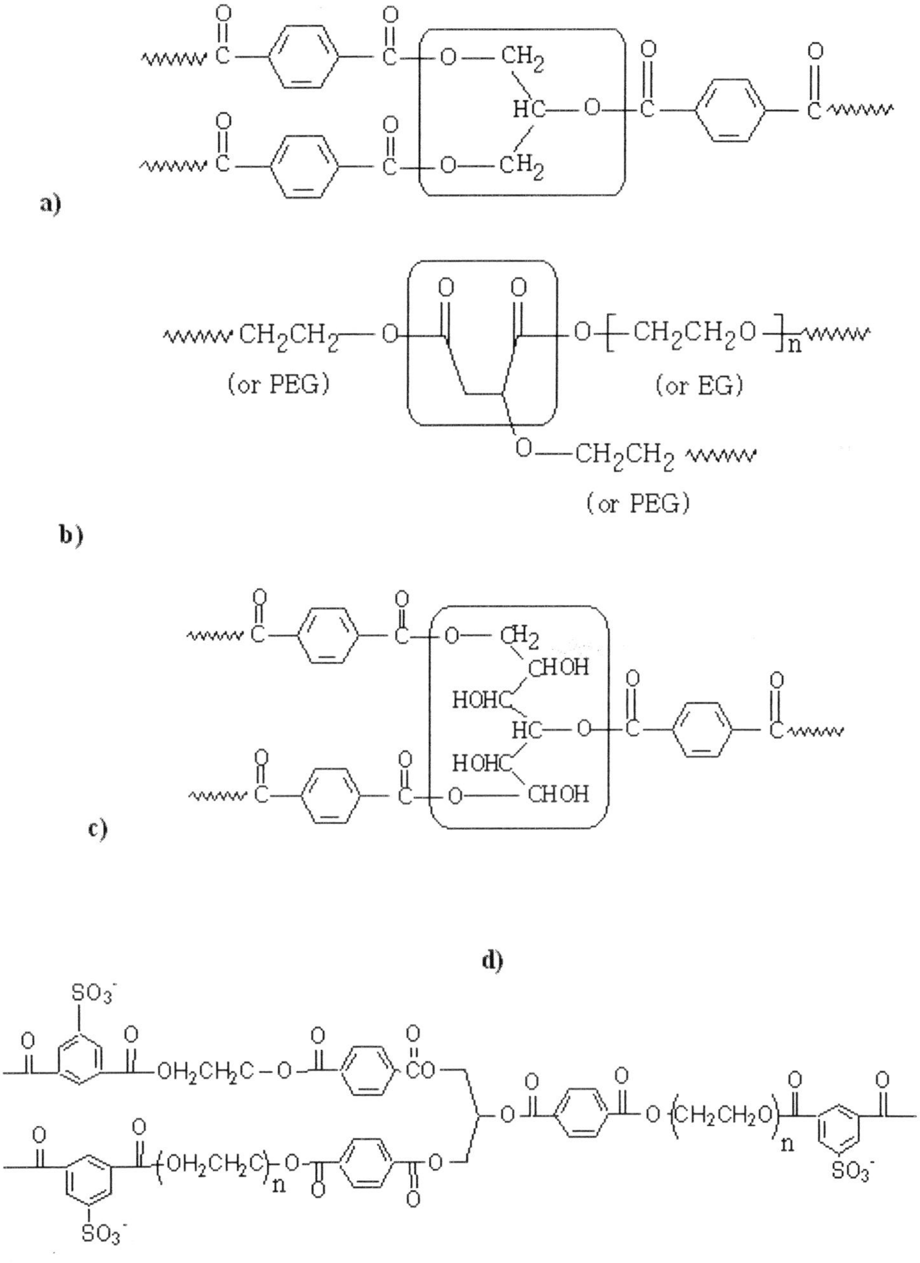

Figure 1. Schematic of PET-PEG copolymer cross-linked by (a) glycerine, (b) sorbitol, (c) maleic anhydride, and (d) glycerine and sulfoisophthalate

2. EXPERIMENTAL

2.1. Materials

DMT, EG, and PEG-200 were commercially obtained. Glycerol, sorbitol, maleic anhydride, and dimethyl 5-sulfoisophthalate sodium salt were used as obtained. Epoxy laminate beam with 1.0 mm thickness was supplied by Korea Fiber Co. and used to prepare unidirectional glass- fiber-reinforced composite with 2.7 mm total thickness.

2.2. Preparation of Cross-Linked Copolymer

PET-PEG copolymers were synthesized by melt-condensation method with a custom made reactor [13-15]. Polymerization was carried out in two steps; oligomer was prepared in the first step with DMT, EG, and PEG-200, and the oligomer made in first step was condensed and cross-linked with the agent in the second step at high temperature and high vacuum to shift reaction equilibrium further to product. Detailed synthetic procedure for PET-PEG copolymer can be found in our previous paper [13-15]. Synthetic scheme and characterization of copolymers are shown in the results and discussion section.

2.3. Intrinsic Viscosity

Intrinsic viscosity [η] of copolymer dissolved in 1,1,2,2-tetrachloroethane/phenol (4/6, w/w) mixture was measured with Ubbelohde viscometer at 35°C and 0.5 g/dl of concentration.

2.4. Thermal Analysis

T_g and T_m were measured by differential scanning calorimeter (DSC, Perkin Elmer Diamond 6). Specimen was heated to 200°C at 10°C/min of heating rate, and kept at that temperature for 3 min, and then cooled to –30°C at –10°C/min. T_g and T_m were determined from the second heating scan at 10°C/min heating rate.

2.5. Spectroscopic Analysis

NMR of the copolymer dissolved in $CDCl_3/CD_3OD$ mixture or trifluoroacetic acid (TFA) for cross-linked one was scanned by 600 MHz Bruker FT-NMR (Avance 600) at the National Instrument Center for Environmental Management, Seoul National University. FT-IR spectra were taken by Bomen MB series 104.

2.6. Preparation of Laminate Beam

Copolymer was preheated at 60°C for 24 hour to prevent any hydrolysis from moisture, and compression-molded to a sheet with 1 mm thickness and 5 mm width at 30 °C above its T_m. Sandwich-type laminate composite was prepared by compressing two 1.0 mm epoxy laminate beams and the above PET-PEG copolymer sheet in middle layer at 210 °C in a heating press.

2.7. Mechanical Properties and Shape Memory Analysis

All specimens for the mechanical property measurement were preheated at 60°C for 24 hours in order to prevent reverse reaction (hydrolysis) from the moisture during the processing, and compression molded to a sheet having 1mm thickness and 5 mm width at the temperature of 30°C above its T_m. Tensile tests were performed using universal testing machine (UTM, Lloyd LR 50K), and dumbbell type specimens according to ASTM D-638 were used with a crosshead speed of 100 mm/min. For dynamic mechanical property measurements, dynamic mechanical thermal analyzer (DMTA, Rheometric Scientific, Mark IV) was employed and specimen with 1mm thickness and 5 mm width sheet was used. All tests were conducted at a 3°C/min heating rate and 1.1 Hz.

Shape memory effect was tested using the identical specimen for a tensile test, and UTM equipped with temperature controlled thermal cabinet was used. For the shape retention rate measurement, following procedures were used. First, specimen with a length L_0 was strained to 100 % at the temperature above its T_g but below T_m, and stayed at that condition for 1min. Secondly, with the strain fixed, specimen was cooled back to the temperature below its T_g, and the load was removed and stayed at this condition for 30min, then the deformed length L_1 was measured. For the shape recovery measurement, after measuring L_1, specimen was heated again to the temperature above its T_g but below T_m, and stayed at this temperature for 10 min. Then the specimen was cooled back to the temperature below its T_g and stayed at this temperature for 30 min. After this, final specimen length L_2 was measured. Thus, shape retention rate and shape recovery rate can be defined as follows.

$$\text{Shape retention rate}(\%) = (L_1 - L_0)/L_0 \times 100$$
$$\text{Shape recovery rate}(\%) = (2L_0 - L_2)/L_0 \times 100$$

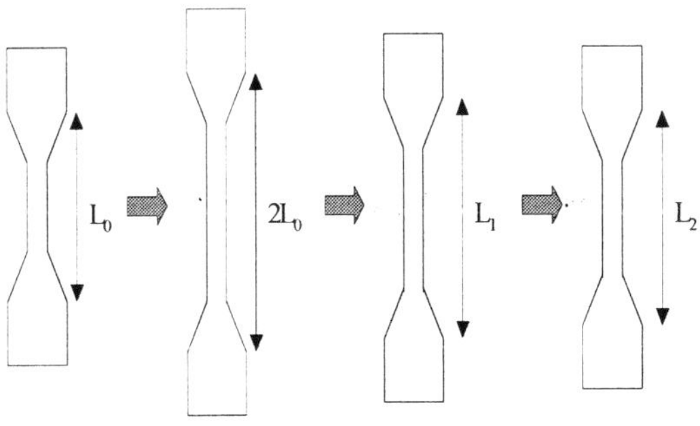

where

L_0 = initial specimen length

$2L_0$ = length of L_0 strained 100% at above T_g, below T_m

L_1 = deformed length at below T_g after load removal

L_2 = final specimen length at above T_g, below T_m

Figure 2. Specimen for testing shape retention and shape recovery rate

This whole cycle represent 1 shape memory test, and the copolymers were treated 3 shape memory tests.

3. Result and Discussion

3.1. Preparation of Copolymers

PET-PEG copolymers were prepared with DMT by melt-condensation method using the polymerization reactor custom made by Go Do Engineering Co. The copolymers were prepared in two different ways by either changing MW or composition of PEG. In addition, cross-linking agent was added in various ratios into the copolymer composed of 4 mol% of EG and 1 mol% of PEG. The polymerization reaction was carried out in two steps; in the first step, oligomers were prepared from DMT, EG, and PEG, and in the second step, the oligomers made in first one was condensed to polymers by esterification reaction which required high temperature, constant stirring, and high vacuum to shift the reaction equilibrium further to the products. The mole ratio of reactants was 50:80:20 (DMT:EG:PEG).

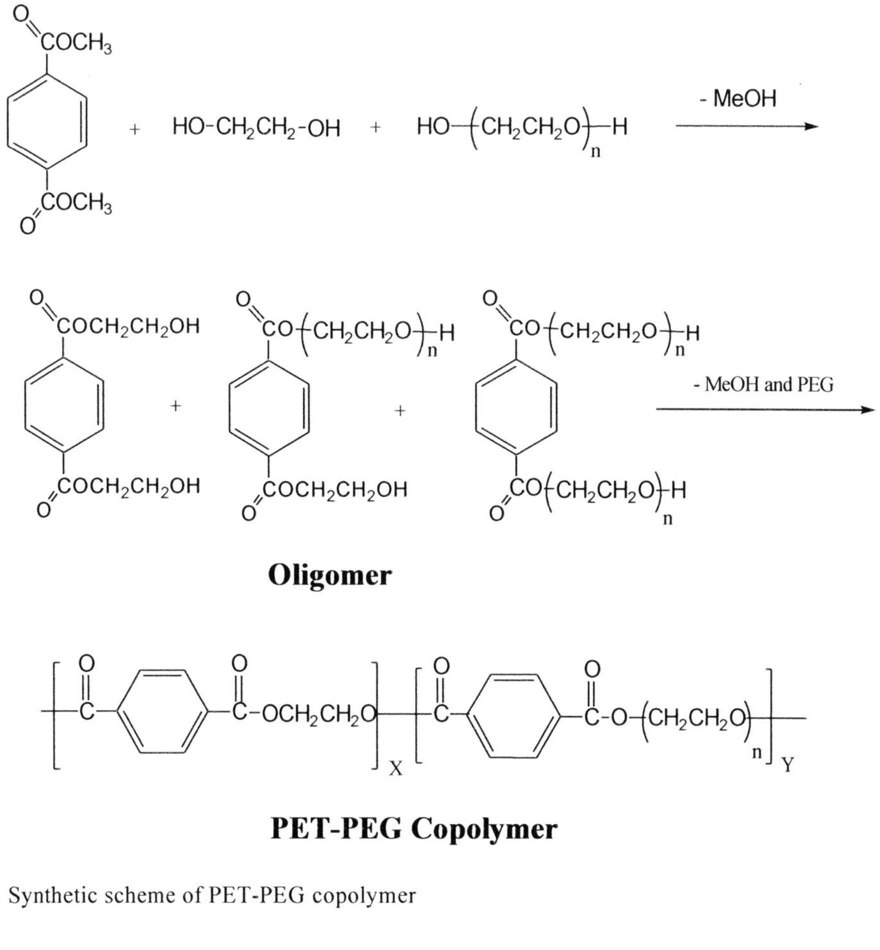

Figure 3. Synthetic scheme of PET-PEG copolymer

Into the mixture containing EG, PEG (200 g/mol), and DMT that was heated to 145°C was added calcium acetate as catalyst, and the mixture was further heated to 200 °C. Methanol was removed from the reaction mixture by distillation and the esterification was continued until the methanol was not produced any more, followed by the addition of phosphorous acid as stabilizer with an additional 10 minute stirring. After transferring the oligomer mixture into the polymerization reactor that was heated to 245°C and addition of antimony oxide as catalyst, the mixture was heated to 270°C and kept under vacuum (45 mmHg) for 30 minute to remove unreacted EG and PEG. Following the addition of cross-linking agent, the reaction was carried out for polymer condensation under high vacuum (1 mmHg) for 3 hrs. The synthetic scheme of copolymers synthesized is shown in Figure 3.

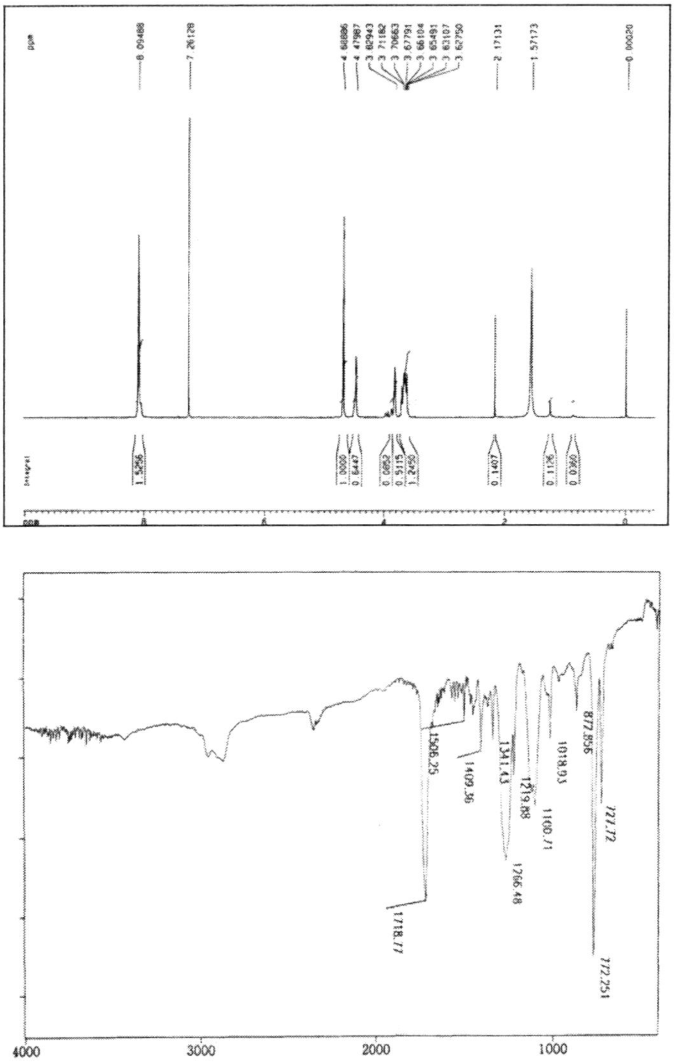

Figure 4. Spectra (NMR and FT-IR) of PET-PEG copolymers

3.2. Spectroscopic Analysis of the Copolymers

In Figure 4, the proton NMR spectrum of a cross-linked polymer showed that chemical shift of EG moved from 3.4 ppm of free EG to 4.4~4.7 ppm of the polymer, originating from the electron-withdrawing ester formation during polymerization, and the composition of EG and PEG calculated by dividing integration ratio between EG and PEG with their number of protons was found to be 4.05:1.00 which was in excellent agreement with the theoretical ratio of 4:1. Proton NMR peaks of PEG and DMT were observed at 3.5-3.8, and 8.0-8.2 ppm, respectively, and inclusion of cross-linking agent generally shifted the above peaks downfield by 0.2-0.3 ppm. The peaks appearing at 1.57 ppm and 2.17 ppm were coming from the trace amount of methanol in CD_3OD solvent and cross-linking agent, respectively. In the FT-IR spectrum of Figure 4 was shown the complete disappearance of the –OH stretching at 3300 cm^{-1} which was found as broad band for the free EG and PEG. Judging from the NMR and FT-IR data, the copolymerization proceeded as expected.

3.3. Tensile Property

Figure 5 shows the maximum stress of PET-PEG copolymers prepared in this study. As can be seen, the maximum stress increased with the addition of PEG, but started to decrease afterwards. Especially, when the PEG with molecular weight 200 g/mol was used, highest maximum stress was obtained at 10 mol% of PEG. However, at more than 10 mol% PEG, the maximum stress started to decrease. Also, as the molecular weight of PEG increased, maximum stress generally decreased. Strain at break increased very slightly in the case of PET-PEG copolymers with the PEG molecular weight of 400, 600 and 1000 g/mol. However, in the case of PEG having 200 g/mol of molecular weight, the percent strain at break increased very slightly up to 10 mol% PEG. However, at more than 10 mol% PEG, percent strain at beak increased abruptly, and at 20 mol% of PEG, it reached more than 1000 %. Considering that the minimum percent strain at break that can test the shape memory effect is 100 %, shape memory effect can only be tested for the PET-PEG copolymers where the molecular weight of PEG is 200 g/mol and the composition is 15 to 20 mol%.

Maximum stress and percent strain at break of PET-PEG copolymers cross-linked by maleic anhydride (MAH) were shown in figure 6. Maximum stress showed an abrupt increase from about 3 MPa to 8.8 MPa at 1.5 mol% of MAH addition, and decreased to about 6 MPa at 2.5 mol% of MAH. But regardless of MA content, all PET-PEG-MAH copolymers showed higher maximum stress compared to PET-PEG copolymer. This is due to the increased resistance to deformation resulting from the formation of cross-links within the polymer chains. In the case of percent strain at break, it decreased from about 1100 % to about 750 % as the MAH is introduced. Even though there was a slight increase in percent strain at break between 1.5 and 2.5 mol% of MAH, the overall trend is obviously a decrease. This is an expected result considering that the added MAH increases the number of cross-links in the polymer chains, thus reducing the extensibility of chains.

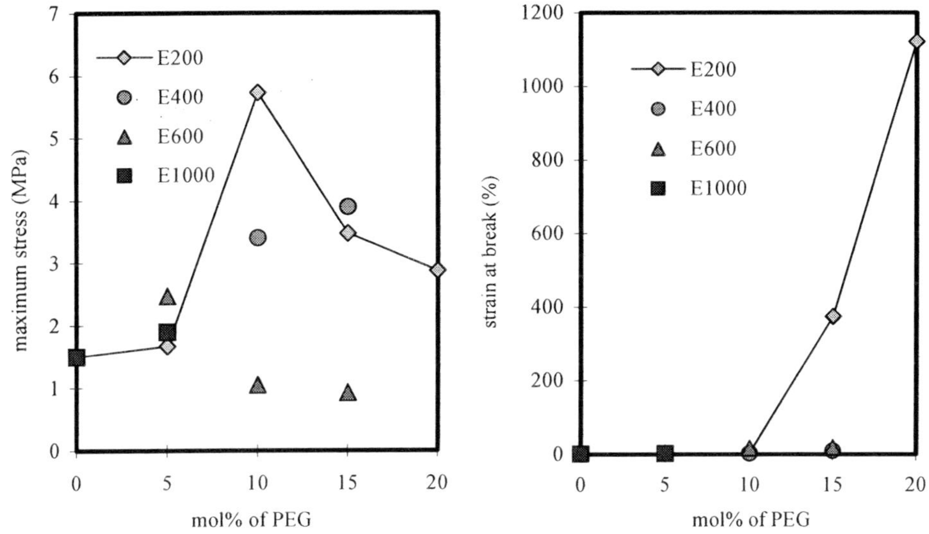

Figure 5. Maxmum stress and strain at break profiles of PET-PEG copolymers

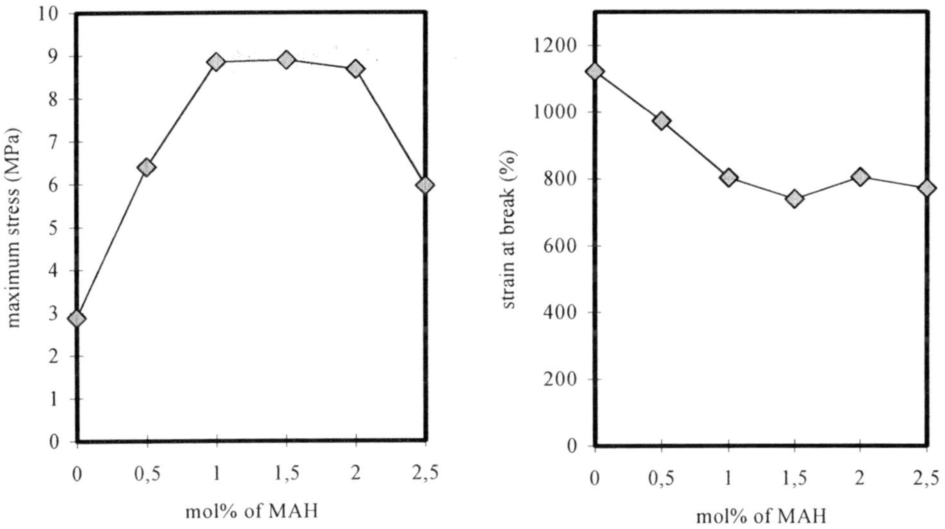

Figure 6. Maximum stress and strain at break profiles of PET-PEG-MAH copolymers

Because PET-PEG copolymer with 20 mol% of PEG 200 (E200-20) showed the best tensile test result, the composition was used for preparation of copolymer with cross-linking agent (glycerine or sorbitol) and related shape memory test (Table 1). Tensile test results of cross-linked copolymers indicated that tensile strength generally increased, but strain at break decreased as more cross-linking agent was included (Table 2).

Table 1. Properties of PET-PEG Copolymers

Sample Code	PEG(mol%)	PEG MW(g/mol)	T_g (°C)	T_m (°C)	[η](dl/g)
PET	0	-	79.1	252.0	0.68
E200-10	10	200	33.5	213.5	0.62
E200-20	20	200	8.1	167.3	0.60
E400-10	10	400	-12.9	200.0	0.70
E600-10	10	600	-32.4	194.6	0.67
E1000-5	5	1000	-42.5	221.2	0.55

Table 2. Comparison of Cross-linked PET-PEG Copolymers

Sample Code	PEG (mol%)	PEG MW (g/mol)	Cross-link Agent (mol%)	Maximum Stress (N/mm^2)	Strain at Break (%)	T_g (°C)
E200-20	20	200	-	5.9	1122	11.2
G-15	20	200	Glycerine (1.5)	16.5	317	14.3
G-25	20	200	Glycerine (2.5)	19.6	48	23.7
M-15	20	200	Maleic anhydride (1.5)	9.1	832	13.2
M-25	20	200	Maleic anhydride (2.5)	9.1	770	14.1
S-15	20	200	Sorbitol (1.5)	14.4	805	9.2
S-25	20	200	Sorbitol (2.5)	14.3	632	11.7

Such result undoubtedly originated from higher resistance to deformation by introduction of cross-linking in copolymer chain, and the concomitant cross-linked solid structure resulted in high tensile strength and low strain at break. Among cross-linked PET-PEG copolymers, one cross-linked by glycerine showed the highest tensile strength and the lowest strain at break.

Maximum stress of glycerol (1.0 and 1.5 mol%) cross-linked copolymers showed initial decrease with SP content and sharp increase at SP content higher than 1.5 %; copolymers with 2.5 mol% glycerol content had gradual increase with SP content. Strain at break of copolymers (1.0 and 1.5 mol% of glycerol) increased with SP contents up to 1.5 % and decreased at higher content, but 2.5 mol% glycerol copolymer showed slow decrease with inclusion of SP. Both maximum stress and strain at break results are similar to Tg data in the following section, and explanation for the combined results will appear there.

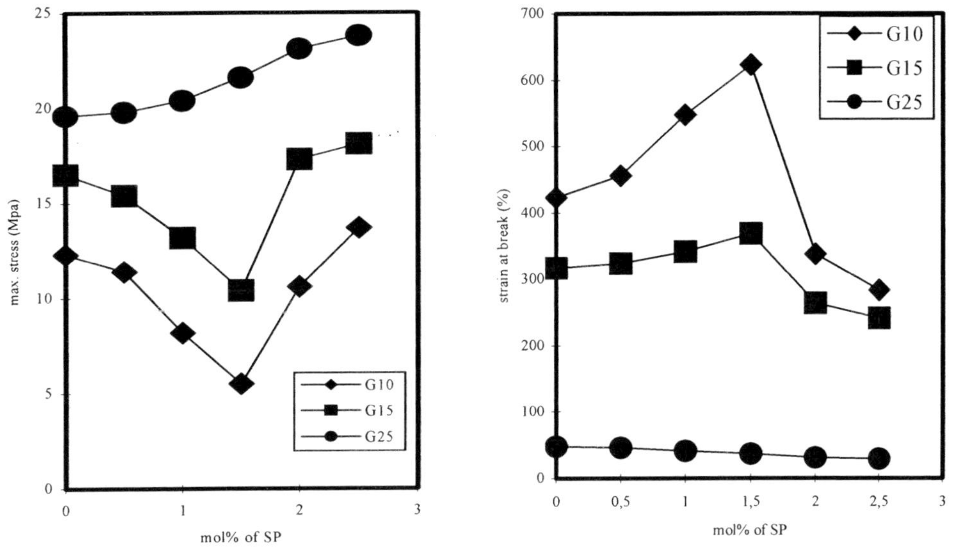

Figure 7. Maximum stress and strain at break profiles of PET-PEG-Glycerol-SP copolymers

3.4. Thermal Property

As shown in Figure 8, T_g decreases rapidly as the PEG is introduced, but this tendency is weakened as the PEG content increases. At the same PEG content, PEG with higher molecular weight showed a larger decrease in T_g values. T_m also decreases linearly with the PEG addition. Compared to T_g behavior, T_m change with respect to molecular weight of PEG is small, but as the molecular weight of PEG decreases, T_m decreases less. Thus, T_g and T_m decrease behavior is due to the increase of chain flexibility and decrease of the chain regularity with the addition of PEG.

Table 3 shows the similar data of PET-PEG copolymers cross-linked by MAH. T_g was affected by the MAH addition, but T_m was not. In general, T_g increased from 8.1°C to 17.7°C when 2.5 mol% of MAH was introduced. The observed T_g increase with the MAH addition is due to the restricted molecular motion resulting from the increased cross-linking [16].

Table 3. Characterization of PET-PEG copolymers cross-linked by MAH

Sample Code	PEG MW (g/mole)	MAH (mole%)	T_g (°C)	T_m (°C)	[η] (dl/g)
E200-20	200	0	8.1	167.3	0.60
MAH05	200	0.5	8.3	165.1	0.62
MAH10	200	1.0	9.8	164.2	0.71
MAH15	200	1.5	11.6	166.3	0.65
MAH20	200	2.0	13.6	164.3	0.79
MAH25	200	2.5	17.7	164.2	0.67

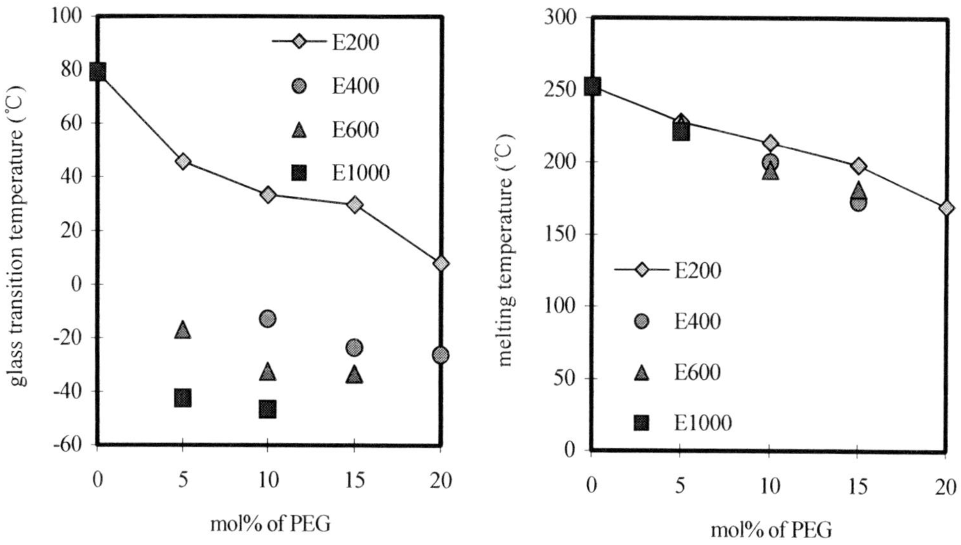

Figure 8. Phase transition temperatures of PET-PEG copolymers

T_g of the copolymers with different combination of glycerol and SP was compared in Figure 9.

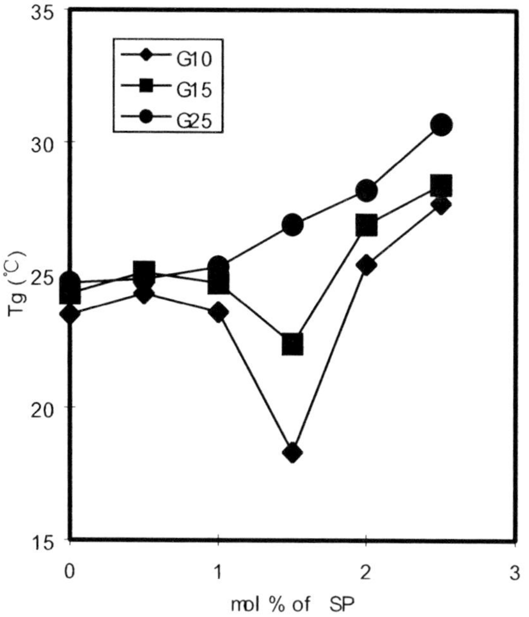

Figure 9. Glass transition temperatures of PET-PEG-Glycerol-SP copolymers

For the copolymers with 1.0 and 1.5 mol% of glycerol, addition of SP decreased the transition temperature up to 1.5 mol%, and steep increase of Tg was observed from 1.5 to 2.5

mol% region. Unlike the above series, the copolymers with 2.5 mol% of glycerol showed constant increase of T_g with the SP content. Such result can be understood from the fact that the copolymer chain is not tightly cross-linked at lower mol% of glycerol (1.0 and 1.5 mol%) and the presence of SP along the chain hinders ordered chain packing, thus resulting in reduced T_g. But, as the SP content increases, inter-chain ionic interaction among sulfonyl sodium salt groups additionally improves inter-chain attraction and, therefore, T_g value increases at SP content higher than 1.5 mol%. Highest inclusion of glycerol (2.5 mol%) dominates the copolymer cross-linking, and the decrease of T_g at lower mol% of SP observed in the above case do not happen because covalent cross-linking of glycerol can subdue the disrupting effect of SP. As more SP was included, T_g of 2.5 mol% glycerol copolymer also increased due to the additional ionic inter-chain interaction originating from sulfonyl groups. The above explanation can be applied to the tensile test results. Such trend can be consistently observed in the following mechanical and shape recovery data, and combination of covalent cross-linking by glycerol and ionic interaction by SP can manipulate the desired transition temperature of copolymers.

3.5. Shape Memory Effect

Shape memory experiments were performed on the PET-PEG-MAH copolymers. As shown Figure 10, PET-PEG-MAH copolymers generally exhibited a slightly lower shape retention value compared to PET-PEG copolymer, but all specimens showed an excellent shape retention of more than 90 %. The overall trend is a decrease in shape retention with the MA addition, but shape retention increased again at more than 2 mol% of MAH. Maximum shape recovery rate is obtained at 1.5 to 2.0 mol% of MAH, which is due to the effect of cross-links formed by the MAH on the recovery of the soft segment: PET acts as physical cross-link point and MAH does as chemical cross-link point. And it is dispersed evenly within the matrix as physical cross-link point, and results in the improvement of shape memory effect in PET-PEG-MAH copolymers. This behavior can be observed in the Figure 10, which shows the shape recovery rate of PET-PEG-MAH specimens after 3 cycles of shape memory tests. Best shape recovery effect is observed for the PET-PEG-MAH copolymer with 2mol% of MAH, which is due to the high resistance to plastic deformation of chain segment to the stress by chemical cross-linking of MAH. In contrast, PET-PEG copolymer fractured after only 2 cycles of shape memory tests due to the severe plastic deformation of chains to the stress.

As for glycerol or sorbitol cross-linked copolymers, shape retention rate slightly decreased, but shape recovery significantly increased after cross-linking. Especially, PET-PEG copolymer cross-linked by glycerol showed the highest shape recovery rate (Figure 11). For example, shape recovery rate of E200-20 decreased from 50.5 % to 21.5 % after two cycles of shape memory test, but G-15 and G-25 maintained 51.0 % and 79.9 %, respectively, under the same condition. PET-PEG copolymer cross-linked by sorbitol also exhibited high shape recovery rate. The specific shape memory test results are summarized in table 4. As for all cross-linked copolymers shape retention rate was more than 95 %, which told us that deformed shape of copolymer could be almost perfectly kept below transition temperature.

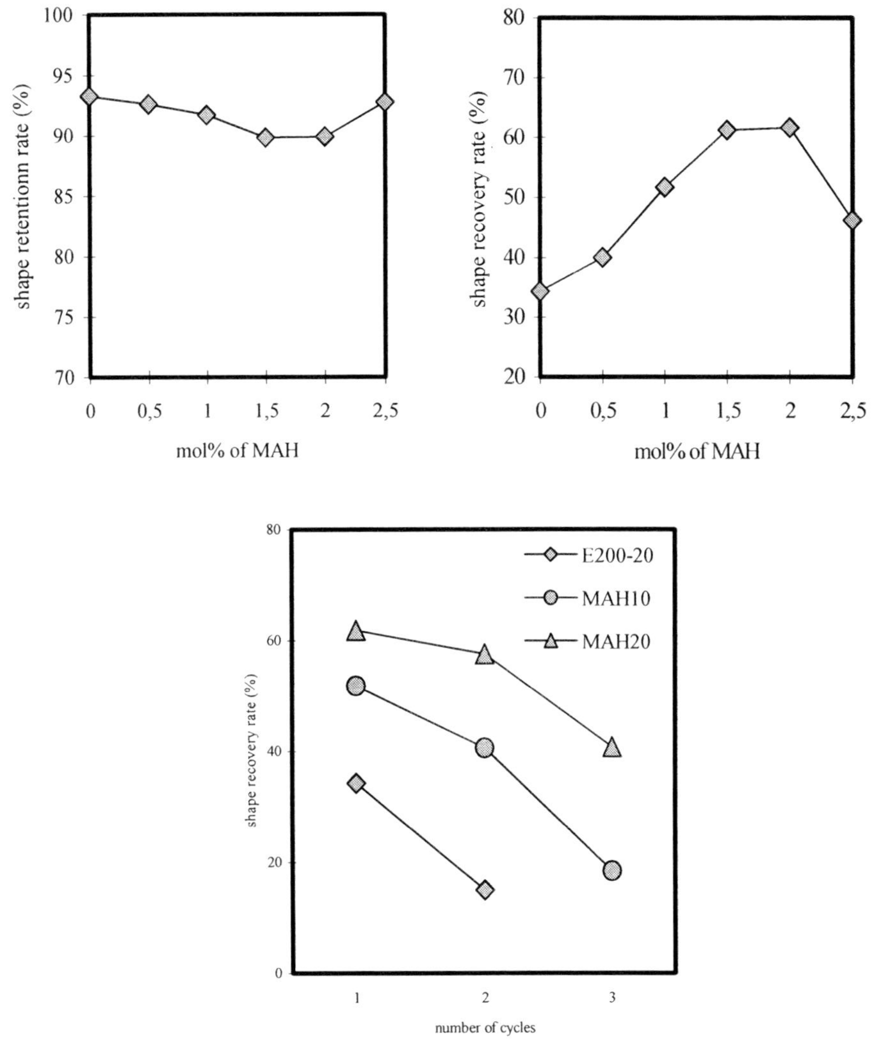

Figure 10. Shape memory profiles of PET-PEG-MAH copolymers

Table 4. Shape Memory Test Results of Copolymers

Sample Code	First Cycle		Second Cycle		Third Cycle	
	Recovery (%)	Max Stress (Mpa)	Recovery (%)	Max Stress (Mpa)	Recovery[1] (%)	Max Stress (Mpa)
E200-20	51.8	0.85	23.5	0.56	-	0.37
G-15	69.4	0.90	53.7	0.75	-	0.60
G-25	85.8	1.00	82.2	0.85	-	0.52
M-15	70.3	0.97	41.7	0.49	-	0.46
M-25	76.1	1.01	66.0	0.77	-	0.33
S-15	52.3	0.76	36.1	0.68	-	0.42
S-25	68.7	0.92	34.7	0.78	-	0.53

[1]Recovery for the third cycle was not proceeded because of the very low value of E200-20

Shape retention rates of the glycerol and SP cross-linked copolymers generally maintain 97-99 % of the original shape, but shape recovery rates are significantly dependent on cross-linking agents. In Figure 11, cyclic shape memory test results of 1.0 mol% glycerol copolymers with various SP contents were compared; shape recovery was getting worse with inclusion of SP up to 1.5 % and was substantially improved at 2.5 mol% of SP to a level that was better than the starting one without SP. Cyclic tests were done with 1.5 mol% glycerol copolymers with various SP contents; shape recovery again decreased up to 1.5 mol% of SP and went up at 2.5 mol% of SP, a result showing similar trend as 1.0 mol% case. Finally, 2.5 mol% glycerol copolymers were not much affected by the content of SP: there are some drops in shape recovery for the cases of 0.5 and 1.0 mol% of SP, but 1.5, 2.0, and 2.5 mol% of SP all showed higher shape recovery. From above results, shape recovery rates of the copolymers at low mol% of glycerol can be controlled by the content of SP, and high glycerol content dominates the shape recovery effect with very minor dependence on SP content. Again, shape memory results remind us of the differences in inter-chain interactions coming from glycerol and SP as explained above.

In addition to chemical cross-linking, following characteristics can be responsible for high shape recovery of cross-linked copolymer: (1) molecular interactions among hard segments (PET part), such as hydrogen bonding, dipole-dipole interaction of carbonyl groups of PET, and induced dipole-dipole interaction of aromatic rings, work as pivoting point for shape memory after several deformation-recovery processes; (2) soft segment (PEG part) effectively and entirely absorb the applied stress so that relative position of hard segment or original shape of copolymer can be safely preserved under severe experimental conditions; (3) glycerol, shorter cross-linking agent than sorbitol, is more effective for shape memory, because, although both are connecting hard segment, short and rigid structure of glycerol is more favored for shape retention of hard segment: (4) MAH which cross-links soft segment is inferior to glycerol or sorbitol, because soft segment's role in shape memory copolymer is to absorb external stress and cross-linking by MAH hinders flexible movement of soft segment. The mechanism for the higher shape memory properties of glycerol-cross-linked copolymers can be schematically explained as in Figure 12: hard segments connected through cross-linking and inter-chain interactions maintains the original shape, while soft segments reversibly absorb external stress.

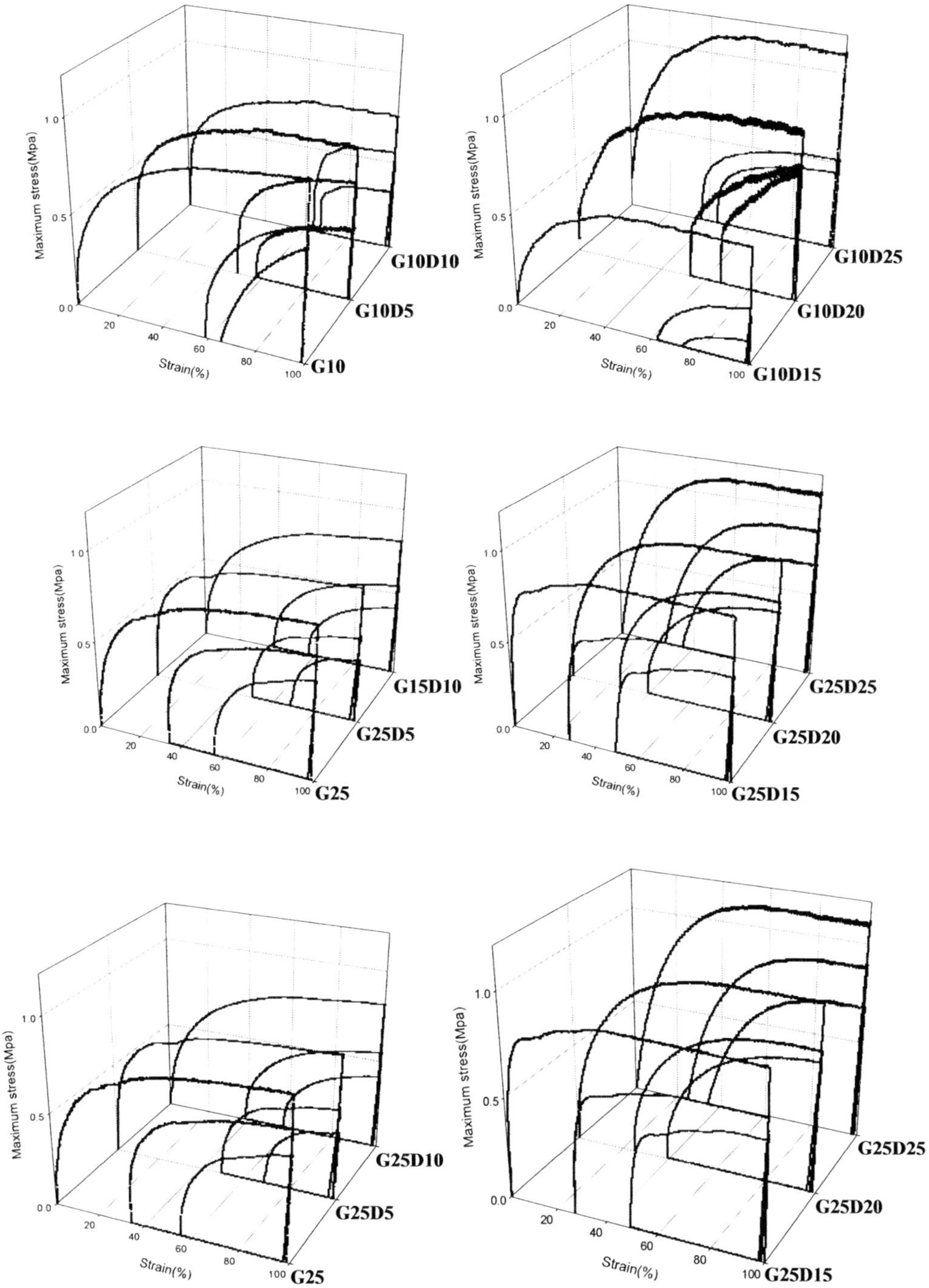

Figure 11. Shape memory profiles of PET-PEG-Glycerol-SP copolymers: G and D designate the mole % of glycerol and sulfoisophthalate, respectively

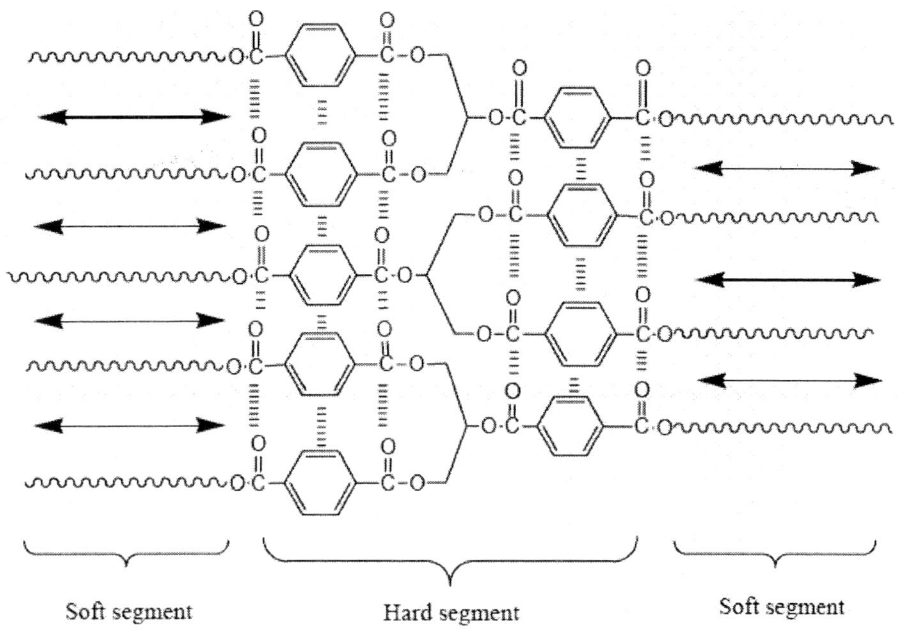

Figure 12. Schematic of shape memory effect of glycerol cross-linked copolymers

3.6. Charpy Impact Strength Test

Unnotched Charpy impact strength of composite laminate was compared in Figure 13. Impact strength normal to the laminate layer of sandwich structure direction was measured with unnotched specimens [17]. As compared to epoxy beam laminate by itself, composite laminate showed higher impact strength; laminate with uncross-linked copolymer (E200-20) or cross-linked one by sorbitol (S-25) or maleic anhydride (M-25) had more than 4 times higher impact strength than epoxy laminate; laminate with glycerol (G-25) was the lowest in impact strength, but still 2 times higher than epoxy laminate. Above results are closely related to the stiffness of copolymer which differentiates overall impact strength of laminate; from the transition temperature of copolymer (E200-20: 8.1 °C, G-25: 23.7 °C, S-25: 11.7 °C, and M-25: 14.1 °C), relative softness at room temperature where impact strength is measured should be reversely proportional to transition temperature. Therefore, G-25 itself and the laminate made of G-25 get the most solid structure among compared copolymers, but such rigid structure is very weak in absorbing external impact and shows low impact strength. Compromise of shape memory and impact strength of copolymer is necessary depending on what is the required property under various surrounding conditions.

Laminates made from the copolymers with glycerol and SP showed 1.3-2.6 times higher impact strength than epoxy laminate. Inclusion of glycerol lowered impact strength of the laminate and presence of SP additionally decreased the value; the higher rigidity of cross-linked copolymer is mainly responsible for the decrease of impact strength and the desired impact strength can be adjusted by control of the composition of glycerol and SP.

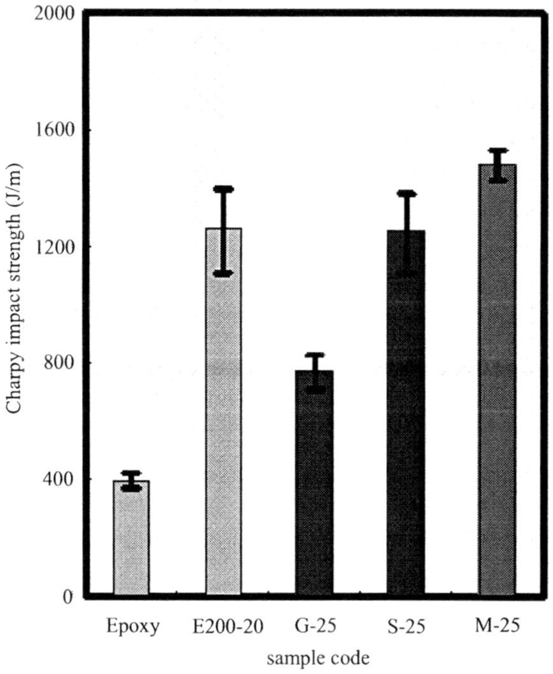

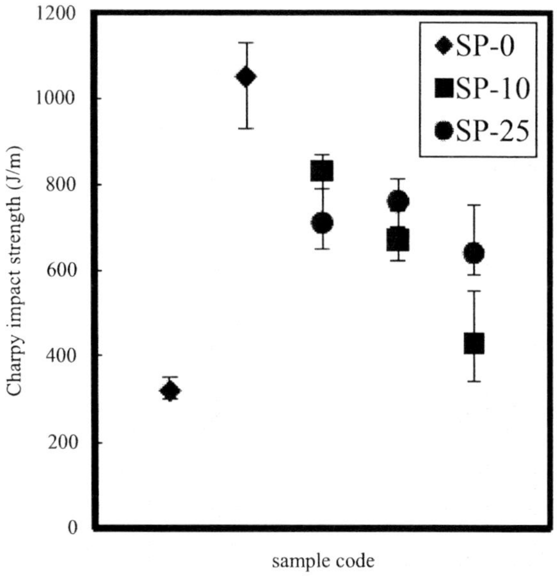

Figure 13. Charpy impact strength of cross-linked copolymers

The impact strength results are closely related to the transition temperature of copolymers (E200-20: 8.1 °C , G10D10: 23.6 °C, G10D25: 27.7 °C, G15D10: 24.7 °C, G15D25: 28.4 °C, G25D10: 25.3 °C, G25D25: 30.7 °C); relative softness at room temperature where impact strength is measured should be reversely proportional to transition temperature. Therefore, laminates made of G-25 series get the most solid structure among compared ones, but such rigid structure is very weak in absorbing external impact and shows low impact strength. Compromise of shape memory and impact strength of copolymer is necessary depending on what is the required property under various surrounding conditions.

3.7. Dynamic Mechanical Property of Sandwich Laminate Beam Structure Composite

Storage modulus and tan δ of sandwich laminate beam composites were compared in Figure 14. Storage modulus of epoxy laminate beam was in high contrast with all other copolymer laminates; all of copolymer laminates except G-25 showed similar shape over the temperature range, significant drop at transition temperature and immediate recovery at higher temperature. Laminate with G-25 with its rigid and well cross-linked copolymer is hard to rearrange its chains at high temperature once organized structure is dismantled. Deviation of G-25 laminate from other copolymer laminates is in accord with the above Charpy impact strength result. Tan δ sharply changed at transition temperature for all of copolymer laminate with no exception of G-25 laminate; their peak temperatures are in good order with transition temperatures of each copolymers. But epoxy laminate was not responsive over temperature ranges as expected from the storage modulus results. Because tan δ indicates damping ability, high tan δ of copolymer laminates (23 times as high as epoxy laminate for the best case) around room temperature range suggests the possibility of developing highly vibration-controlling composite material.

Storage modulus and tan δ of sandwich laminate beam of glycerol and SP copolymer composites of G25 series were compared in Figure 15. Storage modulus of epoxy laminate beam was in high contrast with all other copolymer laminates; G25 copolymer laminates showed similar shape over the temperature range, significant drop at transition temperature and immediate recovery at higher temperature. Additional ionic interaction from SP, together with glycerol cross-linking, increased transition temperature and the increase was proportional to SP content. Tan δ sharply changed at transition temperature for all of copolymer laminate. But epoxy laminate was not responsive over temperature ranges as seen in the above copolymers. There are still more rooms in fine-tuning of copolymer structure through modification of cross-linking method, PEG chain length, and hard segment structure for the control of transition temperature, vibration absorption, and mechanical properties.

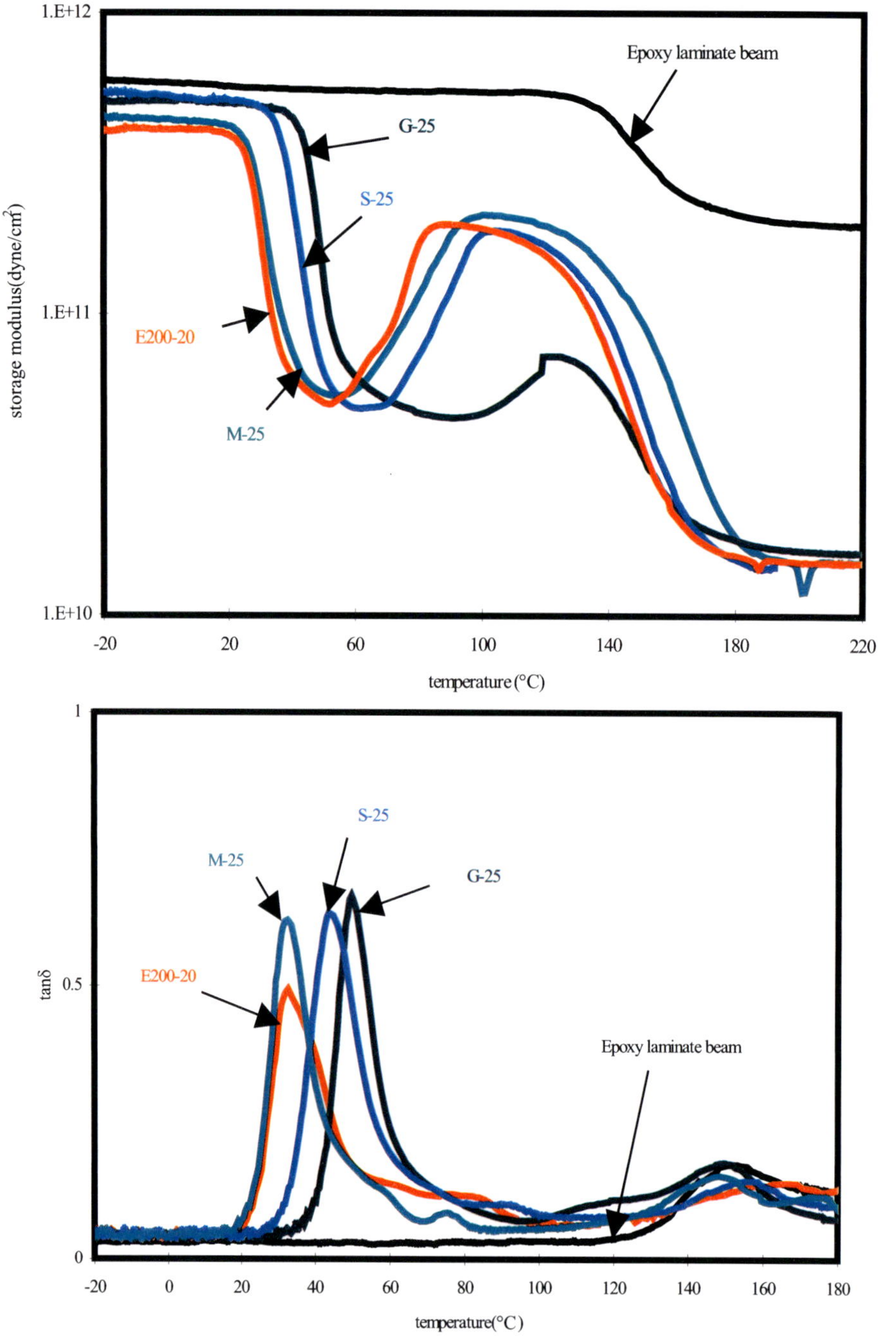

Figure 14. Storage modulus and tan δ of composite laminates with cross-linked copolymers

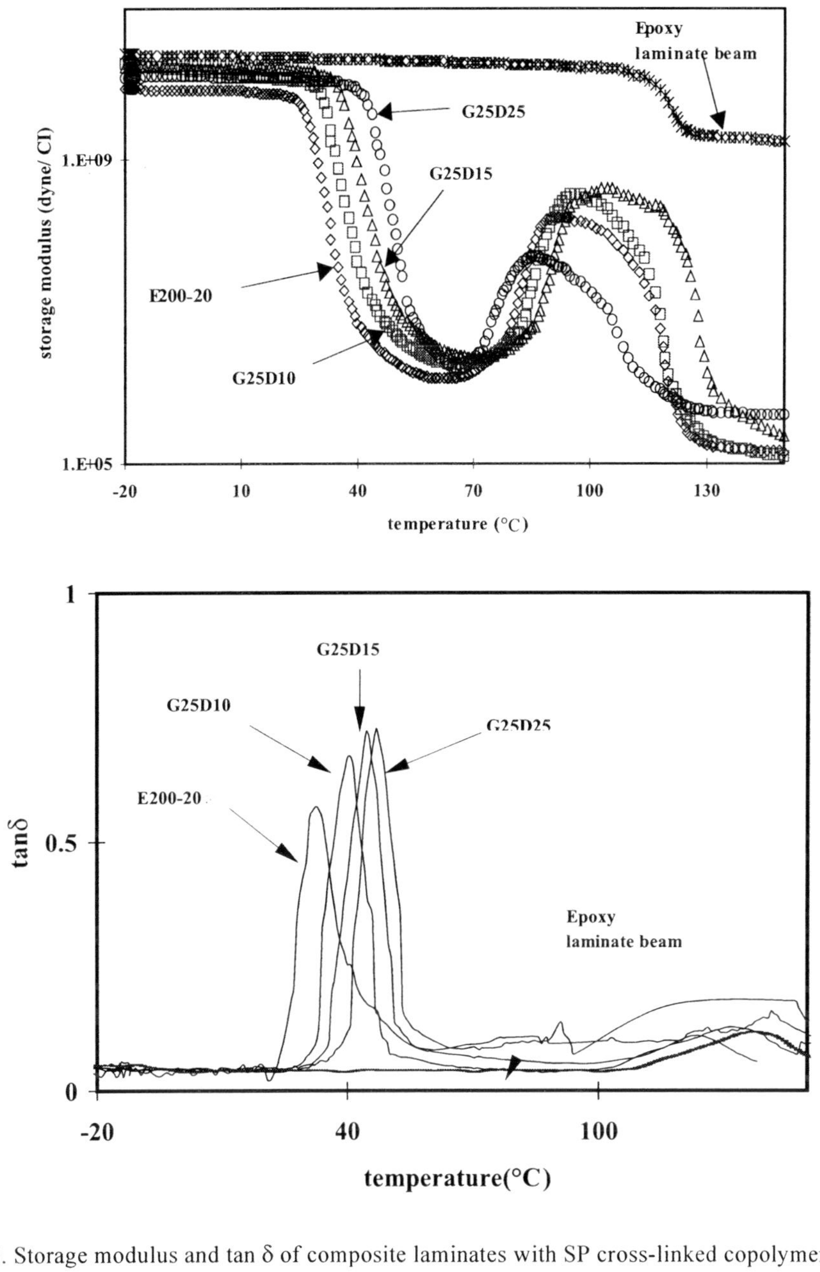

Figure 15. Storage modulus and tan δ of composite laminates with SP cross-linked copolymers

3.8. Dynamic Mechanical Thermal Property

Figure 16 shows loss tan δ and storage modulus E' of PET-PEG (PEG MW= 200 g/mol) and PET-PEG-MAH copolymers (1.0 and 2.0 mol% of MAH). All specimens exhibit a order of 2 or 3 decrease in storage modulus around their T_g. Storage modulus below T_g becomes higher as MAH is introduced into PET-PEG copolymer, originating from the reduced chain mobility by MAH cross-linking. As a shape memory copolymer, PET-PEG copolymer will retain the deformed shape when it is deformed in the rubbery plateau region and cooled below

T_g. And it will recover its original shape, after heating to the rubbery plateau region. At the temperature below T_g, copolymer with higher E′ shows better shape retention, and the copolymer with smaller E′ becomes easily deformed at the temperature above T_g. In general, modulus drop at T_g is about 10~1000 [18]. The storage modulus drop width (transition width) is narrower for the PET-PEG-MAH copolymers compared to PET-PEG copolymer, indicating that MAH cross-linking conferred temperature sensitivity to the copolymer by inducing more organized structures [19]. Higher tan δ was observed as MAH is introduced, and also the peak was found at the higher temperature. It is known that better shape recovery is achieved as tan δ becomes higher since it can deform like an elastomer. Thus, it can be predicted that PET-PEG-MAH copolymer will get better shape recovery since it has higher tan δ value. Another point to note is that as MAH is introduced, the width of rubbery plateau region above T_g increases. Because it is well known that the width increases with molecular weight, the above result confirms that MAH cross-linking increases molecular weight of the PET-PEG copolymer. Similar behavior is also observed in the intrinsic viscosity of PET-PEG-MAH copolymer (Table 3).

Storage modulus and tan δ of three types of cross-linked PET-PEG copolymers (glycerol, sorbitol, and MAH) were compared. Storage modulus abruptly decreased around T_g, and increased later at about 60 °C above their T_gs for all three kinds of cross-linked copolymers, which could be coming from disorganzation at T_g and recrystallization at higher temperature of PET-PEG copolymer chains during heating process in dynamic mechanical property measurement, and such result was not uncommon as previously observed for poly(trimethyleneterephthalate) [20-23]. As more cross-linking agent was included in the copolymer chain, storage modulus profile generally shifted toward higher temperature, and thus recrystallization of copolymer chains occurred at higher temperature, which originated from combined interactions from additional chemical cross-linking, dipole-dipole interaction of carboxyl groups of PET, and induced dipole-dipole interaction between phenyl rings [24,25]. Based on the result, it can be suggested that variation of dynamic mechanical property over temperature can be overcome by drawing process of specimen before heating.

Similar thermomechanical trend was observed in the tan δ profile at around T_g and small relaxation shoulder immediately after T_g was probably coming from temporary disruption of chains during recrystallization. Addition of cross-linking agent resulted in broader tan δ profile, because restricted chain movement by cross-linking agent made chain rearrangement slower and more energy-consuming process than uncross-linked copolymer; therefore, both disorganization and recrystallization were observed at a broader temperature range. In addition to broadening of transition area in tan δ vs. temperature profile, and temperature at peak transition shifted toward higher region, which could be interpreted by the same mechanism. As high tan δ means better deformation and shape recovery, cross-linked PET-PEG copolymers are promising candidates for shape recovery material based on tan δ results. Rubbery plateau region observed at higher temperature narrowed with introduction of cross-linking agent such as glycerine or sorbitol, but MAH-cross-linked copolymer had similar width of rubbery plateau region as uncross-linked one; such difference which is presumed to be coming from the cross-linking point, hard-segment linking (glycerine and sorbitiol) or soft-segment linking (MAH), tell us once again the importance of selection of cross-linking agent in developing highly reliable shape memory copolymer.

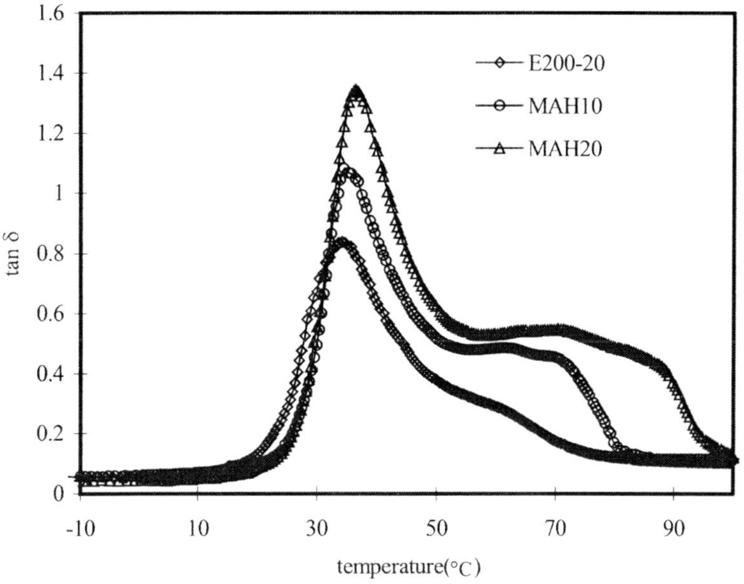

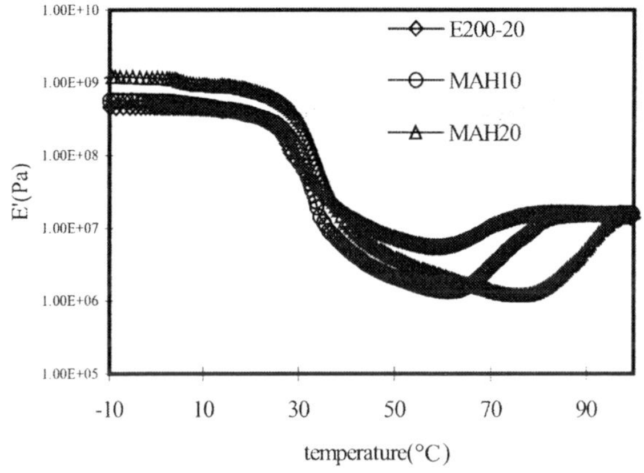

Figure 16. Storage modulus and tan δ of PET-PEG copolymers

It is generally known that shape memory polymer retains its deformed shape, if stressed at rubbery plateau region that is above T_g and then cooled to a temperature below T_g, and it recovers original shape if heated to rubbery plateau region again. Therefore, shape memory polymer has better shape retention property at temperature below T_g with the resulting high storage modulus, and its deformation becomes easier at temperature above T_g because of low storage modulus. From a different perspective, cross-linked PET-PEG copolymer, especially one by glycerine or sorbitol, showed significant modulus change around T_g which was above the necessary modulus change (order of three) as a temperature sensitive material, resulting in high hope for application in that area. Among the copolymers, comparatively sharper transition in glycerine-cross-linked one suggests that cross-linking at hard segment (PET), and short and rigid cross-linking agent are also advantageous in developing highly thermosensitive material [26].

Dynamic mechanical test results of copolymers with three different cross-linking agents are summarized as follows: (1) cross-linking by glycerine provides higher T_g, storage modulus, tan δ, and apparent change of storage modulus plus sharp transition of tan δ around T_g; (2) cross-linking by sorbitol also shows similar characteristics as glycerine, but transitions for both storage modulus and tan δ are blunt, which comes from the fact that flexible cross-linking of hard segment by sorbitol may mitigate strong interaction and shape retention among hard segments as observed in shorter glycerine; (3) cross-linking by MAH is least successful in obtaining improved thermomechanical properties, and interconnection of soft segments like MAH cross-linking is not desirable for our purpose, because soft segment should absorb external stress instead of blocking free movement of PEG chains.

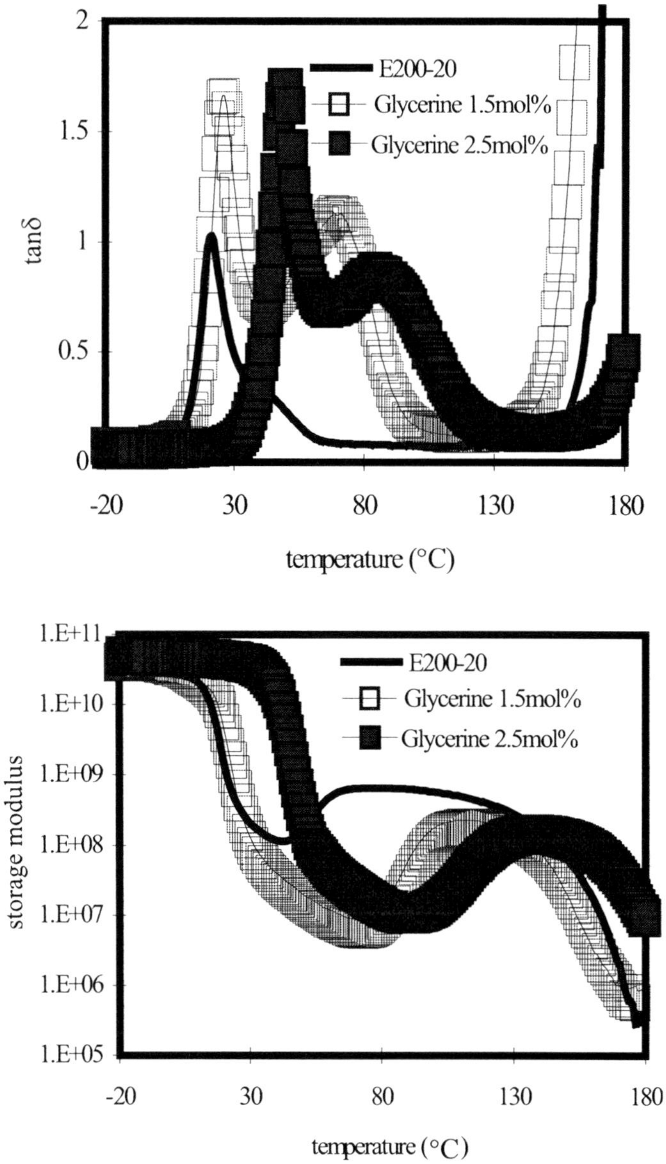

Figure 17. Storage modulus and tan δ of glycerine cross-linked copolymers

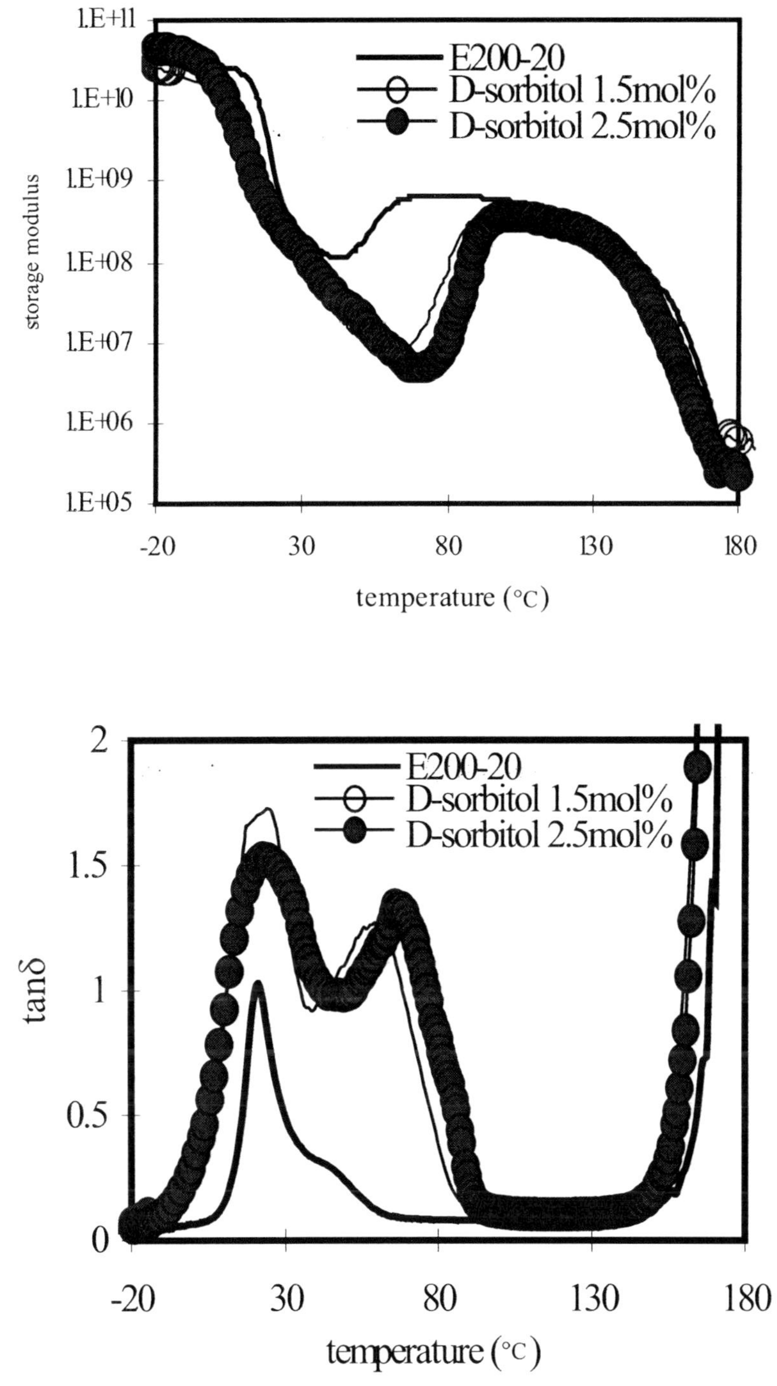

Figure 18. Storage modulus and tan δ of sorbitol cross-linked copolymers

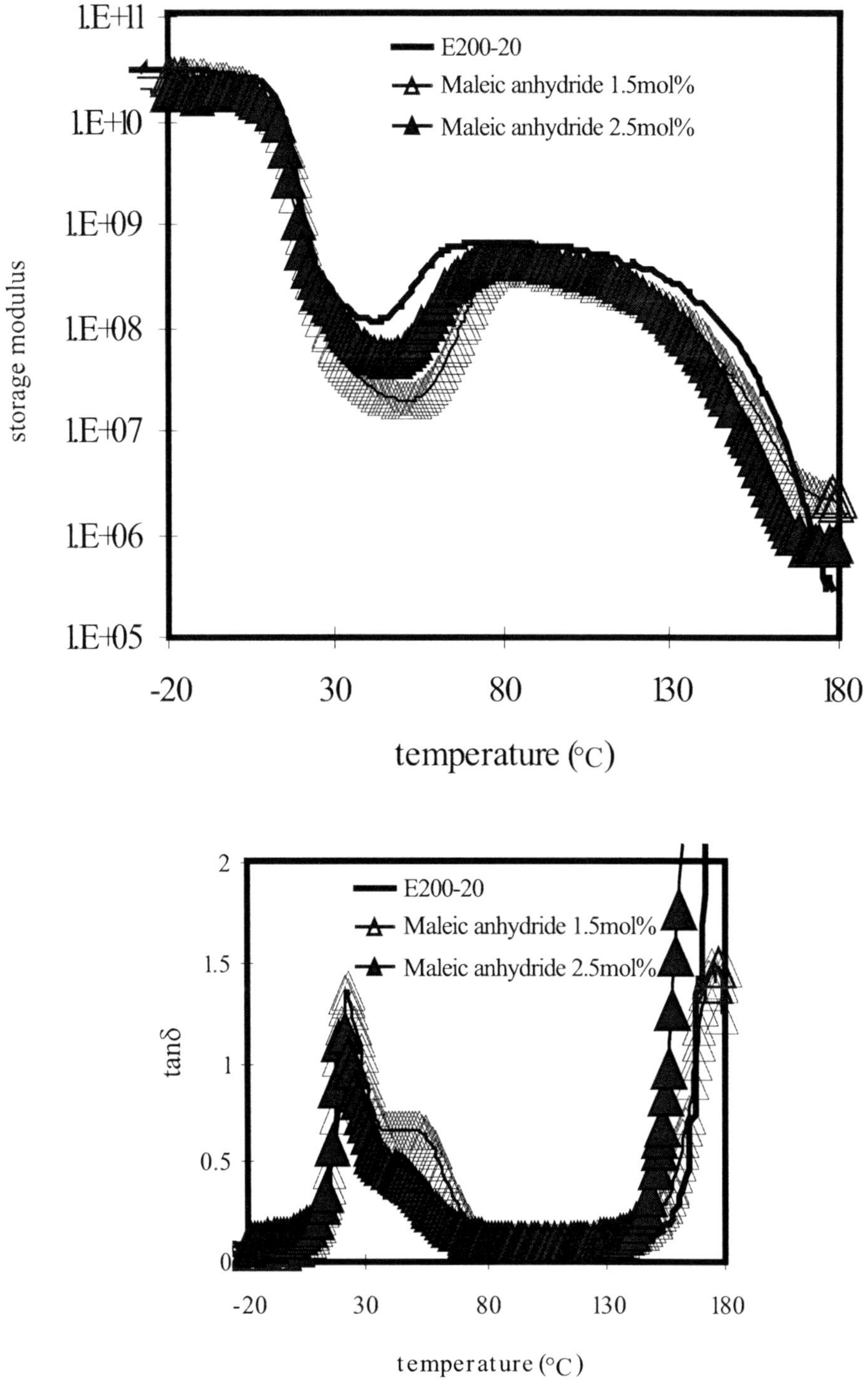

Figure 19. Storage modulus and tan δ of MAH cross-linked copolymers

4. CONCLUSION

As a preliminary search for developing composite laminate material with vibration control ability, various cross-linked shape memory copolymers were investigated for damping effect. PET-PEG copolymers consisting of PEG soft segment and PET hard segment was synthesized along with PET-PEG copolymers cross-linked by maleic anhydride (MAH). Experimental results showed that highest tensile properties were obtained in the case of PET-PEG copolymer with the mol ratio of 80 mol% EG and 20 mol% PEG having 200 g/mol of molecular weight. Addition of MAH resulted in decrease of strain at break and increase of maximum stress, together with increase of shape memory effect and damping (1.5 to 2.0 mol% of MAH). Cross-linking agents such as glycerine and sorbitol were also tried, and each cross-linked copolymer showed its own merits in different tests. Addition of glycerine, or sorbitol as a cross-linking agent also improved both shape memory and damping effect. Highest shape recovery rate was obtained with 2.5 mol% of glycerine, and highest tan δ was from 2.5 mol% of sorbitol. When sandwich-type composite laminate was prepared with glass-fiber-reinforced epoxy beam and cross-linked copolymer, Charpy impact strength increased 1.9 ~ 3.7 times and damping effect also increased 23 times for the best case at room temperature as compared to pure epoxy beam. Thus, cross-linked PET-PEG copolymer sandwiched to laminate composite, enables developing composite materials with controlled T_g and high damping effect. Copolymers having covalent cross-linking by glycerol and ionic interaction from SP group were compared with the above cross-linked copolymers. Shape retention rate of the copolymers showed very high value (97-99 %); shape recovery at low mol% of glycerol showed initial decrease and sharp increase with SP content, but 2.5 mol% glycerol copolymer series had gradual increase with SP content. Such results originated from the balance between strong covalent cross-linking by glycerol and weak ionic-interaction from SP, and the additional fine control of shape recovery by ionic interaction helps in developing custom-made smart material. Similar trend was observed for the glass transition temperatures of copolymers. Composite laminates prepared with the copolymer (glycerol and SP) showed high impact strength and loss tangent δ, and their value can be adjusted with SP content. Finally, physical properties of the sandwich-type laminate with shape memory copolymer can be controlled by proper selection of cross-linking agents, which is very helpful in developing composite materials with controlled T_g and high damping effect.

REFERENCES

[1] Gandhi, M. V.; Thompson, B. S. *Smart Materials and Structures*, Chapman and Hall, New York, 1992.

[2] Takagil, T. *J. Intelligent Mater. Sys*. 1990, 1, 149.

[3] Varadan, V. K. *Smart Structures and Materials*, 1994, 2189.

[4] Riley, P. J.; Wallace, G. G. *J. Intell. Mater. Syst. And Struct*. 1991, 2, 228.

[5] Wallace, G. G.; *Materials Forum*, 1992, 16, 111.

[6] Granqvist, C. G.; Azwns, A.; Isidorsson, J.; Ronnow, D.; Mattsson, M. S.; Veszelei, M. *J. Non-crystalline Solids*, 1997, 218, 273.

[7] Shirai, Y.; Hayashi, S. *Development of Polymeric Shape Memory Material*, MTB184, Mistubishi Heavy Industry, Inc., Japan, December, 1988.

[8] Gandhi, M. V.; Thompson, B. S. *Smart Materials and Structures*, New York: Chapman and Hall, 1992.
[9] Takahashi, T.; Hayashi, N.; Hayashi, S. *J. Applied Polym. Sci.* 1996, 60, 1061.
[10] Wei, Z. G.; Sandstrom, R. *J. Material Science*. 1998, 33, 3743.
[11] Liang, C.; Rogers, C. A.; Malafeew, E, *J. Intellingent Material Systems and Structures,* 8, 380.
[12] Takagil, T. *J. Intelligent Mater. Sys*. 1990, 1,149.
[13] Chun, B. C.; Cha, S. H.; Chung, Y. C.; Cho, J. W. *J. Applied Polym. Sci*. 2002, 83, 27.
[14] Oh, P. R.; Kim, K. J.; Kim, Y. H. *J. Korean Fiber Soc.* 1999, 36, 132.
[15] Chun, B. C.; Cha, S. H.; Park, C.; Park, M. J.; Chung, Y. C., Cho, J. W. *J. of Applied Polym. Sci.* 2002, 90, 3141.
[16] Spathis, G.; Niaounakis, M.; Kontou, E.; Apekis, L.; Pissis, P.; Christodoulides, C. *J. Appl. Polym. Sci.* 1994, 54, 831.
[17] Dutra, D.; Soares, B.; Campos, E.; Silva, J. *Polymer* 2000, 41, 3841.
[18] Lee, S. Y.; Kim, B. K.; Xu, M., *Polymer*, 1996, 37, 5781.
[19] Choi, Y. J.; Kim, B. K.; Jeong, H. M. *Polymer (Korea)*, 1988, 22, 1.
[20] Oh, P. R.; Kim, K. J.; Kim, Y. H. *J. Korean Fiber Soc.* 1999, 36, 132.
[21] Hoffman, D. M. *J. Applied Polym. Sci*. 2002, 83, 1009.
[22] Wu, C.; Yanagishi, T.; Nakamoto, Y.;Ishida,S.;Nitta,K.;Kubota, S. *J. Polym. Sci*. 2000, 38, 2285.
[23] Cascone, M. G. *Polym. Intnl*. 1997, 43, 55.
[24] Cho, D.; Choi, Y.; Drzal, L. T. *Polymer* 2001, 42, 4611.
[25] Krumova, M.; Lopez, D.; Benavente, R.; Mijangos, C. *Polymer* 2000, 41, 9265.
[26] Lee, S. Y.; Kim, B. K., Xu, M. Polymer 1996, 37, 5781.

In: Recent Developments in Polymer Research
Editor: Antony V. Hopper, pp. 163-191
ISBN 1-60021-346-4

***Chapter** 7*

THE POTENTIAL OF CHEMILUMINESCENCE IN THE RESEARCH OF THERMAL OXIDATION OF POLYMERS

Lyda Matisová-Rychlá and Jozef Rychlý
Polymer Institute, Centre of Excellence of Slovak Academy of Sciences CEDEBIPO,
842 36 Bratislava, Slovakia

ABSTRACT

The oxidation of polymers is accompanied by more or less intense chemiluminescence which intensity is determined by the quality of a polymer, temperature, concentration of oxygen in surrounding medium, presence of additives and stabilizers and by the initial molar mass of the substrate. The potential of the measurements of the light emission which appears during oxidation has not been explored yet fully and the method is mostly used in scientific institutions either as the main technique or it complements other methods used for the study of oxidative degradation of polymers. The review is mostly based on personal experience of the authors with the method and may serve as fast introduction for those who intend to apply the chemiluminescence technique in their own research.

INTRODUCTION

A combined effect of heat and oxygen on polymers leads to a more or less intense light emission (chemiluminescence) which is related to the oxidation of substrate in a way that is still the subject of discussion. It is of interest that the chemiluminescence accompanying the oxidation of low molar mass hydrocarbons (ethyl benzene, tetralin, decalin and cumene) and its relation with the kinetics of their oxidation mostly initiated by N,N'-azo(bisisobutyronitril) were investigated by the school of Russian scientists in the past [1] almost in parallel with the first papers on chemiluminescence from polymers. Such papers appeared in 1961 and 1964 when Ashby [2] and Schard [3] reported several examples of the phenomenon from polymer chemistry. They underlined the necessity of better understanding the chemiluminescence from polymers but, at the same time, they also indicated an unexplored potential of the study of the light emission from the oxidized polymers in better comprehension of the oxidation process. The essential difference in interpretation of chemiluminescence during oxidation of low molar mass substrates consists in a well defined rate of initiator decomposition and the oxidation usually occurs in kinetic region of oxygen surplus. In the case of polymer oxidation, however, a continuous accumulation of initiator (hydroperoxides) occurs while the concentration of oxygen in the condensed system of oxidized polymer is determined by the considerably lower rate of diffusion. For oxidation of low molar mass hydrocarbons there were no doubts that the light emission originates from the disproportionation of peroxyl radicals. For oxidation of polymers, however, such mechanisms performs one of the possible alternatives. Interlinks between oxidation of low molar mass model compounds and that of polymers originate from a classical Bolland-Gee scheme hydrocarbon oxidation [4] the validity of which is largely accepted by many scientists even today. Liquid hydrocarbons and solid polymers usually have the similar solubility of oxygen but the rate of oxygen diffusion is different. This may lead to the appearance of oxidation profiles, non-uniformity of the process due to higher local concentrations of hydroperoxides and heterogeneous distribution of stabilizers in the case of polymer oxidation.

Since the pioneering papers by Ashby and Schard some new knowledge accumulated related to chemiluminescence from polymers which were described in several review papers [5-11] from which it follows that chemiluminescence accompanies free radical reactions and that its observation is a function of oxygen concentration in surrounding medium. In the case of polymer oxidation free radicals formed in a different initiating routes ultimately lead to peroxyl radicals which similarly as in the case of low molar mass substrates may provide potential light emitters in recombination reactions. The spectrum of free radicals which in their reactions give light emission is, however, broader, and in conditions of the solid state oxidation various elementary reactions may provide a sufficient exothermicity for excitation of potential light emitters.

MECHANISMS LEADING TO THE LIGHT EMISSION FROM THERMALLY OXIDIZED POLYMERS

The heat gain higher than 400 kJ/mol in an elementary reaction of oxidation is reported as the necessary pre-requisite of the light emission appearance if the light emission comes from excited triplet carbonyl groups. It may be significantly lower when the light emitter is singlet oxygen. This is also the reason why the most preferred elementary reaction considered as the potential light emitting process is disproportionation of secondary peroxy radicals which occurs as follows:

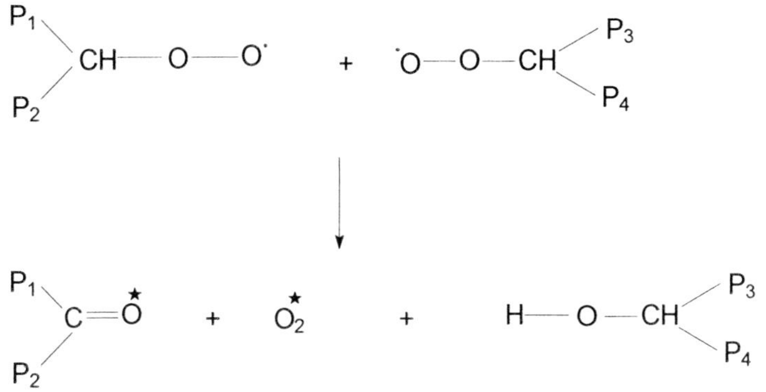

Scheme 1

In the above reaction, the excited triplet ketones and singlet oxygen may be formed due to synchronously released heat (450 kJ/mol). This is an optimum configuration for the highest yield of the light emitted. One has to keep in mind, however, that deactivation of excited states occurs via the heat dissipated in collisions and that one photon [$h\nu$] is formed per several millions or even billions of such deactivating procedures. Regardless of this fact the sensitive photomultiplier connected with photon counting unit is able to register it. This also pre-determines the extreme sensitivity of the chemiluminescence method. $6 x 10^{23}$ molecules or mers in one mole of a polymer indicates that the method is capable of successful monitoring of nano or even pico moles of respective compounds which either oxidized or decomposed may provide the excited particles.

The maximum of emission from excited triplet carbonyls is situated at 460 nm while that from singlet oxygen at wavelengths at 634 nm (dimol emission) and 1270 nm (monomol emission) [12].

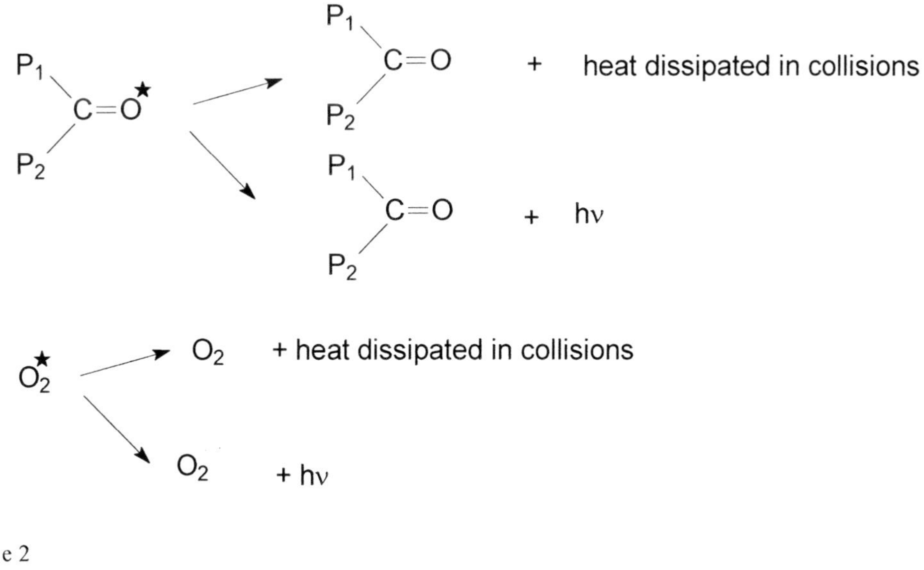

Scheme 2

Singlet oxygen is formed in the decomposition of hydrogen peroxide which appears during oxidation of polymers carrying hydroxyl groups like cellulose according to the following stoichiometric equations:

$$-C(OH)H- + O_2 \rightarrow -CO- + HOOH$$
$$2\ HOOH \rightarrow 2\ H_2O + O_2^*$$

The singlet oxygen ($^1\Delta_g$ state) appears in a reaction of superoxide anion radicals ($^-O_2^\cdot$) which are the reaction intermediates of the above oxidation being formed in alkaline medium. In the presence of water they are converted to hydrogen peroxyl radicals HOO· as follows:

$$^-O_2^\cdot + H_2O \rightarrow {}^-OH + HOO^\cdot$$

The exothermicity required for excitation of oxygen comes from recombination of hydrogen peroxyl radicals

$$HOO^\cdot + {}^\cdot OOH \rightarrow HOOH + O_2^*$$

According to [13] the sequence of subsequent reactions may lead either to ozone or to dimols of singlet oxygen:

$$HOOH + O_2^* \rightarrow O_3 + H_2O$$
$$O_3 + HOOH \rightarrow {}^*(O_2)_2^* + H_2O$$

Emission from dimols of singlet oxygen may be detected with photomultipliers used for measurement of chemiluminescence from hydrocarbon polymers with a maximum spectral sensitivity at 460 nm. The above scheme, however, requires the presence of at least one molecule of hydrogen peroxide in the close vicinity of two recombining peroxyl radicals and assumes a large heterogeneity of the oxidation process.

As it was already indicated, the Russel scheme is not a unique mechanism which was considered as the origin of the light emission in the case of polymer oxidation.

Quinga and Mendenhall [14] proposed that luminescent process in oxidation of polymers involves the disproportionation of two alkoxyl radicals while Audouin-Jirackova and Verdu [15] expressed an idea that it might be the monomolecular fragmentation of alkoxyl radicals to alkyl radical and carbonyl groups in which the latter should be formed in excited state. There exists also the proposal by Z. Osawa [16] that the light emission comes from polynuclear aromatics acting as photosensitizers. They are sorbed to the oxidized polymer from surrounding atmosphere during the polymer storage. This opinion is based on a strong reduction of initial chemiluminescence intensity when polymer is extracted by low molar mass solvents.

Blakey and George [17] came with another interesting possibility. They suggested that chemiluminescence from oxidized polymer may involve the so-called CIEEL (chemically initiated electron exchange luminescence) mechanism, that involves the electron exchange between the potential activator and peroxide which leads to a radical cation of an activator and a radical anion of a peroxide. The next step occurs as the rearrangement of the radical anion of a peroxide followed by charge recombination with formation of excited activator molecules. It appears that the role of activator may be played by either isolated carbonyls or carbonyls attached to of C=C unsaturations. Quantum yield for CIEEL mechanism appears to be by three to five orders of magnitude higher than that for the reaction of a pair of peroxy radicals [18].

Other speculative mechanisms [19] may be proposed based on the presence of singlet oxygen and C=C unsaturations in oxidized polymer. The reaction of latter components may lead to the transient formation of dioxetanes the decomposition of which has even higher quantum yield of luminescence than in CIEEL mechanism [20].

In heterogeneous domains of oxidized polypropylene, dioxetanes may be formed from 1,3-peroxyl hydroperoxides radicals in the sequence of reactions:

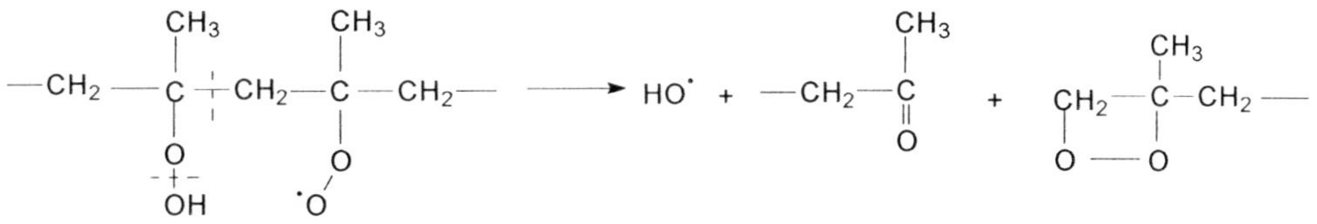

Scheme 3

In the Scheme 3 peroxidic O-O bond of hydroperoxyl group is splitted together with β-scission of formed alkoxyl radical and ring closure of alkyl peroxyl diradical may occur if the process has some extent of synchronisation. The process gives hydroxyl radical, methylcarbonyl terminal groups, $-CH_2-CO-CH_3$, and dioxetane. The latter is unstable and decomposes into an excited triplet state of formaldehyde and/or excited triplet state of methylcarbonyls (Scheme 4).

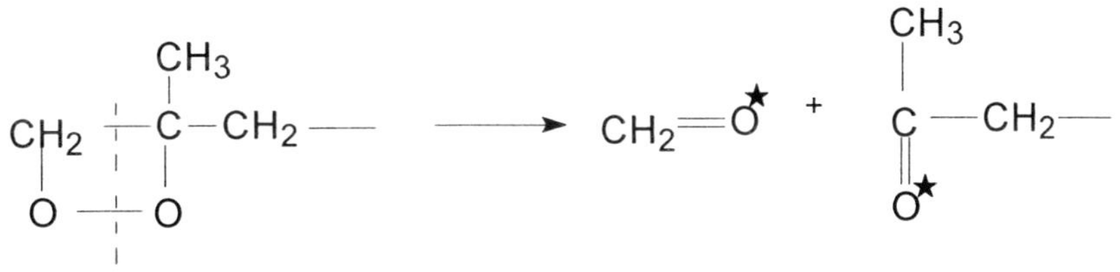

Scheme 4

In the paper by Ptschelintsev and Denisov [21] on oxidation of polyisoprene it was shown that one recombination event of free radicals leads to scissions of about 30 C-C bonds, approximately. The authors expressed an idea that the chain scission is a matter of propagation of peroxyl radicals and not of β-scission of alkoxy radicals. One of the possible pathways, which are at the same time strongly exothermic is also the scission of 1,3-hydroperoxy alkyl radicals which are formed by intramolecular transfer of peroxyl radicals within six-membered transition state (Scheme 5).

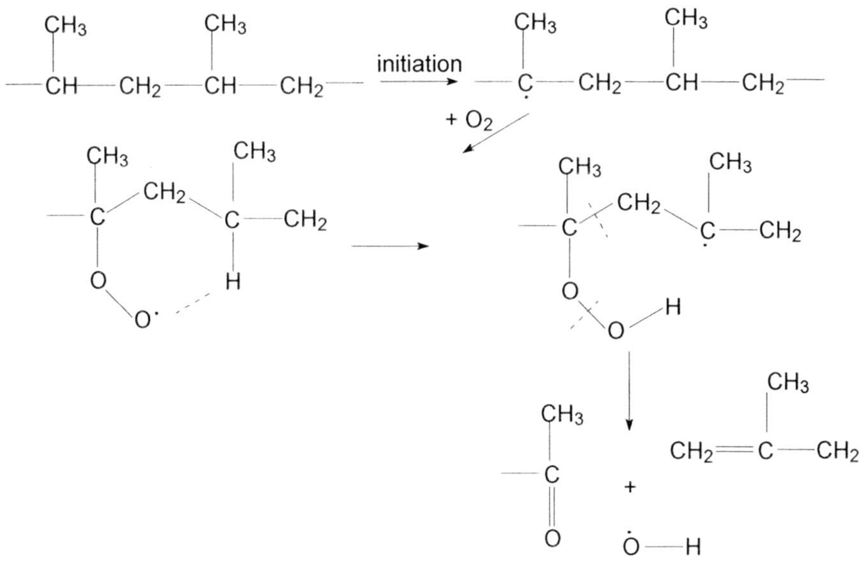

Scheme 5

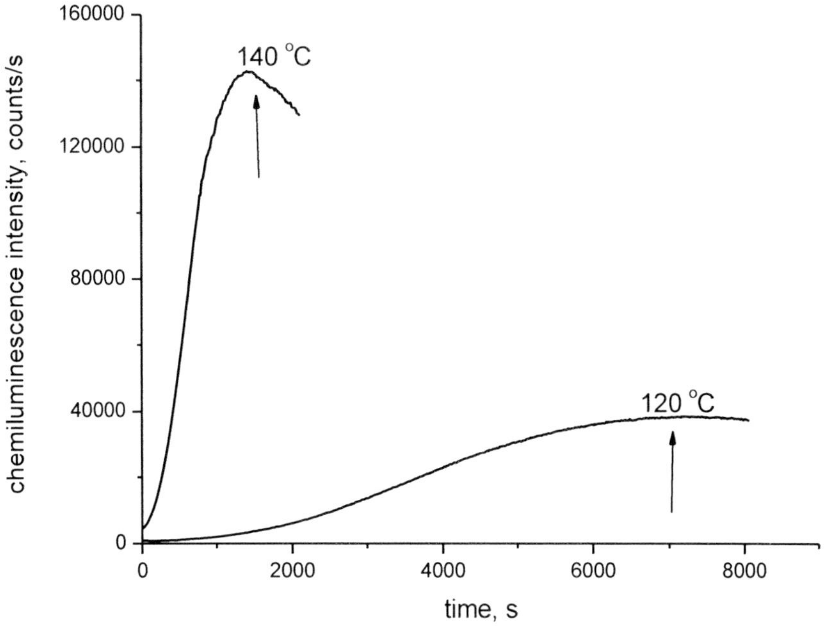

Figure 1. Chemiluminescence at the oxidation of polypropylene films, typical isothermal records (3 mg, oxygen)

The heat gain of the last sequence of above reaction is surprisingly high (about 450 kJ/mol) and under conditions of reduced mobility of the polymer chain it may compete successfully with any mechanism of the light emission. Moreover, the above mechanism is directly linked with main chain scissions. Schard [3] suggested that the intensity of chemiluminescence is directly proportional to the number of tertiary carbons in a macromolecular chain which thus tends more to chain scissions. This rule has a limited validity as oxidized polystyrene in an advanced stage of oxidation gives considerably lower chemiluminescence than polyethylene. In typical isothermal runs for chemiluminescence accompanying the oxidation of polypropylene films, the state when a complete loss of mechanical properties of polypropylene has been observed is designated by an arrow which is just after the maximum of the light emission [Fig. 1].

Recently, we have shown that non-isothermal chemiluminescence runs for oxidized cellulose provide the same rate constants of cellulose degradation, approximately as experiments determining the decay of polymerization degree. This may be taken as an indirect evidence that the light emission is linked with the scission of polymer chains [22].

The intensity of chemiluminescence signal from the heated polymer and kinetics of its change in time/ or with temperature is determined by [23]:

(1) the quality of the polymer, its photo-oxidation and thermo-oxidation history expressed in concentration of hydroperoxides, carbonyl groups or of other oxidized structures including terminal groups. The rate of an oxidative attack may then be related to the average molar mass and to its distribution and to the ratio of amorphous/crystalline structures. Polymers cannot be simply ordered according to the intensity of light emission at a given temperature. The chemiluminescence-time patterns are related with the rate of sample oxidation but they may differ from one to the next polymer.
(2) the quality of the polymer surface; for one polymer the intensity of the signal will be different for film and for powder.
(3) temperature and concentration of oxygen in atmosphere surrounding the oxidized sample.
(4) the extent and quality of polymer stabilization.
(5) a direct oxidation of polymer additive, which may occassionaly give a much stronger signal than the oxidation of polymer itself. This is vey important as it may lead to erroneous relation between the rate of polymer oxidation and chemiluminescence intensity. Qualitatively two different patterns of chemiluminescence intensity in time may be observed for polymer materials, namely:
   a. autoaccelerating increase from the very low values to a maximum followed by the subsequent decay. Such runs show oxidative induction time and may be observed for oxidation of polyolefins, polydienes and polyamides. All of them are polymers which have a relatively long length of kinetic oxidation chains [Fig. 2]. Initial level of chemiluminescence intensity at time t=0 usually corresponds to the extent of polymer pre-oxidation (Fig. 2, see the case of polyamide 66).
   b. steady decay from an initial value of chemiluminescence intensity.

Such courses are typical for oxidation of polymers with heteroatoms in the main chain, like poly(2,6-dimethyl-1,4-phenylene oxide) [Fig.3], etc. having rather short length of oxidation chains and also for polymers oxidized below their glass transition temperature.

In a glassy state, polymer segments, the mobility of which is restricted, attain a limited reaction zone and the rate of the process becomes decays. The chemiluminescence from the decay of peroxidic structures in pre-oxidized polymer samples occurring in an inert atmosphere belongs to this category as well.

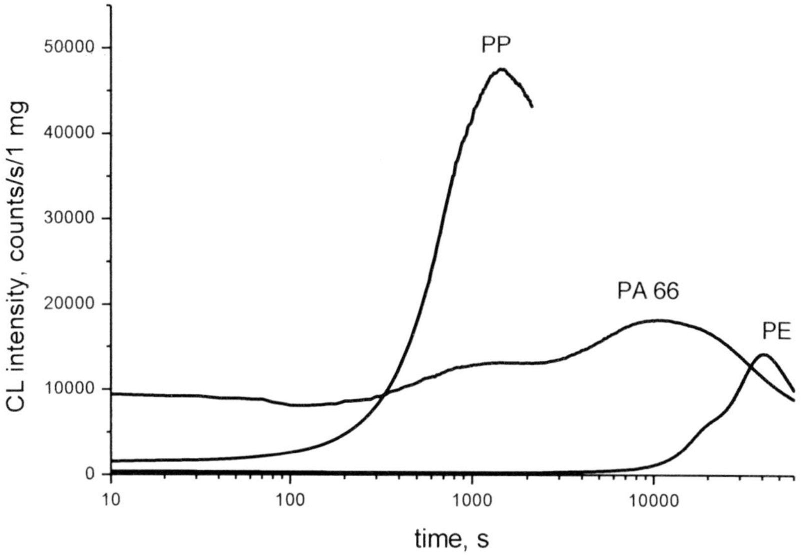

Figure 2. Chemiluminescence oxidation runs at 140 °C in oxygen for films of polypropylene (PP), polyamide 66 (PA 66) and polyethylene (PE)

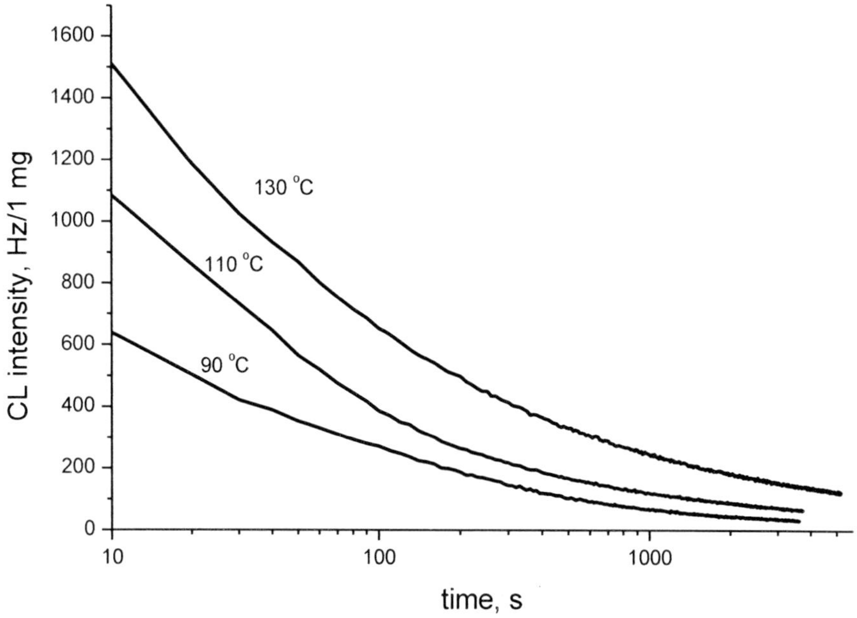

Figure 3. Chemiluminescence oxidation runs for powder of poly(2,6-dimethyl-1,4-phenylene oxide) at different temperatures, oxygen

Some polymers show quite a complex pattern composed of both cases a) and b) (oxidation of pre-oxidized polypropylene) or from two autoaccelerating increases of the case a) (polyamides, polyethylene [Fig.2]) which may be related to the change in oxygen diffusion to the system. The first wave may correspond to the oxidation under surplus of oxygen concentration, the second to the oxidation in oxygen starving systems.

Thermo-oxidative stabilizers in the polymer shift the increase of chemiluminescence to higher times in the case a) and reduce the initial chemiluminescence intensity to lower values in the case b).

Bolland-Gee scheme of oxidation [4] involves all three essential stages of free radical reaction, namely that of initiation, propagation and termination.

$$PH \xrightarrow{w_i} P^{\cdot}$$
$$POOH \xrightarrow{k_{mono}} PO^{\cdot} + {}^{\cdot}OH$$
$$2POOH \xrightarrow{k_{bi}} PO_2^{\cdot} + PO^{\cdot} + H_2O$$
$$PO^{\cdot}(HO^{\cdot}) + PH \xrightarrow{k_{tr}} POH(H_2O) + P^{\cdot}$$
$$P^{\cdot} + O_2 \xrightarrow{k_3} PO_2^{\cdot}$$
$$PO_2^{\cdot} + PH \xrightarrow{k_4} POOH + P^{\cdot}$$
$$2PO_2^{\cdot} \xrightarrow{k_6} products$$
$$PO_2^{\cdot} + P^{\cdot} \xrightarrow{k_7} products$$
$$2P^{\cdot} \xrightarrow{k_8} products$$
$$PO_2^{\cdot} + InH \xrightarrow{k_{inh}} products$$
$$POOH + D \xrightarrow{k_d} products$$

Scheme 6

Initiation is formation of free radicals which may occur either as some adventitious event from polymer deffect structures, catalysts residues and impurities or as the monomolecular and bimolecular decomposition of hydroperoxides. In the presence of oxygen, alkoxyl and alkyl radicals are converted fast to peroxyl radicals which are the main chain carriers and react with substrate giving new alkyl radicals. The propagation process ends by termination of alkyl and peroxyl radicals. In the presence of an inhibitor, peroxyl radicals are either scavenged to give non-reactive products or hydroperoxides are decomposed in a non-radical route. The oxidation process is more or less efficiently inhibited. The inhibitor efficiency depends on the length of kinetic chains, i.e. on the number of cycles of peroxyl radicals regeneration before the termination takes place. The higher the length of kinetic chains, the more efficient inhibitors InH or D.

Provided that chemiluminescence intensity I_{CL} is proportional to the rate of peroxyl radicals termination, i.e. $I_{CL} \approx k_6 [PO_2^{\cdot}]^2$, which is one of the assumption of the chemiluminescence mechanism, from the above scheme of non-inhibited oxidation ([InH], [D]=0) occurring in stationary conditions, it is valid:

$$w_i + k_{mono}[POOH] + k_{bi}[POOH]^2 = k_6 [PO_2^{\cdot}]^2$$

For the case a) of chemiluminescence intensity – time pattern, $k_{mono}[POOH] + k_{bi}[POOH]^2 >> w_i$ after a relatively short oxidation time and the process is controlled by the decomposition of hydroperoxides.The case b) $k_{mono}[POOH] + k_{bi}[POOH]^2 << w_i$ is in fact the consumption of oxidisable defect sites in a polymer and the level of hydroperoxides is low. The main oxidation channel goes through the initiation represented by the first elementary step in the above scheme 6 which is not easy to quantify.

Analytical Expression for Isothermal Kinetics of Oxidation and Its Relation to the Concentration of Oxygen in Surrounding Atmosphere

The chemiluminescence intensity accompanying the oxidation of polymers is usually dependent on concentration of oxygen in surrounding atmosphere (Figs. 4 and 5). This is the confirmation of restricted diffusion of oxygen to the corresponding reactive sites but the correct interpretation is still missing. In the Fig. 4, which shows the chemiluminescence from polybutadiene, the maximum values of chemiluminescence intensity decrease with decreasing concentration of oxygen in the mixture with nitrogen; the similar observation was made with polyisoprene. In the case of oxidation of Whatman cellulose, when we reduce the concentration of oxygen in the mixture with nitrogen stepwisely we receive a stepwise reduction of equilibrium chemiluminescence intensity.

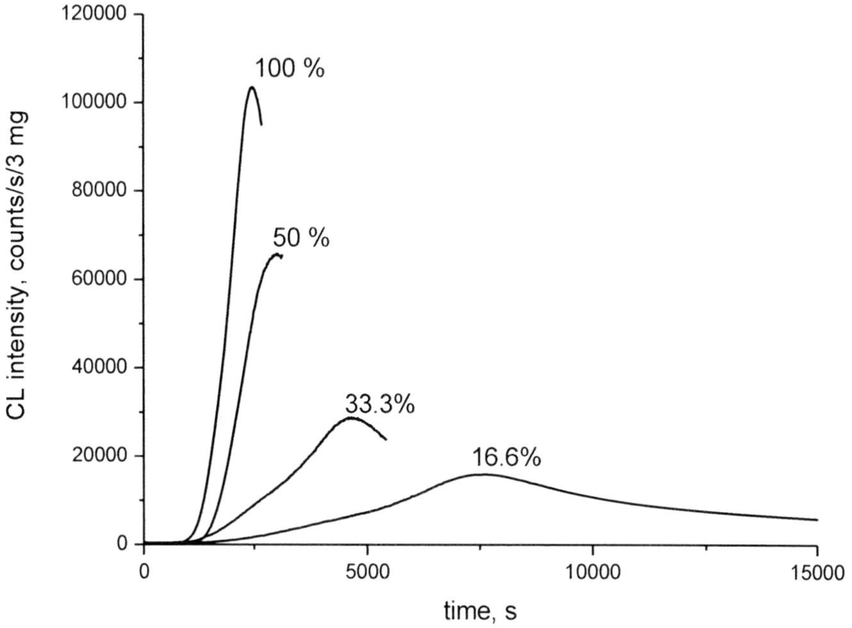

Figure 4. Chemiluminescence from polybutadiene, the effect of oxygen concentration in the mixture with nitrogen, 130 °C, total flow of gas 3.6 l/h, the numbers denote the volume percents of oxygen in the mixture with nitrogen

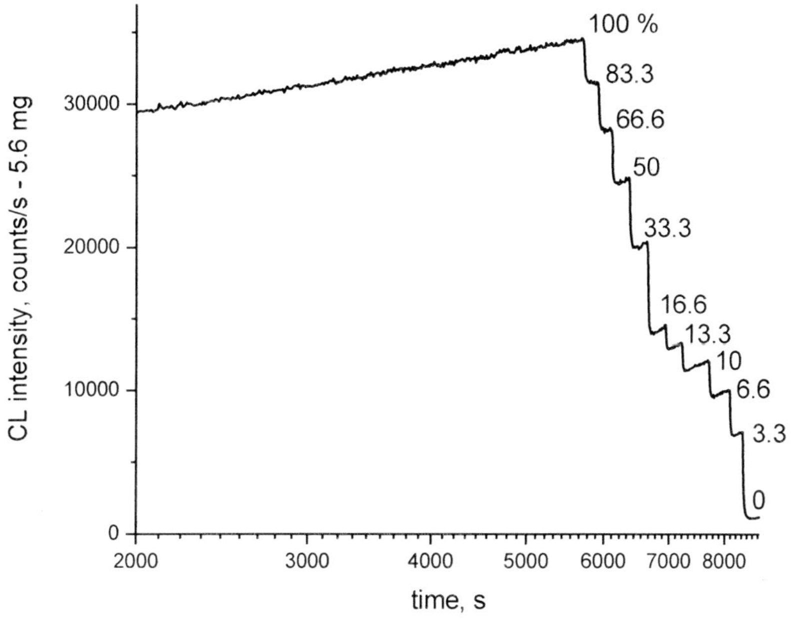

Figure 5. Continual changes of the intensity of chemiluminescence signal with concentration of oxygen in surrounding atmosphere, 180 °C, Whatman paper. Numbers are % vol. of oxygen in the mixture with nitrogen

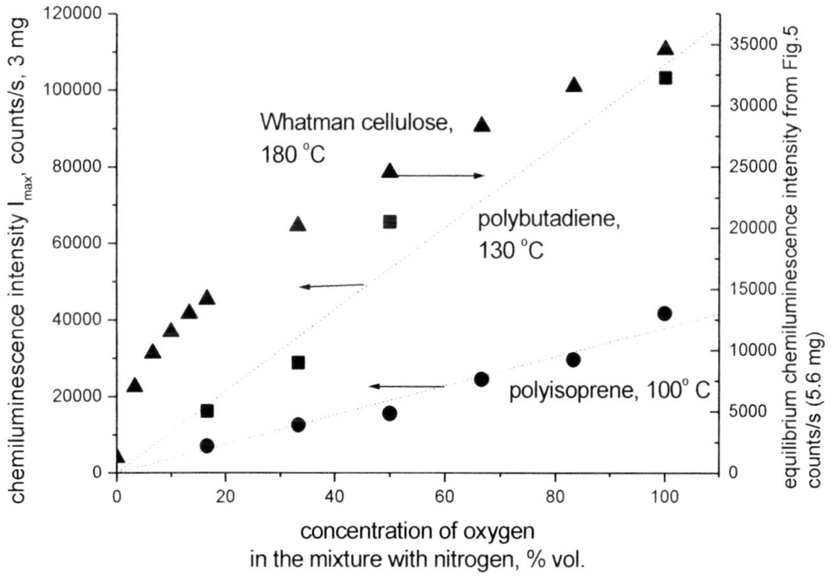

Figure 6. The plot of maximum and equilibrium chemiluminescence intensity in dependence on concentration of oxygen in the mixture with nitrogen

The plot of chemiluminescence intensity vs. concentration of oxygen in surrounding atmosphere is either linear or curved in dependence on the quality of the polymer examined (Fig. 6). Assuming that the reaction occurs in homogeneous conditions, it appears quite clear that in the case of sufficiently high length of kinetic chains, the primary appearance of free radicals by the rate w_i is very soon replaced by the decomposition of hydroperoxides and the process has the character of "closed loop" [24]. From the shape of kinetic runs for oxidation of polypropylene it was deduced that predominant decomposition of hydroperoxides is bimolecular reaction after very short time of the oxidation.

The implementation of oxygen into the kinetic scheme which for the case of polypropylene was attempted in paper [25] may be accounted for via equations that follow from the subtraction and addition of terms for time changes of alkyl and peroxyl radicals $\frac{d[P^{\cdot}]}{dt}$ and $\frac{d[PO_2^{\cdot}]}{dt}$ as follows:

$$\frac{d[P^{\cdot}]}{dt} = k_{bi}[POOH]^2 - k_3[P^{\cdot}][O_2] + k_4[PO_2^{\cdot}][PH] - k_7[PO_2^{\cdot}][P^{\cdot}] - 2k_8[P^{\cdot}]^2 \qquad 1)$$

$$\frac{d[PO_2^{\cdot}]}{dt} = k_{bi}[POOH]^2 + k_3[P^{\cdot}][O_2] - k_4[PO_2^{\cdot}][PH] - 2k_6[PO_2^{\cdot}]^2 - k_7[PO_2^{\cdot}][P^{\cdot}] \qquad 2)$$

As $\frac{d[P^{\cdot}]}{dt}$ and $\frac{d[PO_2^{\cdot}]}{dt}$ are negligible when compared to the hydroperoxide concentration changes we put them to zero and obtain thus the equations 3) and 4).

Subtraction of eqs. 1 and 2:

$$k_3[P^{\cdot}][O_2] + k_8[P^{\cdot}]^2 = k_4[PO_2^{\cdot}][PH] + k_6[PO_2^{\cdot}]^2 \qquad 3)$$

addition of eqs. 1 and 2:

$$k_{bi}[POOH]^2 = k_6[PO_2^{\cdot}]^2 + k_7[PO_2^{\cdot}][P^{\cdot}] + k_8[P^{\cdot}]^2 \qquad 4)$$

Note: such simplification is allowed only when we use eq. 3) and 4) for solution of differential equations involving more stable products, e.g. that for concentration of hydroperoxides, as follows:

$$\frac{d[POOH]}{dt} = k_4[PO_2^{\cdot}][PH] - 2k_{bi}[POOH]^2 \qquad 5)$$

It has nothing to do with the existence of stationary state for which concentrations $[POOH]$, $[PO_2^{\cdot}]$ and $[P^{\cdot}]$ should be constant.

Comparing the probabilities of mutual collisions of $PO_2^{\cdot}$ and $P^{\cdot}$ radicals one has to admit that the rate constants of recombination k_6, k_7 and k_8 are bound by the relation:

$$k_7 = 2\sqrt{k_6 k_8} \qquad 6)$$

which leads to simplification of equation 4):

$$\sqrt{k_{bi}}[POOH] = \sqrt{k_6}[PO_2^{\cdot}] + \sqrt{k_8}[P^{\cdot}] \qquad 7)$$

and by substitution for [P$^{\cdot}$] into eq. 3) we receive:

$$\sqrt{k_6}[PO_2^{\cdot}] = \frac{\sqrt{\frac{k_{bi}}{k_8}k_3[O_2][POOH] + k_{bi}[POOH]^2}}{\frac{k_4}{\sqrt{k_6}}[PH] + \frac{k_3}{\sqrt{k_8}}[O_2] + 2\sqrt{k_{bi}}[POOH]} \qquad 8)$$

After dividing nominator and denominator of the right side of equation 8) by $\frac{k_4}{\sqrt{k_8}}[PH]$ and a small rearrangement we have:

$$\sqrt{k_6}[PO_2^{\cdot}] = \frac{\sqrt{k_{bi}}[POOH](\sqrt{\frac{k_6}{k_8}\frac{k_3[O_2]}{k_4[PH]}} + \frac{\sqrt{k_{bi}k_6}}{k_4[PH]}[POOH])}{1 + \sqrt{\frac{k_6}{k_8}\frac{k_3[O_2]}{k_4[PH]}} + 2\frac{\sqrt{k_{bi}k_6}}{k_4[PH]}[POOH]} \qquad 9)$$

One can see that for certain values of parameters of equation 9) we obtain particular dependences of concentration of peroxyl radicals on concentration of oxygen. Provided that maximum concentration of hydroperoxides in oxidized polymers attains the value 10^{-1} mol/l, $k_4[PH] \approx 10 s^{-1}$ and $\sqrt{k_{bi}k_6}$ is typically 3 $lmol^{-1}s^{-1}$ then the term $\frac{\sqrt{k_{bi}k_6}}{k_4[PH]}[POOH] \rightarrow 0.03$. Provided that the term $\sqrt{\frac{k_6}{k_8}\frac{k_3[O_2]}{k_4[PH]}} \approx 1$, the second term in brackets of nominator and the third in denominator may be neglected and eq. 9) may be simplified:

$$\sqrt{k_6}[PO_2^{\cdot}] = \frac{\sqrt{k_{bi}}[POOH](\sqrt{\frac{k_6}{k_8}\frac{k_3[O_2]}{k_4[PH]}})}{1 + \sqrt{\frac{k_6}{k_8}\frac{k_3[O_2]}{k_4[PH]}}} \qquad 10)$$

or $k_6[PO_2^{\cdot}]^2 = mk_{bi}[POOH]^2$ 11)

$$\text{where } m = \frac{(\sqrt{\frac{k_6}{k_8}}\frac{k_3[O_2]}{k_4[PH]})^2}{(1+\sqrt{\frac{k_6}{k_8}}\frac{k_3[O_2]}{k_4[PH]})^2} \qquad 12)$$

For even lower concentrations of oxygen we have:

$$m = (\sqrt{\frac{k_6}{k_8}}\frac{k_3[O_2]}{k_4[PH]})^2 \qquad 13)$$

Substituting the concentration of peroxyl radicals to eq. 5) we obtain for the time dependence of hydroperoxide concentration [POOH] the same equation (eq.14) as in the paper [24]

$$[POOH] = \frac{X}{1 + Y\exp(-k_2 t)} \qquad 14)$$

where $X = [POOH]_\infty$, $Y = \dfrac{[POOH]_\infty - [POOH]_0}{[POOH]_0}$ and

$$[POOH]_\infty = \frac{k_4[PH]}{2}\sqrt{\frac{m}{k_6 k_{bi}}} \text{ and } k_2 = k_4[PH]\sqrt{\frac{mk_{bi}}{k_6}} \qquad 14a)$$

where $[POOH]_\infty$ and $[POOH]_0$ are stationary and initial concentrations of hydroperoxides in the polymer, respectively.

As it was already pointed out, the chemiluminescence intensity I is related to the rate of polymer oxidation w, however, the unambiguous expression for w is still attempted by various research groups [26]. This is because of the complexity of oxidation mechanism at one side and because of non-specified proportionality constant Φ in the equation

$$I = \Phi w \qquad 15)$$

for different polymers. This constant involves quantum yield of chemiluminescence emission, geometry of the sample, transparency of the sample, filtration and quenching effects of additives or newly formed compounds, etc. and modulates the scale of the coordinate axes on experimental record. The opinion that the higher the intensity of the light emission is, the more easily oxidation occur, need not be, necessarily valid just because of parameter Φ. Deconvolution of the rate of oxidation from equation 15) is not simple when going from one polymer to the next.

One has to take into account that oxidation of polymer may lead to a) degradation, i.e. reduction of molar mass or b) crosslinking with increase of average molar mass, or c) splitting the side groups off with unchanged molar mass when the process of oxidation is focused on the side or terminal macromolecular groups. As a result, not only Φ but also w differs from one polymer to the next depending upon the predominating process in the mechanism in chemiluminescence emission. Let us to focus on two possibilities:

$$w = k_6[PO_2^{\cdot}]^2 \qquad \text{A)}$$

$$w = k_4[PO_2^{\cdot}][PH] \qquad \text{B)}$$

First case A corresponds to already mentioned Russel's mechanism of exothermic disproportionation of peroxyl radicals, the second case B is the rate of propagation of oxidation which for some specific conditions may be significantly exothermic and lead to chemiluminescence. Both possibilities were depicted in the schemes 1 and 5.

From the viewpoint of exothermicity both processes are almost equally probable in the light emission mechanism. It is not possible to determine which process is decisive in chemiluminescence mechanism from a simple kinetic run. There is, however, some chance to distinguish each from the other when we analyse the plots of chemiluminescence intensity on oxygen concentration. In the case of first mechanism A, the chemiluminescence intensity is proportional to m, i.e. to the second power of oxygen concentration (see eq. 13) while in the second case B, it should be proportional to $\sqrt{m}$, i.e. it should be linearly dependent on oxygen concentration. This should be valid throughout the whole oxidation process including the maximum chemiluminescence intensity. It appears that in the case of oxidation of polyisoprene and polybutadiene the second mechanism of chemiluminescence (B) is more probable than that of a Russel's scheme (A). For both polymers the maximum chemiluminescence intensity increases linearly with concentration of oxygen.

The difference between polydienes and cellulose dependence on concentration of oxygen (Fig. 6) is probably due to the fact that the rate constant $k_6 \langle k_8$ for polydiene oxidation while opposite is true for cellulose. The ratio of the above constants follows from relaxation experiments by George [27]. In the case of polydienes when nitrogen atmosphere is replaced by oxygen, chemiluminescence intensity jumps fast to a certain level from which a maximum is attained in the process (Fig. 7). The picture with cellulose is different; when switching to oxygen the chemiluminescence intensity jumps to a some level from which it decays steadily to a some equilibrium level (Fig. 8). The initial increase depends on the time of sample in nitrogen and has a tendency to saturation. This is an indication that peroxyl radicals recombine faster than alkyl radicals, i.e. $k_6 > k_8$.

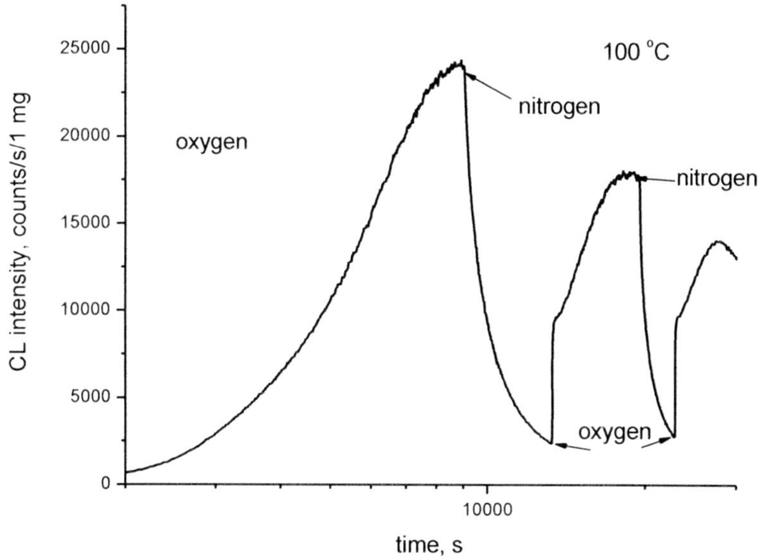

Figure 7. Relaxation changes of chemiluminescence intensity for oxidized polyisoprene during replacement of oxygen by nitrogen and vice versa

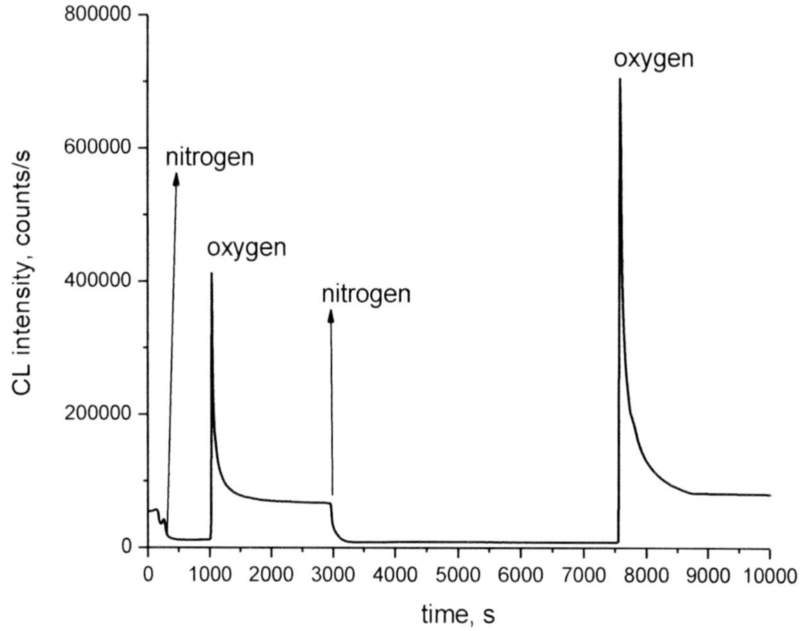

Figure 8. Relaxation changes of chemiluminescence intensity for oxidized cellulose containing magnesium carbonate at 180 °C during replacement of oxygen by nitrogen and vice versa

One should be aware of the fact that in the case of oxidation of polyolefins and polydienes [Case a), see e.g. Fig. 4] after an autocatalytic development of the oxidation, the chemiluminescence intensity passes trough a maximum and then decays. No saturated level of the light emission is achieved as in the case of initiated oxidation of low molar mass models with a constant initiation rate.

COMPARISON OF CHEMILUMINESCENCE ISOTHERMAL RUNS WITH THOSE OBTAINED BY OTHER METHODS

Comparison of chemiluminescence isothermal runs with oxygen absorption measurements for oxidation of polyisoprene indicates that the increase of chemiluminescence precedes that of oxygen uptake (Fig. 9) and that a maximum of chemiluminescence is situated around an inflexion point of oxygen absorption curve.

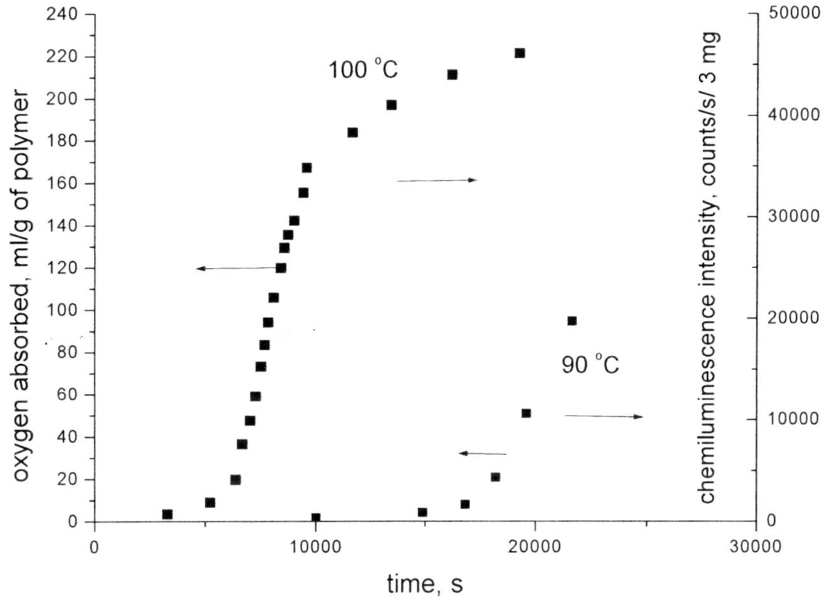

Figure 9. Chemiluminescence intensity and absorption of oxygen [28] runs for oxidation of polyisoprene in oxygen, temperatures 90 and 100 °C

This is in correspondence with an effort encountered in the literature to plot the so-called total chemiluminescence intensity (TLI) vs. amount of oxygen absorbed which in the most cases provides a straight line. A more detailed examination of the equation 3) however reveals that rate of oxygen consumption is in fact involved in its left side and at the sufficient excess of oxygen we have:

$$-\frac{d[O_2]}{dt} = k_3[P^{\cdot}][O_2] = k_4[PO_2^{\cdot}][PH] + k_6[PO_2^{\cdot}]^2 \quad 16)$$

while the right side should be proportional to the rate of reactions leading mostly to the light emission as it is expressed by the cases A and B so that:

$$I = \Phi_1 k_4 [PO_2^{\cdot}][PH] + \Phi_2 k_6 [PO_2^{\cdot}]^2 \qquad 17)$$

In dependence on the mutual ratio of the luminescence proportionality constants Φ_1 and Φ_2 one can postulate the proportionality between the rate of oxygen consumption and chemiluminescence intensity as follows:

$$-\frac{d[O_2]}{dt} = k_3[P^{\cdot}][O_2] \approx I \qquad 18)$$

and after integration

$$\Delta O_2 \approx \int I dt = TLI \qquad 19)$$

where ΔO_2 is the amount of oxygen absorbed. As each polymer system may have different proportionality constant Φ, the correlation 19) is not universal but as expressed in [27], it is polymer-, temperature- and geometry- dependent.

In some laboratories chemiluminescence intensity and DSC signals from samples situated in one oven were registered in parallel. I. Blakey and G. George [17] e.g. presented the correlation between DSC signal and chemiluminescence intensity I during oxidation of polypropylene. The quadratic dependence of I on DSC may be well seen in Fig. 10.

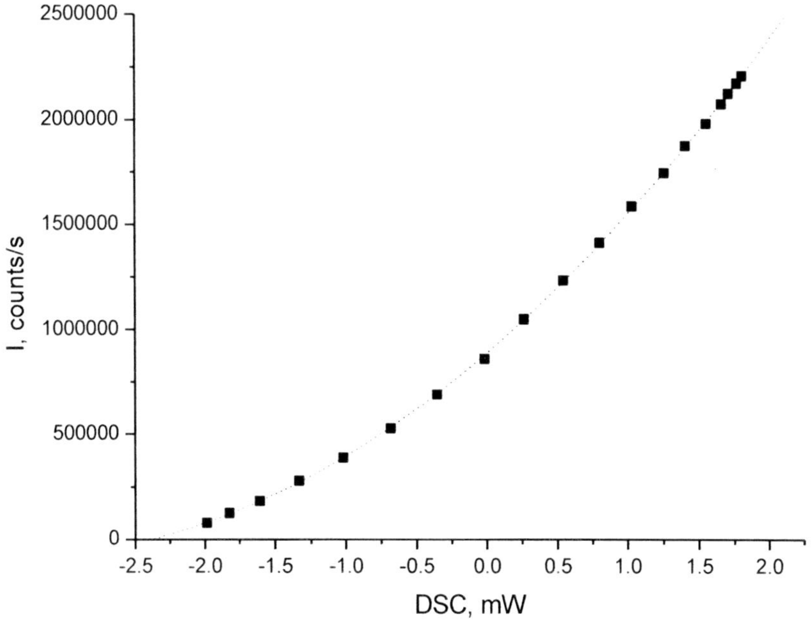

Figure 10. Comparison of DSC and CL signal I monitored paralelly from one oven at oxidation of polypropylene at 145 °C, points are experimental data, the dashed line is theoretical quadratic dependence (the original data taken from the paper [17])

Once we assume that the main portion of heat is released in propagation reaction so that:

$$DSC \approx k_4[PO_2^{\cdot}][PH] \qquad 20)$$

in combination with equation 17) we then receive:

$$I = \alpha DSC + \beta DSC^2 \qquad 21)$$

where α and β are proportionality constants and a quadratic dependence between I and DSC as shown in Fig. 10 is well comprehensible. Whether the equation 21) has the more general validity remains the question of further investigation.

THE EFFECT OF MOLAR MASS ON ISOTHERMAL CHEMILUMINESCENCE RUNS

The effect of molar mass and stereoregularity on thermo-oxidation of polymers and on the chemiluminescence – time patterns was investigated only scarcely [29-31].

There are still some non-explained observations such as that syndiotactic polypropylene was reported as being more stable than isotactic polymer. E.g. at 140 °C, the maximum chemiluminescence intensity was achieved after 2835 minutes for syndiotactic polypropylene while isotactic polymer attained the maximum after 45 minutes, only. In above papers atactic polymer was reported to be more stable than isotactic one. An explanation has been offered that the structure of isotactic polypropylene is much more favourable for autooxidation which proceeds easier via a back biting mechanism where peroxyl radicals abstract adjacent tertiary hydrogens on the same polymer chain.

On the other hand Iring et al. [32] found that the induction period for oxidation of solid isotactic polypropylene is longer than that of atactic polymer. Their result is in accordance with large crystallinity of isotactic polypropylene and with the resistance of its crystalline zones to the oxidation.

On well characterised non-stabilised polypropylene samples [33] having molar mass within 45 – 180 kg/mol and different tacticity and crystallinity we can see that the increasing molar mass (within some interval) leads to the increase of induction time and reduction of maximum chemiluminescence intensity [Fig. 11]. The polymer with higher average molar mass appears to be more stable than that with lower molar mass. This may be ascribed to the effect of increased concentration of more reactive terminal groups which promote initiation of thermal oxidation in the case of lower molar mass polymers.

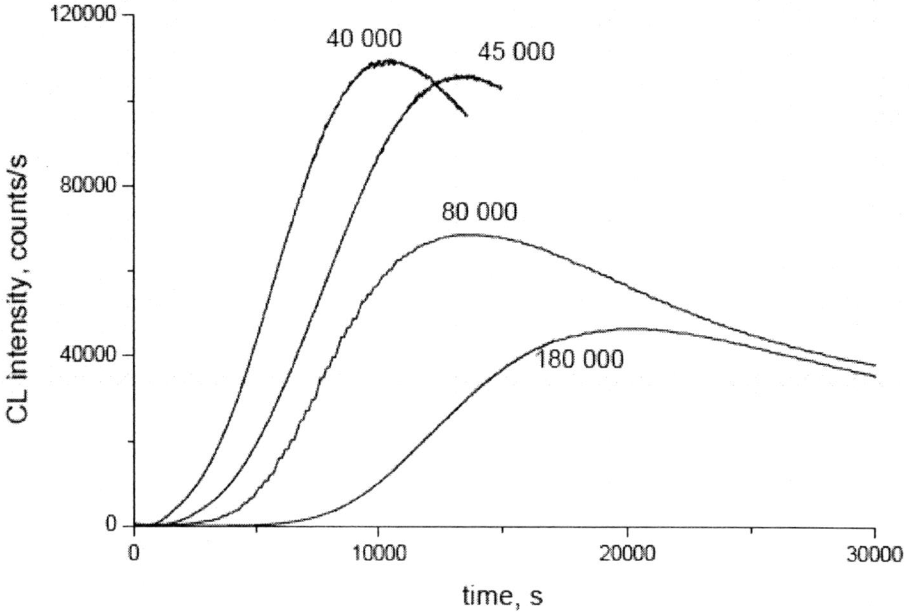

Figure 11. Chemiluminescence oxidation runs at 120 °C in oxygen for powder of polypropylene of different molar mass. The numbers at respective curves express the average molar mass in g/mol

The Effect of Temperature on Chemiluminescence Intensity

The increasing temperature shortens the induction time related to the advanced stages of oxidation and increases the maximum chemiluminescence intensity in the case of chemiluminescence pattern a) (Fig. 12) whereas it increases the initial chemiluminescence intensity in the case of pattern b) (Fig. 3). This is not surprising as the rate of oxidation reaction increases with temperature as well.

For the mechanism of chemiluminescence B) and eq. 10) and 14) the chemiluminescence intensity should be the following function of time:

$$I = \Phi w = \Phi k_4 [PO_2^{\cdot}][PH] = const[POOH] = \frac{const}{1 + B\exp(-k_2 t)} \qquad 16)$$

The rate constant k_2 may thus be determined so that we find the value of $[\frac{d(\frac{I}{I_{max}})}{dt}]_{max}$ which is the slope of chemiluminescence - time records in its inflexion point related to the maximum chemiluminescence intensity. This ratio is function of k_2 as follows:

$$\left[\frac{d\left(\frac{I}{I_{max}}\right)}{dt}\right]_{max} = \frac{1}{2}k_2 \qquad 17)$$

The set of the rate constants k_2 determined for experimental runs of Fig. 12 and their comparison with the rate constants of hydroperoxide decomposition determined by other methods may be seen in Table 1. When we take into account that polypropylenes of different origin were examined, the agreement seems quite satisfactory. This agreement is valid for faster decomposing peroxides which are the species determining the resulting rate of oxidation [34].

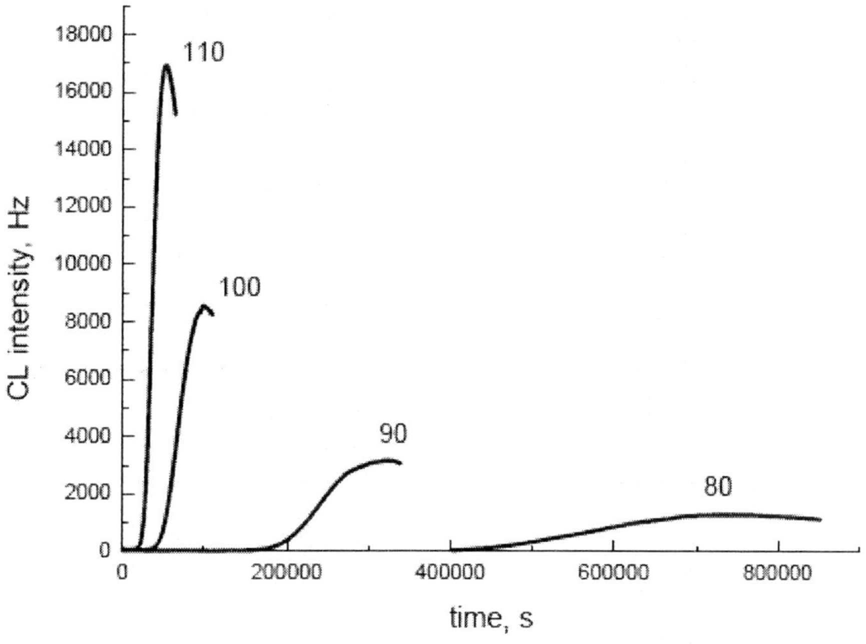

Figure 12. Chemiluminescence oxidation runs for polypropylene powder at different temperatures; the average molar mass of the polymer is 180 kg/mol

Table 1. The rate constants k_2 determined from chemiluminescence experiments of Fig. 11 and eq. 16 and the rate constants of decomposition of hydroperoxides

Temperature, °C	k_2, s^{-1}	k, s^{-1}	Reference
80	$1.7\ 10^{-5}$	-	-
85	-	$3.2\ 10^{-5}$	36
90	$4.7\ 10^{-5}$	$2.8\ 10^{-5}$	35
100	$1.1\ 10^{-4}$	$8.7\ 10^{-5}$	36
110	$2.4\ 10^{-4}$	$3\ 10^{-4}$	37

Eq. 17 appears to be a good approximation for description of isothermal chemiluminescence kinetics for homogeneous systems where oxidation takes place uniformly. However, as it was shown by several authors [38, 39, 40, 41] the different sections of the polymer sample may oxidize with its autonomous kinetics determined by different rates of primary initiation. A chemiluminescence imaging technique revealed that the light emission may be spread from some sites of the polymer film and the isothermal chemiluminescence runs are then modified particularly in the stage of the advanced oxidation reaction [40]. A heterogeneous model of polymer oxidation has been developed and described involving both, the initiation and light emission spreading [41].

Within this model an idea may be adopted that oxidation is initiated faster in regions of lower molar mass and at zones of higher concentrations of eventual structural defects. This might explain how the spreading of the light emission takes place. The initial oxidation attack is focused on the regions of lower molar mass which may be found at the sample edges of cut polymer foils or around the preoxidized domains. In the former case, an increased level of polymer free radicals generated by mechanical degradation may be responsible for the development of the oxidation process, as well. Hydroperoxides formed at the border of such domains will initiate the oxidation of the polymer molecules in the close proximity which will lead to the reduction of the molar mass and the oxidation attack continues within this lower molar mass region and adopts the character of spreading. The mechanism of spreading of oxidation via the gaseous phase in the close proximity to the polymer surface has been proposed as well, however, the molecules responsible for that were not specified.

Kinetics based on the idea of spreading is formally based on the model of development of an infectious disease among human population [42, 43]. The formalism of chemical kinetics, however, should be taken with a care as the similar equation can be derived from the homogeneous model assuming bimolecular decomposition of hydroperoxides as an initiating event.

THE EFFECT OF ANTIOXIDANTS AND POLYMER STABILIZERS

In the case a) antioxidants shift the autoaccelerating increase of chemiluminescence intensity to higher time. This is due to reactions 12) - 14) of the Scheme 2 in which alkyl radicals, peroxyl radicals and hydroperoxides are scavenged until antioxidants InH and D are consumed. The typical example of such a behaviour may be seen on the samples of polypropylene containing 0.1 % wt. of Irganox 1010 (sterically hindered phenol) [Fig. 9]. The presence of antioxidants may even reduce the maximum chemiluminescence intensity [9].

To obtain a reliable induction time from chemiluminescence measurements a homogeneous dispersion of a stabiliser in a polymer is always necessary. If it is not the case, the lower level of chemiluminescence may be due to the start of oxidation in the least stabilised regions. A quenching effect of antioxidant on excited carbonyls may also contribute to the reduction of chemiluminescence intensity after induction time. Induction times increase with increasing concentration of antioxidant in a polymer (Table 2). Nonhomogeneous distribution of Irganox 1010 in polypropylene may be considerably improved provided that the polymer film is thermostated in the atmosphere of oxygen at the temperature lower than it corresponds to the temperature of the oxidation test (Fig. 14). \

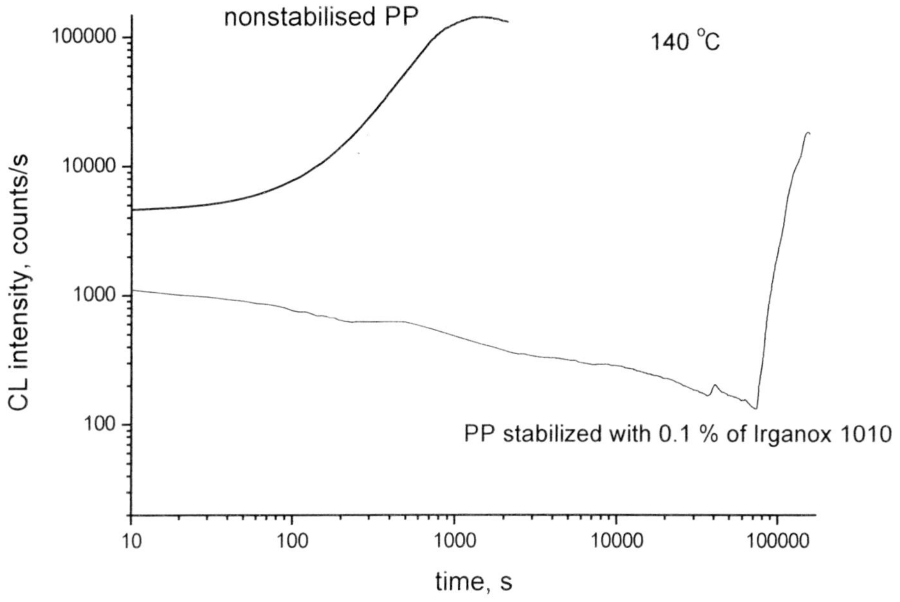

Figure 13. Chemiluminescence runs for oxidation of nonstabilised polypropylene (PP) film and film stabilized with 0.1 % wt. of phenolic antioxidant Irganox 1010, temperature 140 °C, oxygen

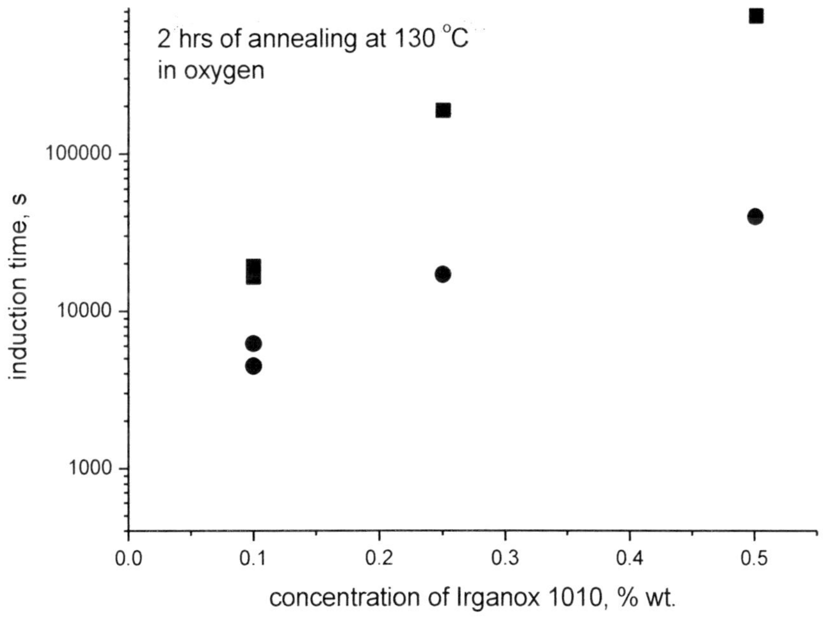

Figure 14. An increase of induction time of oxidation of polypropylene stabilised by Irganox 1010 due to the sample annealing at 130 °C in oxygen. The induction time corresponds to the time of cross-section of the straight line passing the inflexion point and time axis. It was determined for oxygen atmosphere and temperature 150 °C

Table 2. The effect of antioxidants on induction time of oxidation of polypropylene determined at 150 °C from chemiluminescence measurements

Antioxidant, % wt.	t_{ind}, s at 150 °C in oxygen
Irganox 1010, 0..1	6206
Irganox 1010, 0.1	4459
Irganox 1010, 0.25	17018
Irganox 1010, 0.5	39742
Irganox 1076, 0.25	555
Irganox 1076, 0.5	1753

The most vulnerable part of the polymer system with respect to oxidation is amorphous region where macromolecules of the highest molar mass form skeleton of the system responsible for ultimate mechanical properties. The mobility of the chains from one side and the easier access of reaction sites by low molar mass compounds including oxygen brings about that the degradation process starts predominantly there. The domains of higher molar mass are, moreover, less protected by stabilizers which tend to dissolve better in lower molar mass regions. Oxidation of pure polymer starts, therefore, on lower molar mass molecules or on their terminal groups; however, in the case of stabilized polymer these lower molar mass molecules are better protected by higher local concentration of dissolved stabilizer. In such a case, one may expect that degradation starts easier on higher molar mass domains which are less protected [40].

The case of a reduction of the chemiluminescence intensity for stabilised polymers as described by the pattern b) has not been studied as yet.

A Simple Kinetic Model of Polymer Degradation and Its Relation to Non-Isothermal Chemiluminescence Runs

Estimation of kinetic parameters from non-isothermal chemiluminescence runs is more problematic as each rate constant should be described by its own Arrhenius dependence.

The kinetic approach may be, however, simplified taking into account the model developed by Ekenstam [44] and later by Emsley [45] for cellulose degradation.

Such a model quantifies sufficiently the process in time and/or with increasing temperature and becomes thus a necessary pre-requisite of any extrapolations or comparison of the effect of different additives. Polymerisation degree (DP) is understood here as the ratio of concentration of monomer units (N) and polymer molecules (i) as follows:

$$DP = \frac{N}{i} \qquad 18)$$

The concentration of polymer molecules increases in time due to degradation and decays due to crosslinking. Provided that the process is taking place statistically, the kinetics of concentration of macromolecules increase (for the case of degradation) may be described by equation:

$$\frac{di}{dt} = mki^n \qquad 19)$$

where n stands for the order of main chain scissions and m=1 for the case of degradation and m=-1 for the case of crosslinking. The most frequent reaction orders encountered in the literature are n=0 or n=1.

For the former case of n=0 and m=1:

$$\frac{di}{dt} = k \text{ and } i = i_0 + kt \qquad 20)$$

where i_0 is initial concentration of macromolecules in the system. For DP we then have:

$$DP = \frac{N}{i_0 + kt} \qquad 21)$$

or $DP = \dfrac{DP_0}{1 + \dfrac{k}{i_0} t}$, where $DP_0 = \dfrac{N}{i_0}$ and after transformation:

$$\frac{1}{DP} - \frac{1}{DP_0} = \frac{k}{i_0 DP_0} t \qquad 22)$$

$$\text{or } \frac{DP_0}{DP} = 1 + \frac{k}{i_0} t \qquad 23)$$

Formally, the most simple equation which allows for an easy switch to non-isothermal kinetics and to relate it to chemiluminescence intensity is given bellow:

$$I = \Phi[-\frac{dDP}{dt}] \qquad 24)$$

where Φ is already quoted proportionality constant.

According to eq. 6) $-\dfrac{dDP}{dt} = \dfrac{k}{i_0 DP_0} DP^2$ and for non-isothermal conditions

we have:

$$-\frac{dDP}{dT}\frac{dT}{dt} = \frac{A\exp(-E/RT)}{i_0 DP_0} DP^2 \qquad 25)$$

Here T is temperature, A and E is pre-exponential factor and activation energy, respectively and $\frac{dT}{dt} = \beta$ is linear rate of sample heating.

After integration of eq. 25) and substitution into eq. 24) we finally receive for non-isothermal conditions:

$$I = \Phi \frac{A\exp(-E/RT)}{i_0} \frac{DP_0}{[1 + \frac{A}{\beta i_0} \int_{T_{room}}^{T} \exp(-E/RT)dT]^2} \qquad 26)$$

One can see that while the process of the chain scission has the character of the zero order, the chemiluminescence runs formally correspond to the second order scheme.

Provided that we formally consider that chemiluminescence process is governed by two independent mechanisms occurring in parallel, the equation 26) which was derived from the zero order scission of macromolecular chains may be applied as follows:

$$I = \sum_{i=1}^{2} \frac{P_i}{[1 + \frac{A_i}{\beta i_0} \int_{T_{room}}^{T} \exp(-E_i/RT)dT]^2} \qquad 27)$$

Here P_i is the proportionality constant including the corresponding terms from the eq. 26.

The chemiluminescence runs for Whatman (Wh), sulphate (SA) bleached pulp and cotton (C) pulp oxidized under non-isothermal conditions are shown on Fig. 14 and corresponding parameters for two independent steps of degradation (i=2) are summarized in Table 3. Agreement of the fit of experimental measurements by parameters of eq. 27 (points in Fig. 14) is very good, indeed.

Table 3. Parameters of equation 27 found from non-isothermal runs in oxygen for i=2 for Whatman cellulose, SA(sulphate bleached) and C (cotton) pulp sheets

	i=1 Faster process			i=2 Slower process		
Paper	P_1	A_1/i_0	E_1	P_2	A_2/i_0	E_2
	counts/1 mg	s^{-1}	J/mol	counts/1 mg	s^{-1}	J/mol
C	$2.43\ 10^6$	$4.06\ 10^3$	58 141	$1.3\ 10^8$	$1.34\ 10^{10}$	128 409
Wh	$1.13\ 10^6$	$5.35\ 10^4$	64 327	$1.11\ 10^8$	$9.85\ 10^9$	126 951
	$6.31\ 10^7$	$1.19\ 10^3$	60 083	$1.9\ 10^8$	$4.38\ 10^{12}$	149 775

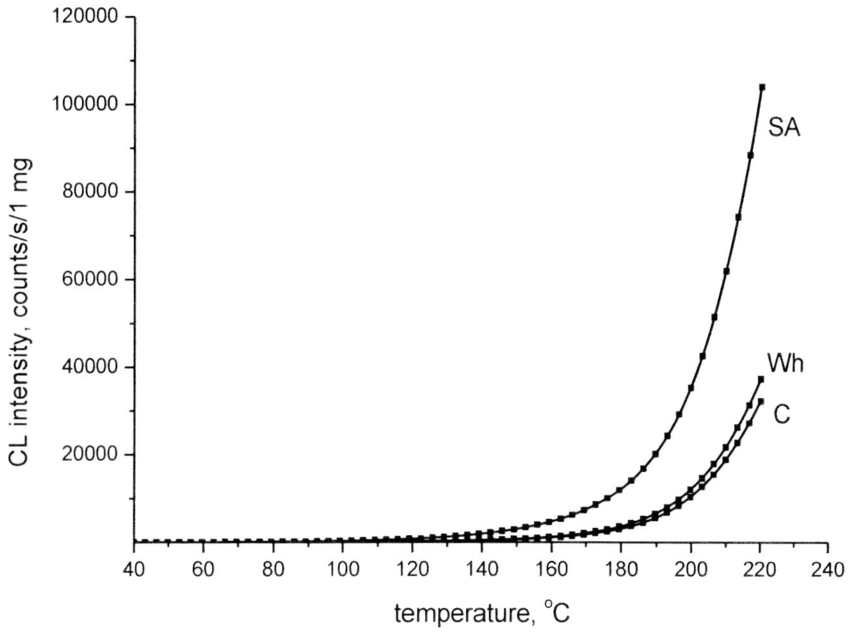

Figure 15. Experimental runs and theoretical fits of chemiluminescence measurements of oxidation Whatman (Wh), sulphate (SA) bleached pulp and cotton C pulp. The rate of sample heating was 3 °C/min, oxygen atmosphere. Points are values fitted by means of eqs. 27 for i=2 and parameters of Table 2

Averaged rate constant k_{av} is defined as:

$$k_{av} = \frac{P_1}{P_1 + P_2} k_{11} + \frac{P_2}{P_1 + P_2} k_{12} \quad 28)$$

Here $k_{1i} = A_{1i} \exp(-\frac{E_{1i}}{RT})$

The above kinetic analysis has some advantages but disadvantages, too. By the use of non-linear regression analysis we may achieve a good fit with experimental chemiluminescence run and get the series of several rate constants which correspond to activation energy and pre-exponential factor determined from an experimental curve. We may thus reveal if the oxidation process is a typical autooxidation with accummulation of hydroperoxides or not. If this is the case, the activation energy obtained from the part of the curve under the largest surface of experimental chemiluminescence course attains abnormally high values due to the contribution of the autoacceleration to the light emission increase. On the other hand, fitting the resulting curve by several first order processes may lead to a certain compensation of parameters which becomes particularly evident when the measured curves are not well developed or slightly distorted.

A question may arise why we put such emphasis on non-isothermal kinetics. The reason lies in the fact that starting from a low temperature and continuously increasing it, a small conversion of the virgin material to the oxidized one is achieved therefore there is a considerably lower probability of the back effect of the reaction products on the rate of the oxidation [6].

REFERENCES

[1] Shlyapintokh, V. Ya., Karpukhin, O. N., Postnikov, L. M., Zakharov, I. V., Vichutinskii, A. A. and Tsepalov, V. F., *Chemiluminescence methods of investigation of slow chemical processes* (in Russian), Publishing House Nauka, Moscow (1966).

[2] Ashby, G. E., *J. Polym. Sci.,* 50, 99, (1961).

[3] Schard, M. P. and Russell, C. A., *J. Appl. Polym. Sci.,* 8, 985, (1964), ibid. 8, 997, (1964)

[4] Bolland, J. L. and Gee, G., *Trans. Farad. Soc.,* 42, 236-243, (1946).

[5] Matisová-Rychlá, L. and Rychlý, J., Inherent relations of chemiluminescence and thermooxidation of polymers, In: Clough, R. L., Billingham, N. C. and Gillen, K. T., editors. Advances in chemistry, Series 249: Polymer durability, degradation, stabilization and lifetime prediction. Am. Chem. Soc. Washington DC, 175, (1996).

[6] Rychlý, J. and Rychlá, L., Chemiluminescence from polymers, In: *Ageing and Stabilisation of Paper*, M. Strlič and J. Kolar, editors, National and University Library, Ljubljana, Slovenia, 2005, pp. 71-90.

[7] Jacobsson, K., Eriksson, P., Reitberger, T. and Stenberg, B., Chemiluminescence as a tool for polyolefin oxidation studies, In: Albertsson, A. C., editor. Long Term Properties of Polyolefins, *Advances in Polymer Sci*., Springer, 169, 151-176, (2004).

[8] Rychlá, L. and Rychlý, J., New concepts in chemiluminescence at the evaluation of thermooxidative stability of polypropylene from isothermal and non-isothermal experiments. In: Jimenez, A. and Zaikov, G. E., editors. *Polymer Analysis and Degradation,* Nova Science Publishers, New York, chapter 7, p.124.

[9] George, G. and Celina, M., Homogeneous and heterogeneous oxidation of polypropylene, In: Halim Hamid, S. and Dekker, M., editors. *Handbook of Polymer Degradation*, Second Ed., Inc. New York, 277, (2000).

[10] Zlatkevich, L., *Luminescence techniques in solid state polymer research*, M. Dekker, New York, chapter 7, (1989).

[11] Rychlá, L. and Rychlý, J., New concepts in chemiluminescence at the evaluation of thermooxidative stability of polypropylene from isothermal and non-isothermal experiments. In: Jimenez, A. and Zaikov, G. E., editors. *Polymer Analysis and Degradation*, Nova Science Publishers, New York, chapter 7, p.124.

[12] Lengfelder, E. Cadenas, E. and Sies, H., *FEBS Letters*, 164, 366, (1983).

[13] Reshetnyak, O. V., Kovalchuk, E. P., Skurski, P., Rak, J. and Blazejowski, J., *J. Luminescence,* 105, 27-34, (2000).

[14] Quinga, E. M. Y. and Mendenhall, D. G., *J. Am. Chem. Soc.* 105, 6520, (1983).

[15] Audouin-Jirackova, L. and Verdu, J., *J. Polym. Sci*., 25,1205, (1987).

[16] Osawa, Z., Kuroda, H. and Kobayashi, Y., *J. Appl. Polym. Sci*., 29, 2843, (1984).

[17] Blakey, I. and George, G. A., *Macromolecules,* 34, 1873, (2001).

[18] Kelogg, R. E., *J. Am. Chem. Soc*., 91, 5433, (1969).

[19] Gugumus, F., *Polym. Degrad. Stab.*, 77, 147-155, (2002).
[20] Broska, R., Rychlý, J. and Matisová-Rychlá, L., *Europ. Polym. J.*, 32, 1251, (1996).
[21] Ptschelintsev, V. V. and Denisov, E. T., *Vysokomol. Soed. (in Russian)*, 27, 1123-1136, (1985).
[22] Rychlý, J., Strlič, M., Matisová-Rychlá, L. and Kolar, J., *Polym. Degrad. Stab.*, 73, 357, (2002).
[23] Matisová-Rychlá, L. and Rychlý, J., *Polym. Degrad. Stab.*, 67, 515, (2000).
[24] Audouin, L., Gueguen, V., Tcharkhtchi, A. and Verdu, J., *J. Polym. Sci. Polym. Chem. Ed.*, 33, 921-927, (1995).
[25] Achimsky, L., Audouin, L., Verdu, J., Rychlá, L. and Rychlý, J., *J. Europ. Polym . J.*, 35, 557-563, (1999).
[26] Matisová-Rychlá, L. and Rychlý, J., *Polym. Degrad. Stab.*, 67, 515-525, (2000).
[27] George, G. A., Use of chemiluminescence to study the kinetics of oxidation of solid polymers, In: Grassie, N., editor. Developments in Polymer Degradation - 3, Applied Sci. Publishers, London , p.173, (1981).
[28] Rosík, L., Ageing and stabilization of elastomers. In: *Synthetic rubber*, J.Pech et al. SNTL Praha, Praha (in Czech), p. 411, (1971).
[29] Mori, H., Hatanaka, T. and Terano, M., *Macromol. Rapid Commun.*, 18, 157, (1997).
[30] Osawa, Z., Kato, M. and Terano, M., *Macromol. Rapid Commun.*, 18, 667, (1997).
[31] Kato, M. and Osawa, Z., *Polym. Degrad. Stab.*, 65, 457, (1999)
[32] Iring, M., Laszlo-Hedvig, Zs., Tudos, F. and Kelen, T., *Polym. Degrad. Stab.*, 467, 5, (1983)
[33] Rychlý, J., Matisová-Rychlá, L., Tiemblo, P. and Gomez-Elvira, J., *Polym. Degrad. Stab.*, 71, 253, (2001)
[34] Gijsman, P. and Verdun, F., *Polym. Degrad. Stab.*, 74, 533-542, (2001).
[35] Gijsman, P., Hennekens, J. and Vincent, J., *Polym. Degrad. Stab.*, 42, 95-105, (1993).
[36] Chien, R. J. C. W. and Jabloner, H., *J. Polym. Sci. A-1,* 6, 393-402, (1968).
[37] Chien, R. J. C. W. and Boss, C. R., *J. Polym. Sci. A-1,* 5, 3091-3101 (1967).
[38] Dudler, V., Lacey, D. J. and Krohnke, Ch., *Polym. Degrad. Stab.*, 51, 115-124, (1996).
[39] Celina, M., George, G. A., Lacey, D. J. and Billingham, N. C., *Polym. Degrad. Stab.*, 311-317, (1995)
[40] Matisová-Rychlá, L. and Rychlý, J., *Polym. Degrad. Stab.*, 73, 393-398, (2001).
[41] Eriksson, P., Reitberger, T., Ahlblad, G. and Stenberg, B., *Polym. Degrad. Stab.*, 73, 177-183, (2001).
[42] Goss, B. G. S., Barry, M. D., Birtwhistle, D. and George, G. A., *Polymer Degradation and Stability*, 74, 271-282, (2001).
[43] Goss, B. G. S., Blakey, I., Barry, M. D. and George, G. A., *Polymer Degradation and Stability,* 74, 523-532, (2001).
[44] Ekenstam, A., *Ber. Deutschen Chem. Gesselschaft,* 69, 553, (1936).
[45] Emsley, A. M., Heywood, R. J., Ali, C. M. M. and Eley, C. M., *Cellulose,* 4, 1-5, (1997).

In: Recent Developments in Polymer Research
Editor: Antony V. Hopper, pp. 193-246
ISBN 1-60021-346-4

Chapter 8

THE APPLICATION OF COUPLING REAGENTS IN POLYMER SYNTHESIS

Anton Blencowe and Wayne Hayes*
School of Chemistry, The University of Reading, Whiteknights,
Reading, Berkshire, RG6 6AD, UK

ABSTRACT

Coupling or condensation reagents are species that increase the rate of condensation between two components, such as an amine and a carboxylic acid. The coupling reagent activates one of the components (generally a carboxylic acid) rendering it more susceptible to nucleophilic attack by the other component. The result is formation of a species arising from the coupled components and a by-product – a hydrated coupling reagent derivative. However, use of coupling reagents can also result in the formation of un-wanted products. As a result of the high efficiency of coupling reagents, it has become commonplace to use them in conjunction with coupling additives that limit side reactions of the activated species. For a coupling reagent to be selected for a particular process, either peptide coupling or polymer synthesis, it must fulfil a number of essential criteria. Ideally a coupling reagent should be stable under the reaction conditions used, work efficiently and with high conversion under mild conditions in stoichiometric quantities. It should produce minimum number of side-products and perhaps most importantly, the hydrated by-product should be removed easily by solvent extraction or other simple isolation procedures.

The aim of this review is to provide the reader with a detailed insight into the use of coupling reagents in polymer chemistries. In addition to providing detailed examples of coupling reagent mediated polymerisations, the article covers the mode of action of coupling reagents and assesses the relative advantages and disadvantages associated with conventional polymerisation methods in comparison to coupling reagent mediated approaches.

* School of Chemistry, The University of Reading, Whiteknights, Reading, Berkshire, RG6 6AD, UK. *Tel: +44 (0)118 378 6491; E-mail: w.c.hayes@reading.ac.uk.*

ABBREVIATIONS

APDE = *N*-(3-aminopropyl)-diethanolamine
AOP = phosphonate*O*-(7-azabenzotriazol-1-yl)-*tris*(dimethylamino) phosphonium hexafluorophosphonate
BEMT = 2-bromo-3-ethyl-4-methyl thiazolium tetrafluoroborate
Boc = *N-tert*-butoxycarbonyl
BOP = benzotriazol-1-yl diethyl
BOP-Cl = *bis*(2-oxo-3-oxazolidinyl)phosphinic chloride
BroP = bromo*tris*(dimethylamino)phosphonium hexafluorophosphate
Cbz = carbobenzyloxy
CDI = 1,1'-carbonyldiimidazole
CDT = 1,1'-Carbonyldi(1,2,4-triazole)
CIP = 2-chloro-1,3-dimethylimidazolidinium hexafluorophosphate
CMC = 1-cyclohexyl-3-(2-morpholinoethyl)carbodiimide methyl-*p*-toluenesulfonate
DB = degree of branching
DBOP = diphenyl (2,3-dihydro-2-thioxo-3-benzoxazolyl)phosphonate
DCC = 1,3-dicyclohexylcarbodiimide
DCU = 1,3-dicyclohexylurea
DIC = 1,3-di*iso*propylcarbodiimide
DMAP = 4-(dimethylamino)pyridine
DMI = 1,3-dimethyl-2-imidazolidinone
DNA = deoxyribonucleic acid
DP = degree of polymerisation
DPCP = diphenyl chlorophosphate
DPPA = diphenylphosphorylazide
Dpp-Cl = diphenylphosphinic chloride
DPTS = 4-(dimethylamino)pyridinium-4-toluene-sulfonate
EDCI = 1-ethyl-3-(3'-dimethylaminopropyl) carbodiimide hydrochloride
Fmoc = *N*-(9-fluorenylmethoxycarbonyl
HATU = *O*-(7-azabenzotriazol-1-yl)-1,1,3,3-tetramethyluronium hexafluorophosphate
HMPA = hexamethylphosphoramide
HOAt = 1-hydroxy-7-azabenzotriazole
HOBt = 1-hydroxybenzotriazole
HODhbt = 3-hydroxy-3,4-dihydro-4-oxo-1,2,3-benzotriazine
MDA = 4,4'-methylenedianiline
MFCA = methano[60]fullerene carboxylic acid
MFPCA = methano[60]fullerene phenylpropyl carboxylic acid
NCA = *N*-carboxylanhydride
NEPIS = *N*-ethyl-5-phenylisoxazolium-3'-sulfonate
NHS = *N*-hydroxylsuccinimide
NMP = *N*-methyl-2-pyrrolidinone
ODA = 4,4'-oxydianiline
PAA = poly(acrylic acid)
PAAME = poly(amic acid methyl ester)

PDI = polydispersity index
PEEK = poly(aryl ether ether ketone)s
PEG = polyethylene glycol
PEI = polyethylenimine
PPEI = poly(propionyl-ethylenimine-*co*-ethylenimine)
PTSA = *p*-toluenesulfonic acid
PyBOP = 1-benzotriazolyoxytris(pyrrolidino) phosphonium hexafluorophosphonate
PyBroP = bromo*tris*(pyrrolydino) phosphonium hexafluorophosphate
PyCloP = chloro*tris*(pyrrolydino) phosphonium hexafluorophosphate
ROP = ring opening polymerisation
$Sn(Oct)_2$ = stannous 2-ethylhexanoate
THF = tetrahydrofuran
TMCFAC = *N,N,N',N'*-tetramethylchloroformamidinium chloride
TOSUO = 1,4,9-trioxaspiro[4,6]-9-undecanone
TPA = terephthalic acid
TPP = triphenyl phosphite

1. Coupling Reagents

1.1. Introduction

Coupling or condensation reagents are species that increase the rate of condensation between two components, such as an amine and a carboxylic acid. The coupling reagent activates one of the components (generally a carboxylic acid) rendering it more susceptible to nucleophilic attack by the other component. The result is formation of a species arising from the coupled components and a by-product – a hydrated coupling reagent derivative. However, use of coupling reagents can also result in the formation of un-wanted products. As a result of the high efficiency of coupling reagents, it has become commonplace to use them in conjunction with coupling additives that limit side reactions of the activated species. For a coupling reagent to be selected for a particular process, either peptide coupling or polymer synthesis, it must fulfil a number of essential criteria. Ideally a coupling reagent should be stable under the reaction conditions used, work efficiently and with high conversion under mild conditions in stoichiometric quantities. It should produce minimum number of side-products and perhaps most importantly, the hydrated by-product should be removed easily by solvent extraction or other simple isolation procedures.

Coupling reagents have been applied extensively to the step-wise synthesis of peptides and for the formation of sensitive esters that require mild reaction conditions. In addition, coupling reagents have been particularly useful for the synthesis of dendrimers and more recently for the synthesis of linear and hyperbranched polymers as they offer a high degree of selectivity and hence excellent control over the polymerisation process. As coupling reagents operate under mild conditions it is also possible to synthesise novel polymers or macromolecular architectures which may prove unstable to more conventional condensation polymerisation methods such as exposure to high temperatures and low pressures.

The aim of this review is to provide the reader with a detailed insight into the use of coupling reagents in polymer chemistries. In addition to providing detailed examples of coupling reagent mediated polymerisations, the article covers the mode of action of coupling reagents and assesses the relative advantages and disadvantages associated with conventional polymerisation methods in comparison to coupling reagent mediated approaches.

1.2. Coupling Reagents

Coupling or condensation reagents have been utilized extensively to construct peptides and proteins. Indeed, since the first report of the use of 1,3-dicyclohexylcarbodiimide (DCC) in 1955 [1] the number of coupling reagents now used commonly exceeds thirty. The coupling reagent used most frequently in peptide synthesis is DCC, a member of the carbodiimide class of coupling reagents. The mode of action for carbodiimide reagents 1 [2] involves nucleophilic attack of a carboxylate group at the carbodiimide carbon to form an *O*-acylisourea 2 (Scheme 1). A nucleophile, such as an amine or an alcohol, can then attack the activated species 2 at the carbonyl carbon thus displacing the hydrated carbodiimide adduct 3 (a urea) and resulting in the formation of an amide or ester group, respectively.

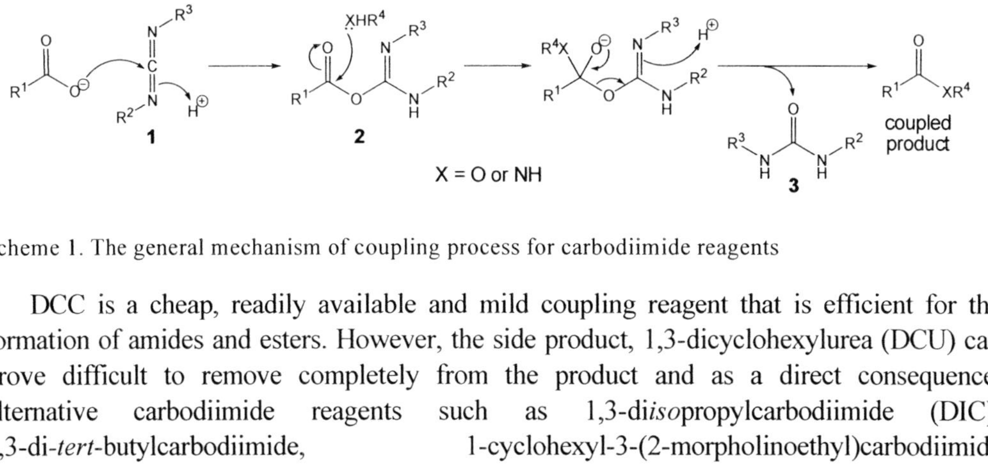

Scheme 1. The general mechanism of coupling process for carbodiimide reagents

DCC is a cheap, readily available and mild coupling reagent that is efficient for the formation of amides and esters. However, the side product, 1,3-dicyclohexylurea (DCU) can prove difficult to remove completely from the product and as a direct consequence, alternative carbodiimide reagents such as 1,3-di*iso*propylcarbodiimide (DIC), 1,3-di-*tert*-butylcarbodiimide, 1-cyclohexyl-3-(2-morpholinoethyl)carbodiimide methyl-*p*-toluenesulfonate (CMC) and 1-ethyl-3-(3'-dimethylaminopropyl)carbodiimide hydrochloride (EDCI) [3] have been developed in order to circumvent this practical issue. EDCI and CMC are water-soluble derivatives designed to obviate the problem of urea contamination of the desired product by rendering the urea by-product water-soluble. In addition, solid supported coupling reagents [4] and coupling additives [5] have been developed recently to aid purification of the desired products through simple isolation procedures. Additional issues associated with carbodiimide couplings include unwanted side reactions such as *N*-acylurea formation and racemisation of chiral substrates, although these problems can be alleviated by the addition of coupling additives. Coupling additives that are used commonly in peptide synthesis include 1-hydroxybenzotriazole (HOBt) [6] and 1-hydroxy-7-azabenzotriazole (HOAt) [7]. HOAt increases the rate of the desired coupling process whilst reducing the loss of chiral integrity and provides a visual indication of the reaction end-point. Alternative coupling additives that are not employed as frequently include

3-hydroxy-3,4-dihydro-4-oxo-1,2,3-benzotriazine (HODhbt) [4], *N*-ethyl-5-phenylisoxazolium-3'-sulfonate (NEPIS) [8] and *N*-hydroxylsuccinimide (NHS) [9].

In recent times carbodiimide coupling reagents (in combination with coupling additives) have been replaced gradually by aminium/uronium salts and phosphonium salts. These salts have been designed carefully so as to incorporate both the coupling reagent and additive in one reagent. Most notably Knorr *et al.* [10] have developed a series of uronium salts based upon 4-oxo-1,2,3-benzotriazine, 5-norbornene-2,3-dicarboxomidine, 2-oxo-pyridyl, and succinimide core structures. The majority of aminium/uronium salts feature HOBt or HOAt at the core. The research groups of Xu [11], Gross [12] and Carpino [7, 13] have reported valuable developments in this field. Uronium coupling reagents function via nucleophilic attack of a carboxylate group at the positive uronium carbon, displacing HOAt or HOBt and forming an *O*-acylisourea cation 4 (Scheme 2). The hydroxyl anion of the HOAt or HOBt unit then attacks this activated species 4 at the carboxyl carbon thus displacing a urea and resulting in the formation of an *O*-acyl-1-hydroxy-7-azabenzotriazole or *O*-acyl-1-hydroxybenzotriazole, respectively. A nucleophile (such as an amine or an alcohol) can then attack at the carbonyl carbon thus displacing HOAt or HOBt and resulting in the formation of an amide or ester group, respectively.

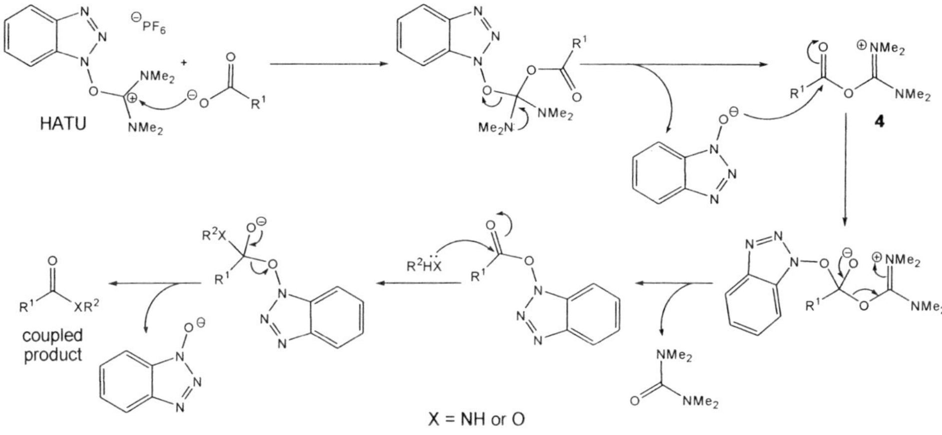

Scheme 2. The mechanism of the coupling process for *O*-(7-azabenzotriazol-1-yl)-1,1,3,3-tetramethyluronium hexafluorophosphate (HATU) (General mechanism shown is applicable to all uronium and amidinium coupling reagents)

Phosphonium coupling reagents include, amongst others, benzotriazol-1-yl diethyl phosphonate (BOP) [14], *O*-(7-azabenzotriazol-1-yl)-*tris*(dimethylamino) phosphonium hexafluorophosphonate (AOP) [15] and 1-benzotriazolyoxy*tris*(pyrrolidino) phosphonium hexafluorophosphonate (PyBOP) [16] and these reagents are now used readily in peptide synthesis as a result of their extremely high coupling efficiencies. Phosphonium coupling reagents act in a very similar manner to uronium and amidinium systems. A carboxylate group reacts with the positive phosphorous atom of the phosphonium unit, displacing HOAt or HOBt thus forming an *O*-acyloxyphosphoramidium cation 5 (Scheme 3). The hydroxyl anion of the HOAt or HOBt then attacks this activated species 5 at the carboxyl carbon thus displacing an amido phosphine oxide and resulting in the formation of an *O*-acyl-1-hydroxy-7-azabenzotriazole or *O*-acyl-1-hydroxybenzotriazole, respectively. A nucleophile, such as an amine or an alcohol, can then attack at the carbonyl carbon thus displacing HOAt or HOBt and resulting in the formation of an amide or ester group, respectively.

Alternative coupling reagents that have been found useful particularly for coupling sterically hindered amino acids include 2-bromo-3-ethyl-4-methyl thiazolium tetrafluoroborate (BEMT) [17], *bis*(2-oxo-3-oxazolidinyl)phosphinic chloride (BOP-Cl) [18], bromo*tris*(dimethylamino)phosphonium hexafluorophosphate (BroP) [19], diphenyl-phosphinic chloride (Dpp-Cl) [20], 2-chloro-1,3-dimethylimidazolidinium hexafluoro-phosphate (CIP) [21], bromo*tris*(pyrrolydino) phosphonium hexafluorophosphate (PyBroP) [22] and chloro*tris*(pyrrolydino) phosphonium hexafluorophosphate (PyCloP) [22].

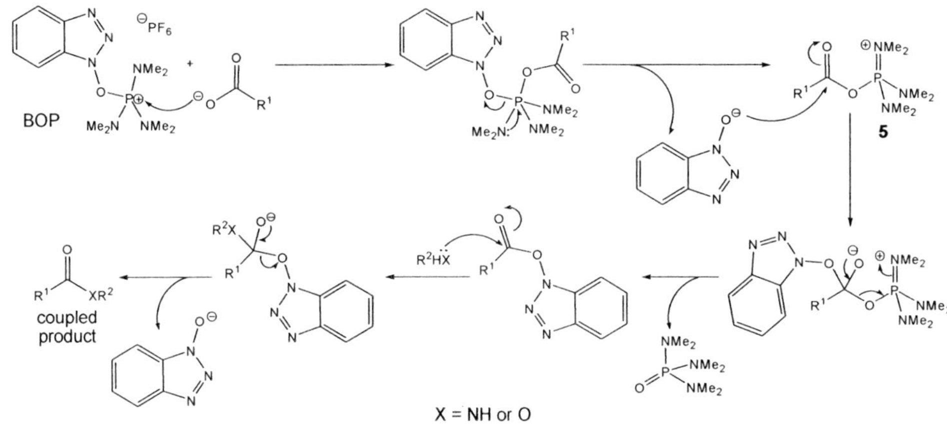

Scheme 3. Mechanism of coupling process for BOP (General mechanism shown is applicable to all phosphonium coupling reagents)

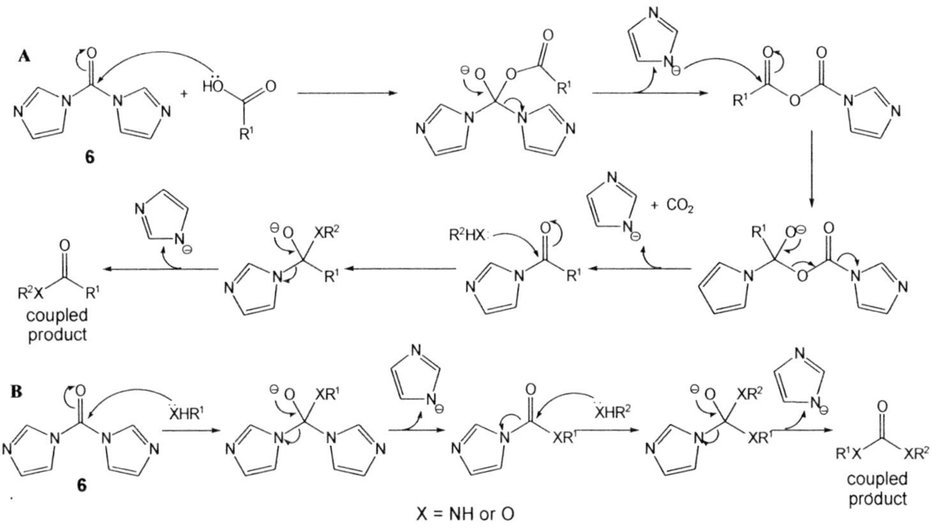

Scheme 4. A; Mechanism for coupling reaction mediated by CDI. B; Mechanism for synthesis of amines, carbonates and urethanes involving CDI carbonyl incorporation

Although 1,1'-carbonyldiimidazole 6 (CDI) [23] is not used regularly in peptide chemistries it remains a particularly useful coupling reagent for the formation of amides and esters (reaction A, Scheme 4) [24]. In addition, CDI can also be used in the synthesis of carbonates, ureas and urethanes [24, 25]. However, in these cases CDI acts as a reagent that couples two nucleophiles (alcohols and/or amines) together through a carbonyl group (reaction B, Scheme 4) as opposed to coupling two functions together directly. 1,1'-Carbonyldi(1,2,4-triazole) (CDT) is a derivative of CDI that was developed recently and incorporates two triazole groups as opposed to imidazole moieties.

1.3. Coupling Reagents and Polymer Chemistry

Direct polycondensation reactions have two major synthetic limitations. Firstly, precise control of molecular weight is difficult to obtain and secondly, forcing conditions are required frequently. As a result it is very difficult to synthesis high molecular weight products from sensitive monomers that are either highly functionalised or chiral in nature. It is common to convert the condensation monomers into activated derivatives, however, this process hinders the formation of isoregic [26] polymers that possess translational polar symmetry.

Carbodiimide reagents, diphenyl (2,3-dihydro-2-thioxo-3-benzoxazolyl) phosphonate (DBOP) [27] and triphenyl phosphite (TPP) [28] are the coupling reagents used most commonly in polymer synthesis. Both carbodiimide reagents and DBOP are used generally at room temperature or below, whereas TPP polycondensations are conducted usually at *ca.* 100 °C. DBOP acts in a similar manner to phosphonium, uronium and amidinium coupling reagents. A carboxyl group attacks the phosphorous atom of DBOP displacing benzo[d]oxazole-2-thione and resulting in the formation of an *O*-acylphosphate. The amine moiety of benzo[d]oxazole-2-thione then attacks this activated intermediate species at the carbonyl carbon thus displacing a phosphate anion and resulting in the formation of an *N*-acyl-benzo[d]oxazole-2-thione. A nucleophile, such as an amine or an alcohol, can then react with this activated intermediate to form an amide or an ester, respectively. In addition to acting as a chain extender in transesterifications [29] and an inhibitor to prevent cross-linking, TPP **7** has been used extensively in the synthesis of polyamides. Although the mechanism associated with the TPP coupling reagent has not been well studied [30] it is conceivable that the formation of an *O*-acylphosphite 8, *P*-acylphosphate 9 or phenylester 10 intermediate is involved (Scheme 5).

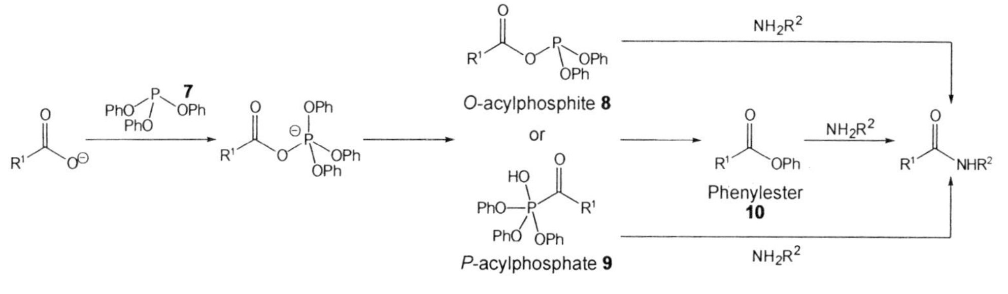

Scheme 5. Possible mechanisms involved in TPP coupling of carboxylic acids and amines

2. LINEAR POLYMERS

2.1. Synthesis of Oligo and Polypeptides using Coupling Reagents

In an attempt to improve the understanding of prebiotic chemistry, Liu and Orgel have studied the oligomerisation of α- and β-amino acids in aqueous solution using CDI and EDCI [31]. Neither EDCI or CDI can be considered as plausible prebiotic coupling systems, however, these reagents are both relevant to prebiotic chemistry since their modes of action are relevant to broad classes of potentially prebiotic reactions. EDCI models closely activating agents that function by reacting directly with the carboxyl group of unprotected amino acids, whilst reactions induced by CDI resemble reactions that involve the activation of an amino acid in the presence of carbon dioxide. Indeed, cyanamide, cyanoguanidine and cyanogens are all plausible prebiotic versions of EDCI.

It was found that when aqueous solutions of α-amino acids such as α-glutamic acid were incubated at appropriate temperatures (25 to 50 °C) with solid CDI, oligomerisation occurred efficiently [32,33]. However, β-amino acids, such as β-glutamic acid could not be oligomerised by using CDI. The CDI-induced oligomerisation of α-amino acids proceeded rapidly at room temperature, with the majority of oligomeric products formed within 2-3 hours. The use of CDI afforded peptides possessing oligomers up to 11 α-amino acids after a period of 24 hours. Longer oligomers (degree of polymerisation (DP) = 11) were obtained when the concentration of CDI was twice that of the amino acid(s) used. Increasing further the concentration of CDI did not improve the polymerisation efficiency. It was hypothesized that the α-carboxyl group of the dipeptide was activated under these conditions and diketopiperazine formation then competes with the desired oligomerisation pathway [34]. The CDI-induced reaction of α-amino acids was found to be more efficient under slightly alkaline conditions (pH 8) and at higher temperatures. For example, CDI induced oligomerisations of α-glutamic acid at 25 °C only yielded oligomers up to 11 units long, in contrast, the same experiment conducted at –20 °C afforded minor amounts of trimer. High performance liquid chromatography (HPLC) elution profiles of the polymeric material obtained from the reaction of L-α-glutamic acid with CDI revealed a series of well-defined single peaks corresponding to oligomers. However, when aspartic acid was used in conjunction with CDI then the resultant HPLC elution profiles featured a second series of subsidiary peaks as a result of the presence of side-products (*iso*aspartate 11) formed by the rearrangement of oligo-α-aspartic acid 12 mediated by excess CDI. The amide nitrogen reacts with the activated β-carboxyl group to afford a succinimide 13 [35], which in turn is ring-opened by water to generate oligomers featuring both α- and β-linked peptides (Scheme 6).

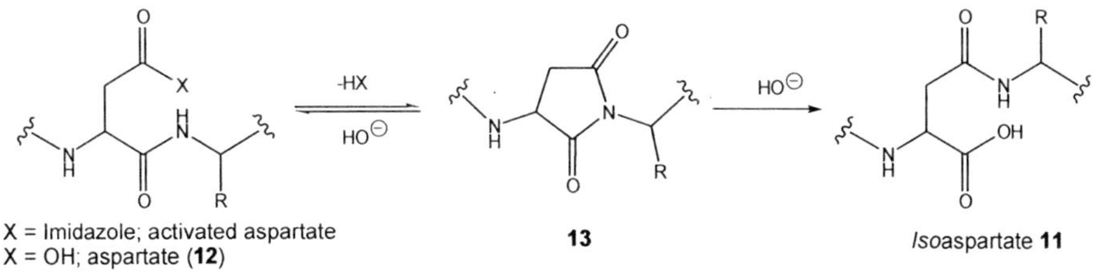

Scheme 6. Formation of α- and β-linked oligo(aspartic acid)

The mechanism of CDI-induced oligomerisations of α-amino acids in aqueous solution proceeds with addition of the amine group of an α-amino acid to CDI to afford a *N*-[imidazolyl-(1)-carbonyl] amino acid 14 which then cyclises to afford a *N*-carboxylanhydride (α-NCA) 15 (Scheme 7) [32].

The amine group of an α-amino acid attacks the carbonyl carbon of an α-NCA derivative resulting in ring opening and the formation of a peptide bond [2, 36]. As a consequence, only individual amino acids are activated and oligomers can be extended only by reacting with α-NCA derivatives but cannot ligate to each other [34]. In addition, since carboxyl-activated dipeptides are not generated under these conditions, formation of 2,5-diketopiperazines does not occur to a significant extent [32]. Basic conditions favour oligomerisations with CDI since the amino acid monomers exist in a deprotonated state and can thus compete efficiently with water and hydroxyl anions in attacking the α-NCA derivatives. However, under acidic conditions the amine groups of the amino acids are protonated and thus disfavour attack on α-NCA derivatives. The CDI-induced oligomerisation of β-amino acids requires the formation of β-NCAs. As six-membered β-NCAs are not formed as readily as their five-membered α-NCA counterparts and are prone to rearrange to form β-isocyanato acids via ring opening [37] and thus polymerisation of β-amino acids via activation with CDI is not observed.

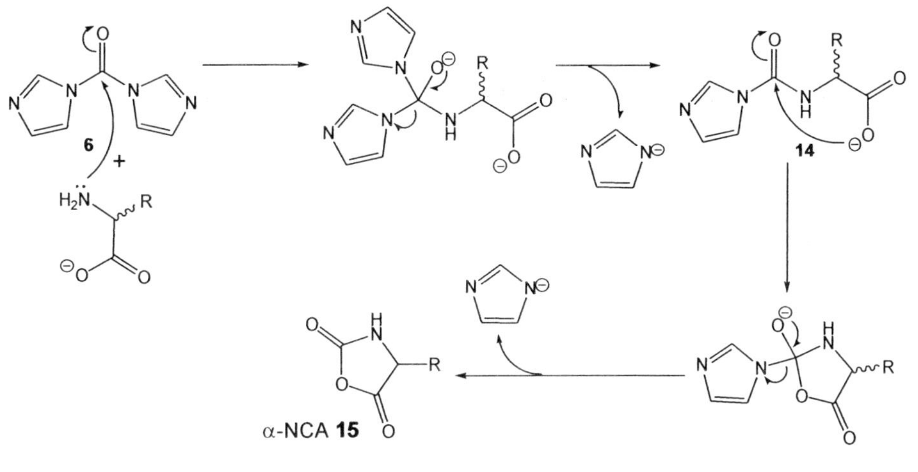

Scheme 7. CDI activation of α-amino acids to afford a *N*-carboxylanhydride

In contrast to the use of CDI in peptide synthesis, when aqueous solutions of α-amino acids were incubated at appropriate temperatures (-25 to 50 °C) with solid EDCI oligomerisation did not occur [32,33], however, β-amino acids were oligomerised efficiently by using EDCI. In the case of oligomerisation of α-amino acids with EDCI, the carbonyl group of the dipeptide can be activated resulting in efficient cyclisation to give a diketopiperazine, thus inhibiting oligomerisation [34]. In contrast, cyclisations of an EDCI activated dimer consisting of β-amino acids are disfavoured because an eight-membered ring does not form readily. Consequently, β-amino acids but not α-amino acids can be oligomerised when EDCI is used as the coupling reagent. The EDCI-induced oligomerisation of β-amino acids proceeded rapidly at room temperature and oligomeric products are formed within 2-3 hours. Oligomers featuring up to 10 β-amino acids were formed after a period of 24 hours. However, under optimised conditions peptides featuring up to 45 units could be obtained [31]. Indeed, when EDCI was used, optimum oligomerisations were obtained when 1 equivalent of EDCI per β-amino acids was used, although high concentrations of coupling reagent led to the formation of branched polymers [38]. Reaction of EDCI with β-amino acids proceeded readily under acidic conditions (pH 4.5-6) as a result of protonation of one of the nitrogen atoms of the carbodiimide, thereby rendering the central carbon atom more electron deficient [39] and thus more susceptible to nucleophilic attack by carboxylic acids. The EDCI-induced oligomerisation of β-amino acids was found to be independent upon the temperature in the range 4-50 °C, although when experiments were conducted at -25 °C [40] oligomers possessing up to 45 repeat units could be detected. Oligomerisation of aspartic acid using EDCI generates peptidic material that possesses a complicated elution profile consisting of a complex mixture of components that result from the ability of EDCI to activate both carbonyl groups in an unselective manner thus affording linear and branched peptides. The side chain carboxyl groups of the oligo-β-aspartic acids were activated by EDCI and coupled with the free amino groups of another amino acid or peptide to form branched peptides [38].

Pascual and co-workers have also studied the CDI and EDCI mediated polycondensation of α- and β-amino acids in aqueous solution [41]. In addition to reinforcing the results obtained by Liu and Orgel [31], Pascual and co-workers studied the effect of the coupling additive NHS. The major advantage of using coupling additives for the aqueous phase coupling reagent mediated polycondensation is that the coupling additive forms an activated ester [42] that is less prone to hydrolysis when compared to EDCI activated esters (*O*-acylureas). Polycondensation of glycine using 1 equivalent of EDCI over a period of 16 hours was conducted both in the absence and presence of the coupling additive NHS. When NHS was not used the resulting material obtained after 16 hours contained considerable amounts of unreacted glycine. In comparison, when 10 mol% of NHS was used, the polymeric material obtained after 16 hours was devoid of glycine. Therefore, NHS reduces the possibility of hydrolysis of activated *O*-acylurea derivatives, thus increasing the coupling efficiency.

In addition to seminal studies of Liu and Orgel upon aqueous phase oligomerisations of α- and β-amino acids, research has also been carried out on the formation of oligopeptides on the surface of minerals (referred to as 'polymerisation on the rocks') in an attempt to mimic the construction of polypeptides on the primitive Earth [43]. It was found that long oligopeptides of β-glutamic acid and aspartic acid accumulate efficiently on the surface of hydroxylapatite using EDCI as the coupling reagent. In addition, hydroxylapatite can facilitate the ligation of tripeptides using EDCI. Oligoarginines were accumulated successfully on the surfaces of illite, kaolmite and montmorillonite using CDI as the coupling reagent and FeS_2 catalyzed their formation, although it does not absorb oligoarginines. The reactions were conducted by adding aqueous solutions containing amino acids or tripeptides and a coupling reagent to an appropriate amount of the mineral substrate. After tumbling the mixtures, the samples were centrifuged, the supernatant removed, and fresh activated amino acid was again added to begin the next coupling cycle [38]. After the required number of coupling cycles had been completed, the mineral was washed with water and the oligopeptides eluted using either aqueous $K_4P_2O_7$ (oligopeptides with negatively charged side-chains) or guanidine (oligopeptides with positively charged side-chains). Reaction of aspartic acid or β-glutamic acid with EDCI on hydroxylapatite using the 'feeding' technique resulted in the formation of oligopeptides consisting of up to 45 units after 25 cycles. If the same procedure was repeated in the absence of the mineral support it was found that successive additions of fresh activated monomer did not lead to the formation of longer oligopeptides. Interestingly, the total yield of oligomers after 1 cycle was greatest when the mineral was not used suggesting that inorganic phosphate groups in hydroxylapatite react with EDCI and reduce the effective concentration of the coupling reagent near the mineral surface. Longer oligomers of β-amino acids accumulated efficiently on hydroxylapatite after repeated 'feeding' with monomer and EDCI since the mineral (a crystalline calcium phosphate) acts as an anion exchange material [38,44]. As the strength of adsorption of an oligopeptide anion to hydroxylapatite increased with the number of negatively-charged side chains [45], shorter oligomers could be desorbed easily, whilst longer oligomers remained for longer on the mineral surface thus permitting further elongation.

The elastic/plastic protein-based polymers are a relatively new class of polymers [46], composed of repeating peptide sequences that occur in all sequenced mammalian elastic proteins [47]. The most striking and longest sequence between cross-links in pig and cows is the polypentapeptide, $(Val^1\text{-}Pro^2\text{-}Gly^3\text{-}Val^4\text{-}Gly^5)_n$ (also abbreviated to $(VPGVG)_n$), where *n* is 11 [48]. Polymers comprised of such naturally occurring peptide sequences can be designed to have a number of medical and non-medical applications [49]. Poly(GVGVP) 16 with molecular weights in excess of 50 kD were synthesised from *p*-nitrophenyl esters of the pentapeptide over a period of 14 days [50]. However, prior to the use of the poly(GVGVP) in medical applications, it is essential to remove any *p*-nitrophenol completely as the presence of even trace amounts may cause toxic side effects. In order to avoid the problems associated with the *p*-nitrophenol approach (high toxicity and long reaction times), Gowda and co-workers investigated the polymerisation of the pentapeptide H_2N-GVGVP-OH 17 using carbodiimide coupling reagents (Scheme 8) in collaboration with the coupling additive HOBt [51]. Polymerisation of H_2N-GVGVP-OH 17 (Scheme 8) using DCC and DIC in the presence or absence of coupling additives was found to produce poly(GVGVP) 16 with molecular weights of *ca.* 15 kD, considerably lower than those obtained with the *p*-nitrophenol approach. Polymerisation of H_2N-GVGVP-OH 17 using EDCI/HOBt over a period of 2 days followed by dialysis using 50 kD molecular weight cut-off dialysis tubing afforded a polymer 16 with molecular weights in excess of 50 kD. Poly(GVGVP) 16 obtained from polymerisation with EDCI was nearly identical to the material obtained via the *p*-nitrophenol approach (i.e. a polymer featuring weight-average molecular weight >50 kD). The synthesis of poly(GVGVP) 16 using a combination of EDCI/HOBt and the subsequent characterisation studies (such as NMR spectroscopy, thermal analysis and FTIR spectroscopy) were in good agreement with the polypentapeptide obtained by the conventional *p*-nitrophenol mediated approach. The results obtained by Gowda and co-workers demonstrate that EDCI/HOBt can be used as an alternative for the synthesis of pure, high molecular weight polypeptides. The water soluble urea by-products can be removed readily using solvent extraction, thus reducing impurities and decreasing the toxicity of the poly(GVGVP) 16.

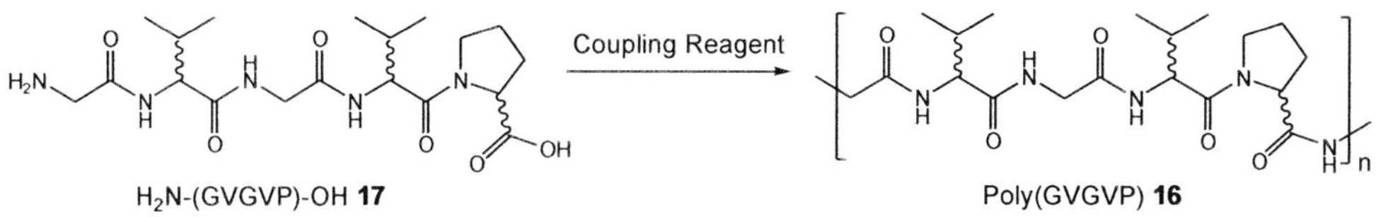

Scheme 8. Synthesis of poly(GVGVP) 16 using coupling reagents. Coupling reagents = DCC, DIC or EDCI [51]

2.2. Synthesis of Synthetic Polyamides Using Coupling Reagents

Aromatic polyamides such as Kevlar® (poly(*p*-phenyleneterephthalamide)) are well known for their high thermal and chemical resistance and, high strength as fibres [52]. The primary disadvantage of these polyamides is that processing is difficult because of the high softening or melting temperatures and the insoluble nature of these materials in most common organic solvents [53]. In order to improve these properties, flexible linkages [54] (e.g. ether and sulfonyl) have been introduced or bulky side groups [55] to the polyamide backbone. For example, Yang *et al.* have synthesised soluble linear aromatic poly(ether amide)s 18 from a diamine 19 and a series of diacids 20 via TPP mediated coupling [56]. The diamine 1,4-*bis*(4-amino-2-trifluoromethylphenoxy)benzene A_2 type monomer 19 was polymerised with a series of aromatic diacid B_2 type monomers 20 (Scheme 9) using an excess of TPP/pyridine in *N*-methyl-2-pyrrolidinone (NMP).

In these polycondensations the molecular weight and inherent viscosities of the polyamides was found to be dependent upon the concentration of the reagents and monomers. When high monomer concentrations were employed, high molecular weight poly(ether amide)s were produced providing that precipitation or gelation of the resulting polymer did not occur [57]. The resulting poly(ether amide)s 18 were soluble in polar amide type solvents as a result of the trifluoromethyl groups present along the polymer backbone and flexible ether linkages, and possessed inherent viscosities in the range 1-1.7 dL g^{-1}.

The coupling reagent *N,N'*-phenylphosphon*bis*[2(3*H*)-benzothiazolone] 21 (Scheme 10) developed by Ueda and Mochizuki [58] was found to be very suitable for the synthesis of polyamides from diamines and diacids, particularly polyamides that could not be prepared successfully by ordinary condensation type approaches.

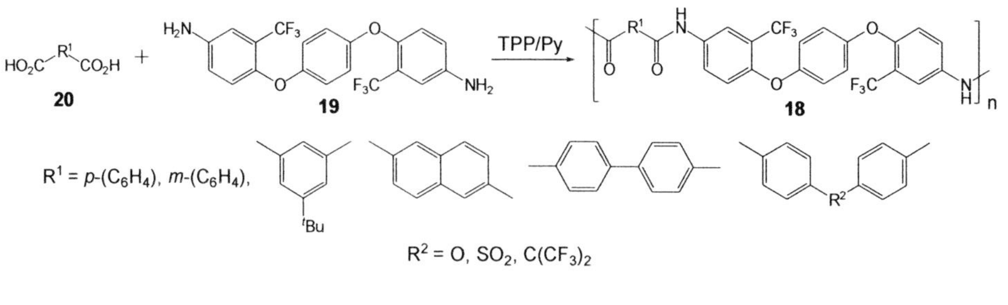

Scheme 9. Synthesis of linear poly(ether amide)s 18 using TPP mediated coupling [56]

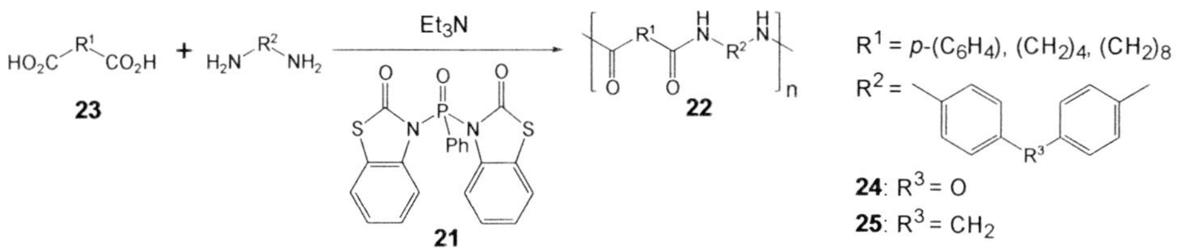

Scheme 10. Synthesis of linear polyamides 22 using *N,N'*-phenylphosphon*bis*[2(3*H*)-benzothiazolone] 21 [58]

Polyamides 22 were synthesised at room temperature from aromatic and aliphatic diacids 23 and, 4,4'-oxydianiline (ODA) 24 and 4,4'-methylenedianiline (MDA) 25 using 2.2 equivalents of 21 in the presence of triethylamine in NMP or hexamethylphosphoramide (HMPA).

The resulting polyamides 22 synthesised from aliphatic diacids possessed inherent viscosities between 0.39-1.00 dL g^{-1}, whereas those produced from aromatic diacids exhibited higher inherent viscosities between 0.44-0.95 dL g^{-1}. Novel poly(keto amide)s [58] having inherent viscosities of *ca.* 0.75 dL g^{-1} were also synthesised from ODA 24 and ketone functionalised diacids 26 and 27 (Figure 1) using 2.2 equivalents of 21 in the presence of triethylamine.

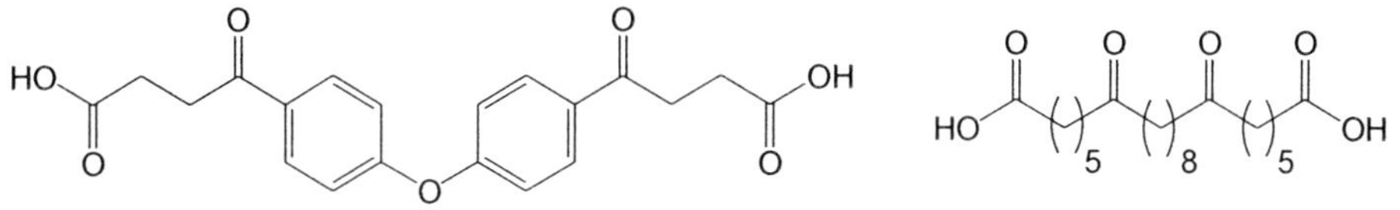

Figure 1. Diketone functionalised diacid monomers [58]

Ueda and co-workers have conducted extensive studies on the DBOP mediated synthesis of polyamides [27,59-63]. Since the late eighties these studies have resulted in the synthesis of a large range of polyamides and block co-polyamides (Figure 2).

The polyamides were synthesised from a variety of symmetrical and/or asymmetrical monomers via both step-wise and random one-pot co-polymerisation methodologies. Initial studies reported that active dithiol esters and diamines derived from 2-benzoxazolethiol reacted with diamines at room temperature to afford high molecular weight polyamides [64]. The development of DBOP [27] prompted Ueda and co-workers to explore further the synthesis of polyamides from activated carboxylic intermediates. The polyamides 29 and 30 and poly(ether amide)s 31 and 32 [27] were synthesised via a step-wise procedure from isophthalic acid or 1,10-decanedioic acid and ODA 24, MDA 25 or 1,6-diaminohexane using 2.2 equivalents of DBOP in the presence of triethylamine in NMP or HMPA at room temperature. The diacids were allowed first to react with DBOP prior to the addition of the diamines in a separate second step.

The polyamides 29-32 were obtained in excellent yields and possessed inherent viscosities of 0.97, 0.63, 0.34 and 0.61 dL g^{-1} (measured at a concentration of 0.5 g dL^{-1} in concentrated H_2SO_4 at 30 °C), respectively. The polyamides 31 and 29 were also synthesized via a one-pot procedure in which all of the reactants were mixed together simultaneously to afford the polyamides possessing inherent viscosities of 1.2 dL g^{-1}. However, the one-pot method was found to be inadequate for the synthesis of the polyamide 30 as a result of the high basicity of the aliphatic diamine which reacts with DBOP to form undesirable products including phosphoramides 33 [27,65] (Scheme 11).

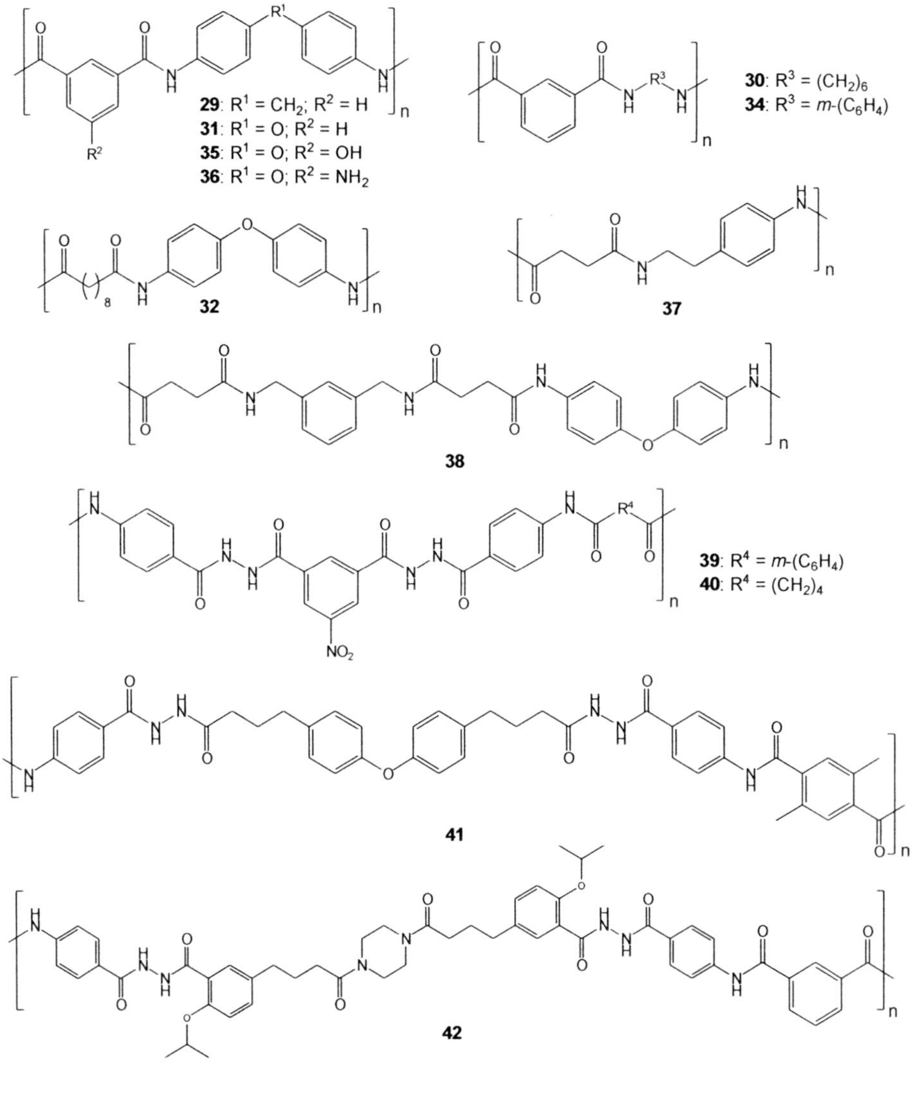

Figure 2. Polyamides and block co-polyamides synthesised using DBOP [27,59-63].

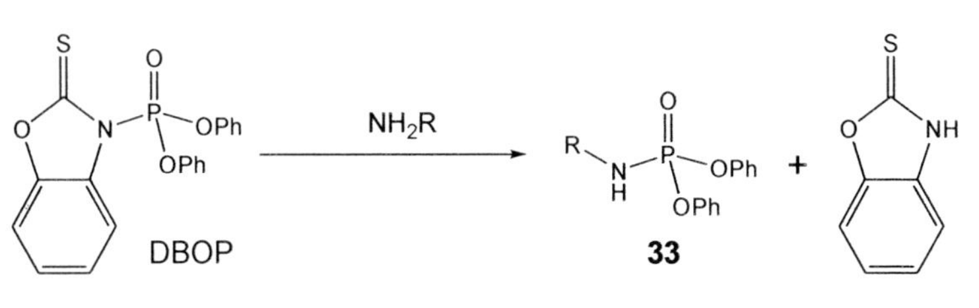

Scheme 11. Phosphoramide 33 formation via the reaction of aliphatic amines with DBOP (where R is aliphatic) [27,65]

In order to further demonstrate the applicability of the DBOP coupling methodology, chemoselective polyamidations were also conducted [27]. The backbone functionalised polyamide 34 and poly(ether amide)s 35 and 36 (Figure 2) featuring inherent viscosities of 0.42 (0.92 dL g^{-1} if the polycondensation was conducted at 80 °C), 0.53 and 0.52 dL g^{-1} were synthesised via the one-pot procedure from *iso*phthalic acid derivatives and ODA 24 or 3,5-diaminobenzoic acid under the conditions described previously [27]. The synthesis of the polyamides 34, 35 and 36 highlighted selectivity issues with the DBOP coupling reagent and the activated diacid intermediates thus formed from it:–

- DBOP activated acids do not react with hydroxyl groups that are less nucleophilic than aromatic and aliphatic amines;
- DBOP activated acids can differentiate between aromatic amines such as aniline and electron poor aromatic amines such as 3-aminobenzoic acid and
- DBOP activated acids can differentiate between aromatic acids such as benzoic acid and electron rich aromatic acids such as 3-aminobenzoic acid.

The polyamide 37 (Figure 2) was synthesized from 2-(4-aminophenyl)ethylamine and succinic anhydride via both stepwise and one-pot methodologies [60] using *ca.* 2.2 equivalents of DBOP in the presence of triethylamine in a mixture of 1,3-dimethyl-2-imidazolidinone (DMI) and HMPA at room temperature. Both step-wise and one-pot methods yielded identical polyamides possessing inherent viscosities of 0.25 and 0.20 dL g^{-1}, respectively. However, one-pot polycondensation of succinic acid and 2-(4-aminophenyl)ethylamine afforded a randomly orientated polyamide consisting of three structural sub-units (Figure 3).

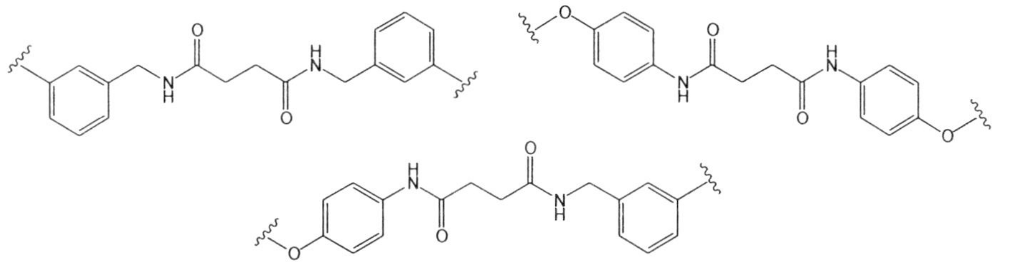

Figure 3. Structural sub-units present in randomly orientated polyamide [60].

The poly(ether amide) 38 (Figure 2) was synthesized from *m*-xylylenediamine, succinic anhydride and ODA 24 [60] via both stepwise and one-pot methodologies using *ca.* 2.2 equivalents of DBOP in the presence of triethylamine at room temperature. The step-wise process involved formation of a diacid via ring opening addition of 2 equivalents of succinic anhydride by *m*-xylylenediamine. DBOP mediated polycondensation of the diacid, *N,N'*-(*m*-xylylene)disuccinamic acid with ODA 24 in the presence of triethylamine in NMP afforded the ordered polyamide 38 with inherent viscosity of 0.23 dL g^{-1}. One-pot polycondensation of the components used to synthesise the polyamide 38 afforded a random polyamide with a random structure and inherent viscosity of 0.22 dL g^{-1}.

Ueda and co-workers have also combined coupling reagent methodologies with activated acid chemistries in order to synthesis a range of co-block polyamides 39-42 (Figure 2). The poly(acylhydrazide amide)s 39 and 40 [61] were synthesized from isophthaloyl chloride (39) or adipoyl chloride (40) and *N,N'-bis*(4-aminobenzoyl)-5-nitroisophthalodihydrazide, which was obtained from 5-nitroisophthalic acid and 4-aminobenzohydrazide [66] using DBOP in the presence of triethylamine. The polyamides 39 and 40 were found to exhibit inherent viscosities of 0.71 and 0.40 dL g^{-1}, respectively. Polyamides of the same structure as 39 and 40 were also synthesised from 5-nitroisophthalic acid, 4-aminobenzohydrazine and *bis*(2,4,6-trichlorophenyl) isophthalate (39) or *bis*(2,4,6-trichlorophenyl) adipate (40) using DBOP. In this case the low reactivity of the trichlorophenyl esters compared to the corresponding acid chlorides allowed the polycondensations to be conducted in a one-pot approach. The components were mixed at room temperature thus forming *N,N'-bis*(4-aminobenzoyl)-5-nitroisophthalodihydrazide and then heated at 70 °C to afford poly(acylhydrazide amide)s 39 and 40 possessing inherent viscosities of 0.25 and 0.30 dL g^{-1}, respectively. The poly(acylhydrazide ether amide) 41 (Figure 2) [62] featuring an inherent viscosity of 0.62 dL g^{-1} was synthesised from 2,5-dimethylterephthaloyl chloride and 4,4'-(oxydi-*p*-phenylene)*bis*[*N'*-(4-aminobenzoyl) butanohydrazide] 43, which was obtained from 4,4'-(oxydi-*p*-phenylene)dibutanoic acid and 4-aminobenzohydrazine using DBOP in the presence of triethylamine. Polyamide 41 was also constructed via direct polycondensation through reaction of 2,5-dimethylterephthaloic acid 44 and 4,4'-(oxydi-*p*-phenylene)dibutanoic acid 45 with DBOP followed by adding 4-aminobenzohydrazine. The initial step in this polycondensation involved the activation of 44 and 45 by DBOP (Scheme 12).

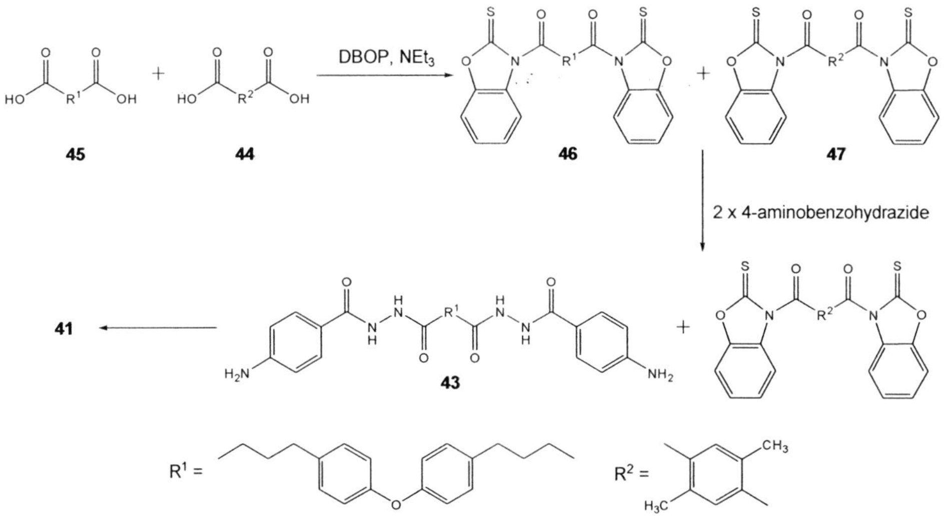

Scheme 12. Synthesis of poly(acylhydrazide ether amide) 41 using DBOP [62]

The resulting activated acids 46 and 47 were then reacted selectively with 4-aminobenzohydrazine. The aliphatic activated acid 46 is more reactive towards nucleophilic substitution than the aromatic activated acid 47 and thus 46 reacts with the more nucleophilic hydrazine component (relative to the aniline component) of 4-aminobenzohydrazine to form 43. The aromatic activated acid 47, which has been present throughout the reaction then reacts with the aniline components of 43 to form the polyamide 41 with an inherent viscosity of 0.40 dL g^{-1}. The ordered poly(amide acylhydrazide amide) 42 (Figure 2) [63] featuring an inherent viscosity of 0.48 dL g^{-1} was synthesised via a three-step process (Scheme 13) involving:– (i) addition of DBOP to 5-(3-carboxypropyl)-2-*iso*propoxybenzoic acid 48 resulting in the formation of the activated acid derivative 49 that in turn reacted selectively with 0.5 equivalents of piperazine to afford the amide 50, (ii) condensation of 50 with 2 equivalents of 4-aminobenzohydrazide to afford 51, and (iii) polycondensation of 51 with isophthaloyl dichloride.

Kitazawa and co-workers have synthesised a novel polyamide incorporating [60]fullerene in the polymer main chain [67]. As a result of the high electron density of [60]fullerene a highly delocalised state can exist on the surface of [60]fullerene [68]. Therefore, co-polymers of this type [69] have been of particular interest because of their potential electron delocalisation properties. The polyamide 52 featuring a weight-average molecular weight and polydispersity of 45 kD and 2.2, respectively was synthesised from [60]fullerene*bis*acetic acid 53 and 4,4'diaminobenzophenone 54 (Scheme 14) using a 20-fold excess of the TPP/pyridine coupling system with respect to the monomer [67]. The resultant polyamide 52 was soluble in common organic solvents, and displayed enhanced extended conjugation in comparison to previously synthesised [60]fullerene polymers.

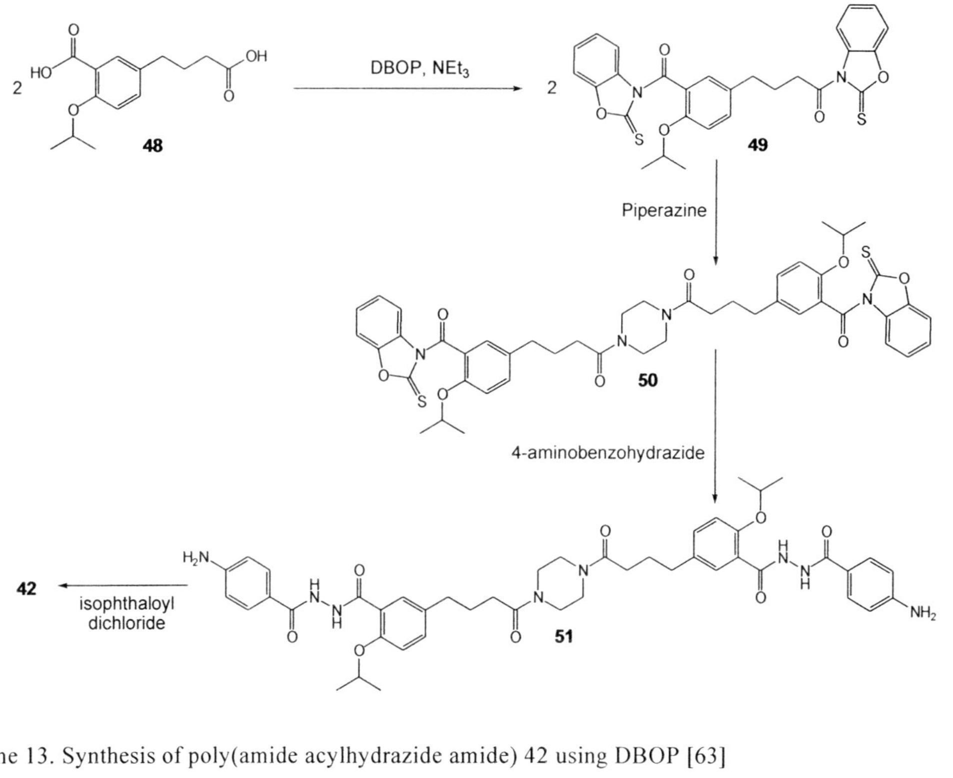

Scheme 13. Synthesis of poly(amide acylhydrazide amide) 42 using DBOP [63]

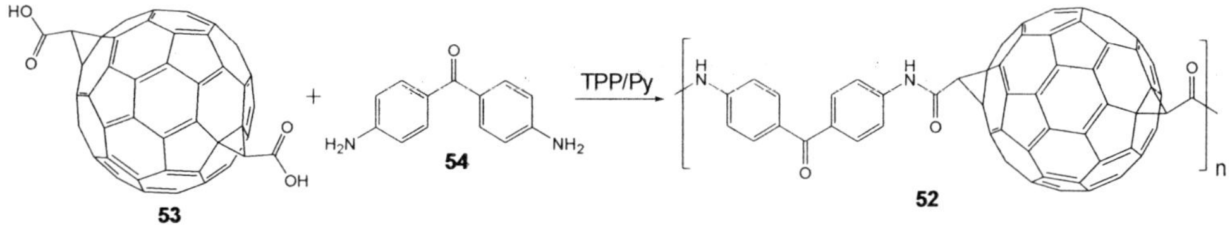

Scheme 14. Synthesis of polyamide 52 incorporating [60]fullerene [67]

Weber and co-workers have synthesised successfully oligo(azo sulfonyl amide)s [70] which contains the non-linear optical (NLO) chromophore 4-dialkylamino-4'-(alkylsulfonyl) azobenzene using DCC in a step-wise manner. The advantage of synthesising main chain NLO polymers [71] as opposed to side chain NLO polymers [72] is that these systems were able to align, thereby enhancing the NLO properties of the material [73]. The monomer 55 (Scheme 15) was deprotected selectively to afford mono-protected monomers with either free carboxylic groups 56 or secondary amines 57.

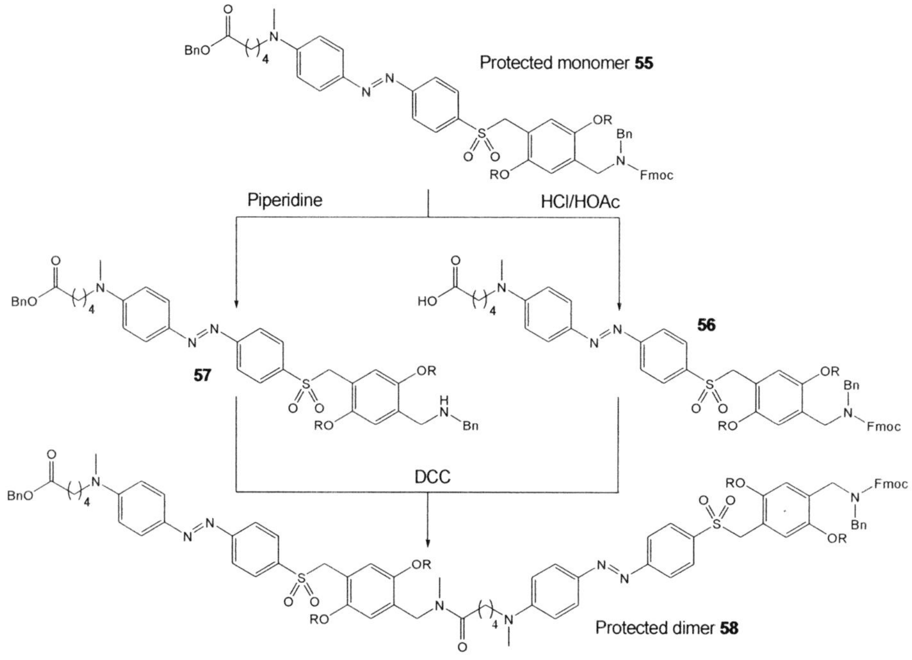

Scheme 15. Stepwise synthesis of oligo(azo sulfonyl amide)s [70]

The two complementary monomers 56 and 57 were then coupled together using DCC to afford a protected dimer 58 (75 %). In the same manner, selective deprotection followed by DCC coupling of the mono-protected dimers yielded a protected tetramer (66 %).

A large variety of linear poly(amide imide)s have been synthesized by Liaw and co-workers [74-77] utilising imide containing diacids and a variety of aromatic amines (Figure 4). The polycondensations were conducted at 100 °C using the TPP/pyridine coupling system in NMP. The diacid A_2 monomers 59-62 consisted of either a 2,2-*bis*(phthalimidyl)hexafluoropropane backbone or two 4-trimellitimidophenoxy moieties joined via a linker with pendant aliphatic groups.

Regardless of the monomers used, all of the resulting poly(amide imide)s were found to be highly soluble in organic solvents, possessed glass transition temperatures in the range 230-290 °C and thermal stabilities up to a minimum of 400 °C. The inherent viscosities of the poly(amide imide)s varied, depending on the diimide diacid A_2 type monomer used. Polyamides consisting of the A_2 type monomers 59 [77] and 60 [74] had inherent viscosities in the range 0.69-1.06 dL g^{-1}, whereas polyamides incorporating the A_2 type monomers 61 [75] and 62 [76] had inherent viscosities in the ranges 0.78-1.63 and 0.73-1.36 dL g^{-1}, respectively. The incorporation of aliphatic and trifluoromethyl groups into to the polymers backbone not only enhanced the solubility of these type of poly(amide imide)s, but also improved the thermal stability of these systems.

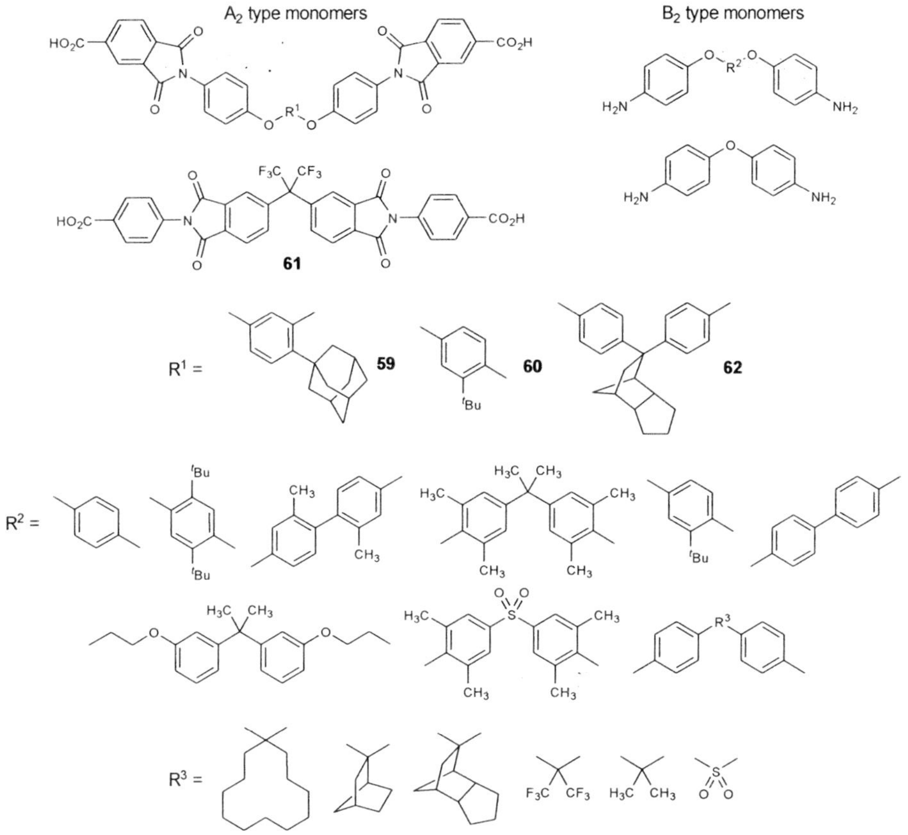

Figure 4. The monomers used by Liaw and co-workers to synthesise a diverse range of poly(amide imide)s [74-77]

2.3. Synthesis of Biodegradable Poly(ester amide)s Using Coupling Reagents

Several biodegradable aliphatic polyesters are employed extensively for agricultural, biomedical, pharmaceutical and environment-related applications [78]. However, low thermal, mechanical and processing performances restrict greatly the practical use of these materials [79,80]. The mechanical properties can be improved by the introduction of amide linkages in the polymer chains, although for optimum biodegradability to be maintained the amide linkages must be sensitive to attack by enzymes and microbes. For example, poly(ester-amide)s containing phenylalanine, leucine, alanine and glycine can be degraded by enzymes [81-84]. Kise *et al.* have produced a series of poly(ester-amide)s 63 containing dipeptide groups composed of phenylalanine and several other amino acids via the EDCI mediated coupling of diacids and diamine monomers in a 1:1 ratio (Scheme 16) [85]. Coupling reagents have proven to be extremely valuable for the synthesis of poly(ester-amide)s as a result of their highly selective and mild coupling conditions, whereas direct polycondensation strategies would lead to cleavage of the ester bonds, resulting in the unwanted formation of amide linkages. Kise *et al.* obtained [85] poly(ester-amide)s 63 with number-average molecular weights ranging from 5.6 kD to 7.7 kD and narrow polydispersities between 1.36 and 1.48. The highest number-average molecular weight polymers were obtained when R = Bn (M_n = 7.7 kD) and decreased in the order R = CH_3 (M_n = 7.2 kD), R = ^{i}Pr (M_n = 6.8 kD), R = ^{i}Bu (M_n = 5.6 kD). The results obtained suggest that the coupling reagent mediated polycondensation was affected by the nucleophilicity of the amino group in monomer 64 (Scheme 16) and not as a result of steric influences of the side-chains present. Biodegradation studies indicated that all of the resultant poly(ester-amide)s 63 were hydrolysed by enzymes such as *α-chymotrypsin*, *subtilisin Carlsberg* and *lipase*, and that the rate of degradation depended strongly upon the nature of the amino acid residues in the polymer chains.

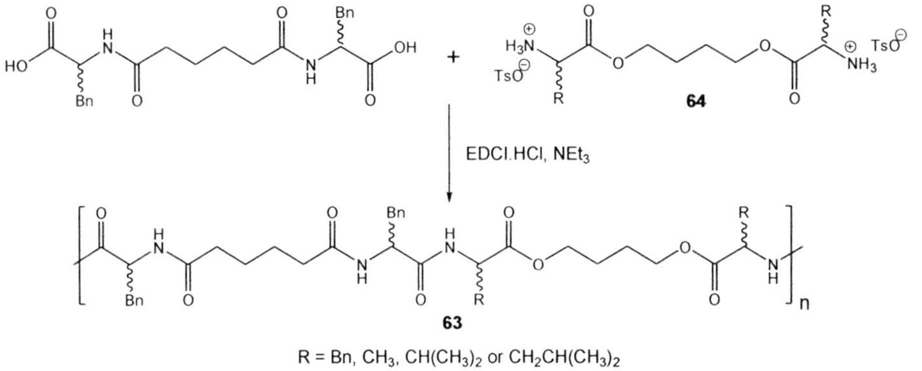

Scheme 16. Synthesis of poly(ester-amide)s 63 consisting of L-phenylalanine, L-leucine, L-alanine or L-glycine [85]

Fey *et al.* have synthesised poly(ester amide)s 65 from glutaric anhydride and aliphatic aminoalcohols using EDCI or esterification catalysts [86] (Scheme 17). To prevent the formation of ammonium salts from the carboxylic acid produced in the ring opening of glutaric anhydride a mixed solvent system (1:1 acetonitrile:diethyl ether) was used that resulted in precipitation of the α-carboxy-ω-hydroxyamides 66 from the reaction mixture.

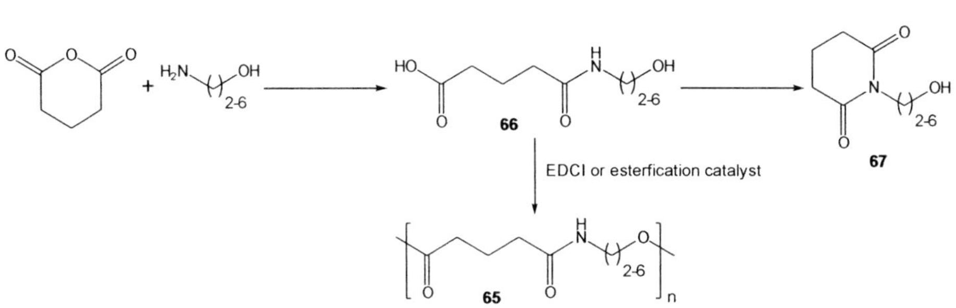

Scheme 17. Synthesis of poly(ester-amide)s 65 from glutaric anhydride and aliphatic aminoalcohols [86]

The AB monomers 66 were polymerised over a period of 24 hours using EDCI in the presence of 4-(dimethylamino)pyridine (DMAP) to afford poly(ester amide)s 65 with number-average molecular weights and polydispersities between 7-11 kD and 1.35-1.58, respectively. Polycondensation of the AB monomer 66 using esterification catalyst (such as $Ti(OBu)_4$, $Bu_2Sn(OMe)_2$, stannous 2-ethylhexanoate ($Sn(Oct)_2$)) at 110 °C under reduced pressure (10^{-2} mbar) afforded poly(ether amide)s 65 with number-average molecular weights in the range 8-13 kD (depending on the catalyst used) after *ca.* 175 hours. However, the harsh reaction conditions employed for the catalyst mediated esterification resulted in considerable formation of cyclic imides 67. In comparison, when EDCI was used only a trace of cyclic imides 67 were detected.

2.4. Synthesis of Linear Polyesters using Coupling Reagents

Melt-phase transesterification has been used extensively to synthesise polyesters, however, this approach is limited as a result of the high temperatures (250 °C) required. Higashi *et al.* used carboxylic acid activating agents, diphenyl chlorophosphate (DPCP) [87] and arylsulfonyl chlorides [88] at 120 °C in polyesterifications, although it has been reported [89] that these activating agents result in considerable by-product formation and lack the ability to selectively activate hydroxyacid monomers. The use of carbodiimide reagents in esterification chemistries was hampered initially as a result of undesired *N*-acylurea formation. For example, early attempts to synthesise polyesters using carbodiimide mediated coupling methodologies only resulted in the formation of low molecular weight products and side-products [90]. However, it was observed that *N*-acylurea formation can be suppressed by the use of hyperacylation catalysts such as DMAP [91] and acid catalysis [92]. In an attempt to develop carbodiimide mediated polyesterification chemistries, Moore and Stupp [89] optimised the polymerisation of the hydroxyacid AB type monomer, 4-(7-(4-hydroxyphenyl) heptyl)benzoic acid 68 [93] using DIC in combination with a range of additives (DMAP, *p*-toluenesulfonic acid (PTSA), pyridine) and solvents. It was concluded that 1:1 a mixture of DMAP and PTSA resulted in high conversions and negligible *N*-acylurea formation. Therefore, a new reagent was prepared, 4-(dimethylamino)pyridinium-4-toluene-sulfonate (DPTS) [89], by combining DMAP and PTSA in a 1:1 ratio. Polycondensation of the monomer 68 over a period of 12 hours using dichloromethane as the solvent, 1.3 equivalents of DIC and 0.2 equivalents of DPTS afforded the polyester 69 (Scheme 18) in a yield of 88 % and with a DP $\geq$ 50 (corresponding to molecular weight in excess of 15 kD).

The purity of the polyester produced was remarkably high and *N*-acylureas were not evident in the material produced. To prove the general applicability of the newly developed method, Moore and Stupp synthesised a small series of polyesters (Figure 5) based upon phenolic-acid AB monomers. All of the polyesters were synthesised using 0.5 equivalents of DPTS and 1.5 equivalents of DIC relative to the monomer and possessed DP values $\geq$ 50.

To illustrate the mildness of the carbodiimide/DPTS mediated polyesterification system a polyester [89] was constructed using the deuterium-labelled chiral hydroxyacid monomer, 4-(2-deutero-2-cyano-7-(4-hydroxyphenyl)heptyl)benzoic acid [93,94].

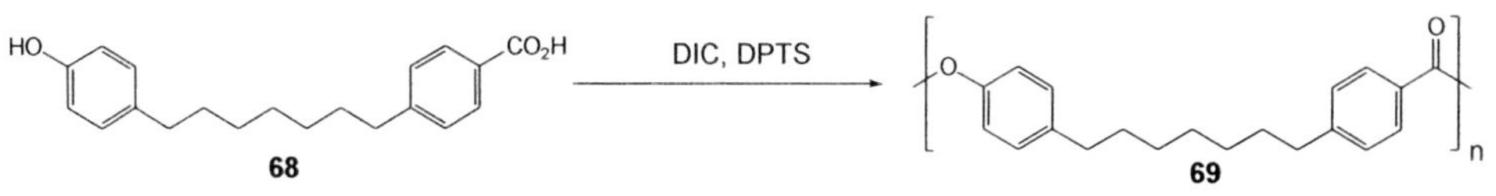

Scheme 18. Synthesis of polyester 69 using DIC and DPTS [89]

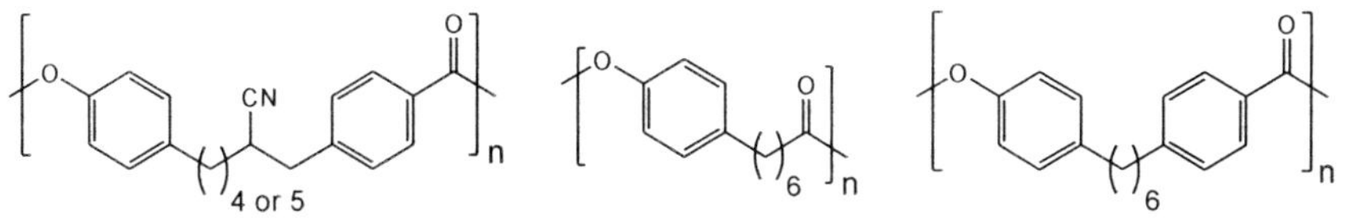

Figure 5. Polyesters synthesised from AB monomers using DIC and DPTS [89].

As a result of the deuterium on the nitrile α-carbon of this monomer it is possible that deuterium-proton exchange could have occurred under the harsh polymerisation conditions. If deuterium-proton exchange did occur, polymerisation of the enantiomerically enriched monomer would result in a polymer with low stereochemical purity. However, when carbodiimide reagents in the presence of DPTS were used less than 3 % of its deuterium label was exchanged. Hence, it was concluded that carbodiimide mediated polyesterifications are very beneficial for the synthesis of polyesters derived from sensitive and optically active monomers.

Fréchet and co-workers have synthesised successfully aromatic linear polyesters 70 with phenolic alcohol groups distributed along the polymer backbone from 3,5-dihydroxybenzoic acid derivatives using DCC as the coupling reagent [95]. The ABB' type monomers were synthesized initially from 3,5-dihydroxybenzoic acid 71 either via monosilylation or monobenzylation with *tert*-butyldimethylsilyl chloride and benzylbromide, respectively. Polymerisation with DCC in the presence of DPTS afforded both silyl ether functionalised 72 and benzyloxy functionalised polyester 73 with weight-average molecular weights of 20 kD and 11 kD and polydispersities of 1.9 and 1.8, respectively. Deprotection of 72 using HCl afforded the hydroxyl functionalised linear polyester 70 (Scheme 19). However, attempts to deprotect the benzyloxy functionalised polyester 73 using standard hydrogenation protocols resulted in recovery of the starting polyester.

However, attempts to synthesise the hydroxyl functionalised linear polyester 70 from 3-benzyloxy-5-(trimethylsilyloxy)benzoyl chloride using procedures developed previously in the synthesis of hyperbranched polyesters from 3,5-*bis*(trimethylsilyloxy)benzoyl chloride [96] resulted in the formation of insoluble gels.

The properties of the linear polyesters were compared with dendritic and hyperbranched analogues. These hyperbranched polyester analogues [97] exhibited high solubility in common organic solvents, whereas in comparison the linear analogue showed a marked decrease in solubility. Interestingly, the thermal properties (such as the glass transition temperature (T_g)) of the three analogues were very similar thus suggesting that the macromolecular architecture had very little effect on the polymers' thermal properties.

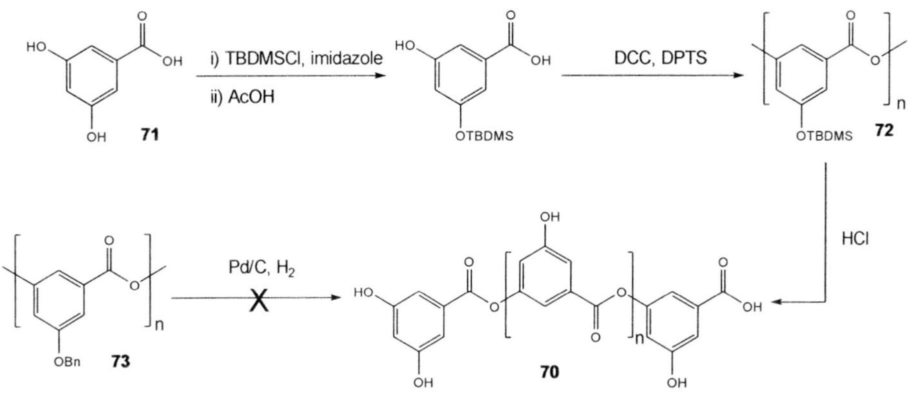

Scheme 19. Synthesis of hydroxyl functionalised aromatic polyester 70 [95]

Higashi and co-workers have synthesised a range of linear aromatic polyesters 74 [98] based upon substituted hydroquinones 75 and diacids 76 (see Scheme 20) using the coupling reagent DPCP in the presence of pyridine [99] at 120 °C.

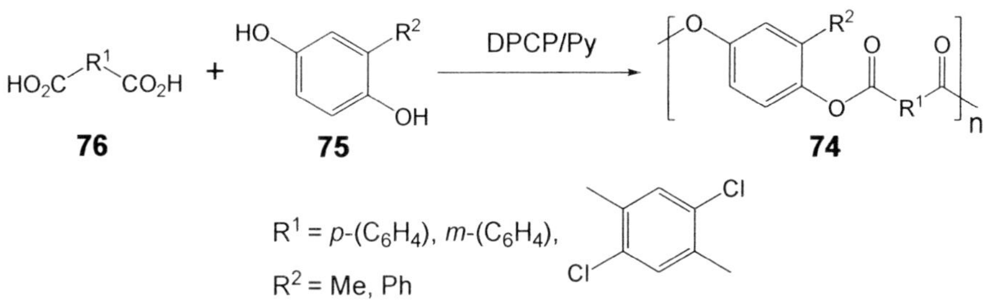

Scheme 20. Synthesis of linear aromatic polyesters from substituted hydroquinones using DPCP [98]

Carboxylic acids react with DPCP 77 by displacing a chlorine anion from DPCP to afford an activated acid intermediate, an *O*-acyldiphenylphosphate 78 (Scheme 21). Nucleophilic attack on this activated species by an amine or alcohol results in displacement of a phosphate anion and the formation of an amide or ester, respectively.

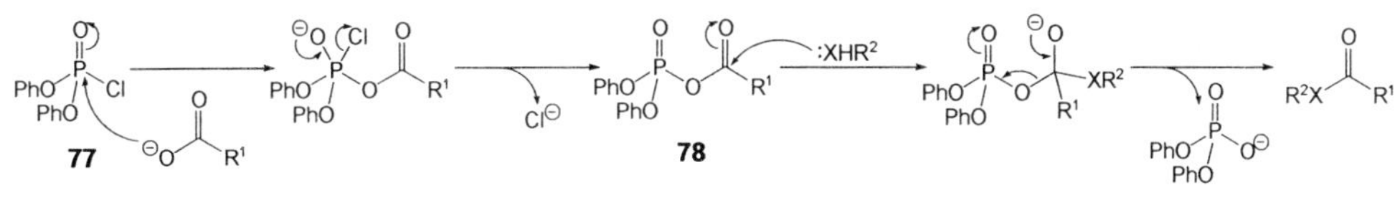

Scheme 21. The mechanism of coupling processes using the coupling reagent DPCP

Polycondensation of 2-phenylhydroquinone and terephthalic acid (TPA) using DPCP afforded polyesters 74 with inherent viscosities in the range 2.89-3.26 dL g^{-1} (measured in *p*-chlorophenol at 50 °C) and that exhibited a nematic mesophase. In comparison, when TPA and 2,5-dichloroterephthalic acid was employed in the polycondensation, the inherent viscosities of the polyesters 74 decreased considerably to 0.76-1.89 dL g^{-1} as a result of different conformations of the polymer chains [100]. Randomly orientated polyesters synthesised from a 1:1 mixture of isophthalic acid and TPA and methylhydroquinone possessed inherent viscosities in the range 0.97-1.34 dL g^{-1} and exhibited lower transition temperatures when compared to the highly ordered polyesters synthesised previously.

Poly(lactic acid) 79 has attracted considerable attention as a biodegradable material [101]. However, it is difficult to synthesize poly(lactic acid) 79 from the direct polycondensation of lactic acid as a result of the instability of oligo(lactic acid) at elevated temperatures and its tendency to depolymerise to lactide [102]. Akutsu and co-workers investigated the direct polycondensation of lactic acid 80 [103] under mild conditions using the coupling reagents, CDI, *N,N,N',N'*-tetramethylchloroformamidinium chloride 81 (TMCFAC) [104] and DCC (Scheme 22).

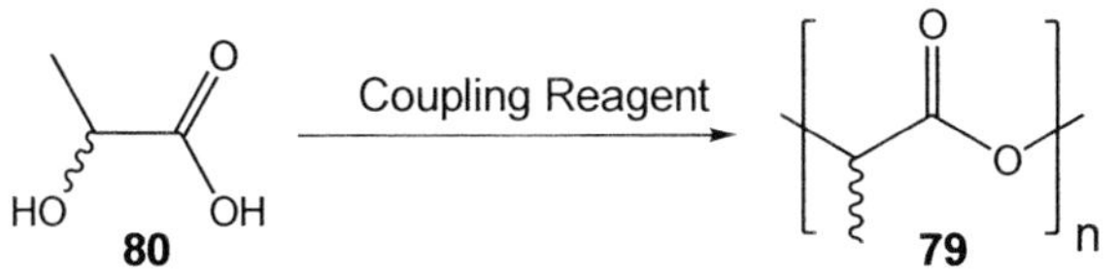

Scheme 22. Polymerisation of lactic acid 80: Coupling reagent = CDI or TMCFAC or DCC/DMAP [103]

These studies [103] revealed that racemic lactic acid could be polymerised successfully using DCC to afford poly(lactic acid) 79 possessing a number-average molecular weight of *ca.* 15 kD. In contrast, the polymerisation of 80 using CDI and TMCFAC 81 only resulted in the formation of low molecular weight oligomers. It was predicted that the poor selectivity of CDI resulted in its reaction with the hydroxyl groups of 80 prior to activation of the carboxylic acid and hence only low number-average molecular weight oligomers (*ca.* 400-670 D) were obtained in a fashion independent of the concentration of the coupling reagent. TMCFAC 81 (Scheme 23) in the presence of pyridine has been shown to activate selectively carboxylic acids that co-exist in a molecule with alcohols [105].

However, when TMAFAC 81 was employed in the polymerisation of 80 only brown viscous oils were obtained that possessed number-average molecular weights of *ca.* 4 kD. Variation of the reaction times and TMCFAC:monomer feed ratio were found to have minimal effect upon the physical appearance or properties of the resulting poly(lactic acid) 79. In comparison, the DCC mediated polycondensation of 80 in the presence of DMAP afforded poly(lactic acid)s 79 with number-average molecular weight between 7.4 kD and 16 kD mass units. The DCU by-product formed in the process was removed by filtration and the filtrate concentrated *in vacuo* to afford the polyester 79 as a white solid. Increasing the reaction time beyond 24 hours did not affect the yield and number-average molecular weight of the resultant poly(lactic acid) 79, however, increasing the monomer concentration from 1 to 3.4 molar equivalents was found to have a positive effect upon these properties. Further increments of the monomer concentration resulted in a decrease in the number-average molecular weight as a consequence of the high viscosity of the reaction mixture as a result of precipitation of DCU.

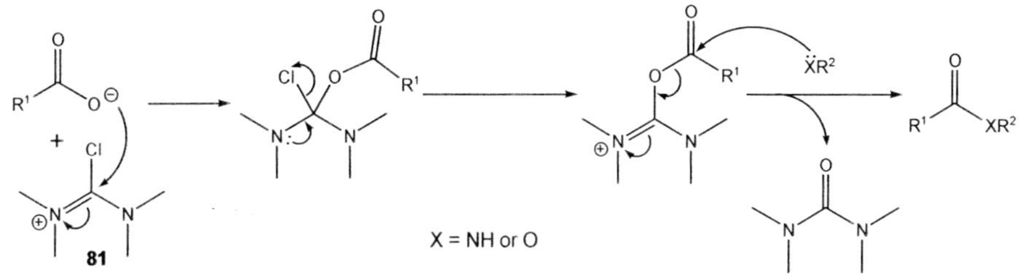

Scheme 23. The mechanism of the coupling process involving TMCFAC 81

Bei *et al.* have also incorporated poly(lactic acid) in multiblock copolymers 82 [106] for use as drug delivery systems [107]. Poly(L-lactide-ethylene oxide) copolymers 82 were synthesized and functionalised with the drug 5-fluorourecil. The drug release was studied and could be tailored by adjusting co-polymer composition. Initially, triblock co-polymers 83 were constructed from polyethylene glycols (PEG) possessing weight-average molecular weights of 2 kD and 4 kD, and lactide in various molar feed ratios using $Sn(Oct)_2$ as a transesterification catalyst. The multiblock co-polymers 82 were then constructed by coupling the triblock co-polymers 83 with succinic anhydride in the presence of DCC and DMAP (Scheme 24).

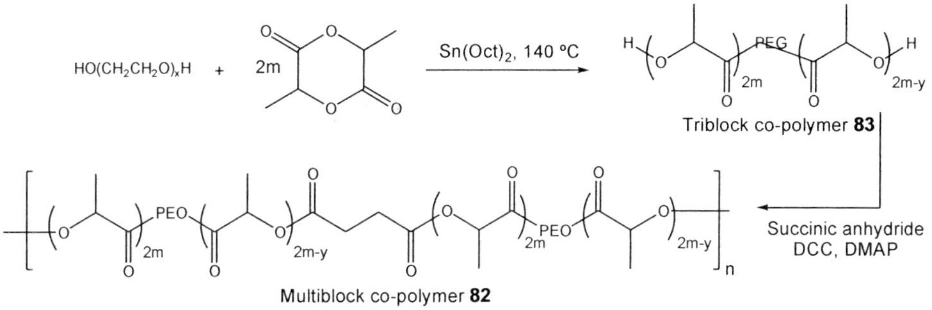

Scheme 24. Synthesis of multiblock co-polymers using $Sn(Oct)_2$ and DCC [106]

Since a carboxylic acid moiety was not present at the outset of this polymerisation process, the reaction proceeded via ring opening of succinic anhydride by DMAP followed by displacement of DMAP by the free hydroxyl groups present on the triblock co-polymer 83. Upon ring opening succinic anhydride, a reactive carboxylic acid is produced that can compete with DCC to form an *O*-acylisourea. Reaction of the activated carboxyl group with another triblock co-polymer 83 results in formation of a multiblock co-polymer 82. The multiblock co-polymers 82 have been synthesized previously via alternative procedures, [108,109] both of which were found to be more complex and less effective as the coupling reagent mediated synthesis presented by Bei and co-workers [106]. Increasing the ratio of PEG to lactide from 1:1 to 1:2 to 1:4 resulted in a linear increase of weight-average molecular weight of 83 from 8.5 kD to 33.4 kD for PEG 2 kD and from 14.3 kD to 49.8 kD for PEG 4 kD. In this approach, the polydispersities of 83 were *ca.* 1.5. DCC mediated coupling of 83 resulted in formation of multiblock co-polymers 82 with weight-average molecular weights between 40.3 kD and 66.8 kD and polydispersities of *ca.* 1.5, although the weight-average molecular weights of the triblock co-polymers used were found to have little effect on the overall properties.

Polycations such as polylysine and polyethylenimine (PEI) have been used within polymer based gene delivery systems that mediate the transfection of cells [110]. PEI is also hypothesized to mediate escape of plasmid deoxyribonucleic acid (DNA) from the endosomal pathway [111,112]. However, polylysine and PEI are associated with a considerable degree of cytotoxicity [113] as a result of the slow degradation, which increases the cellular exposure time. As a consequence, Putnam and Langer investigated the synthesis of a polycation with low cytotoxicity that maintained the capability to electrostatically complex and condense plasmid DNA [114]. The polycation that was developed, poly(L-proline ester) 84, was synthesised via DCC mediated polycondensation of *N*-protected-4-hydroxy-L-prolines 85 (Scheme 25). The polycondensation of *N*-protected-4-hydroxy-L-proline methyl ester (where the amine groups were acylated with long chain fatty acids which could not be removed after polymer synthesis) has been reported previously [115,116] using transesterification catalysts at high temperatures (*ca.* 180 °C).

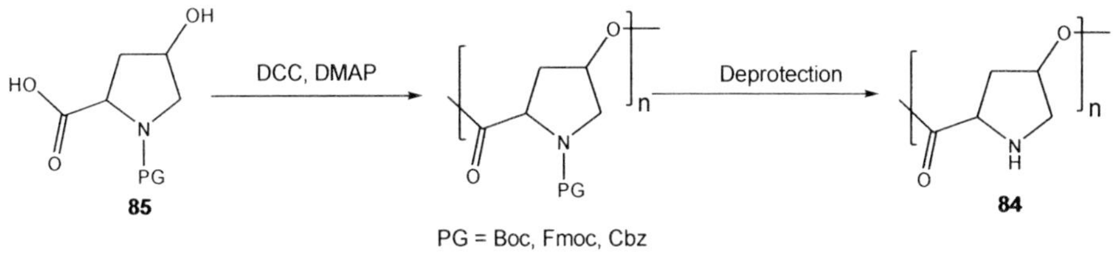

Scheme 25. Synthesis of poly(L-proline ester)s from *N*-protected-4-hydroxy-L-proline [114]

As a result of the harsh high temperatures required for the polycondensation protecting groups such as *N-tert*-butoxycarbonyl (Boc), carbobenzyloxy (Cbz) and *N*-(9-fluorenylmethoxycarbonyl (Fmoc) could not be employed [117]. To avoid this problem Putnam and Langer used the coupling reagent DCC, which is compatible with Boc, Cbz and Fmoc protecting groups [114]. Polymerisation of *N*-Boc and *N*-Fmoc protected 4-hydroxyl-L-proline using DCC resulted in the formation of low weight-average molecular weight poly(proline ester)s 84 (3.5 kD and 1.5 kD, respectively). The conditions required for removal of Boc and Fmoc protecting groups are known to cause ester hydrolysis, therefore deprotection was not attempted. In contrast, poly(proline ester)s possessing weight-average molecular weights up to 7.8 kD and polydispersities of 1.8 were synthesized from *N*-Cbz-4-hydroxyl-L-proline using an optimised procedure. The optimised polymerisation process involved stirring a 2 M DCC solution in DMI, one mole equivalent of DCC relative to monomer and DMAP (10 mol%) at 0 °C for a period ranging between 40 and 72 hours. The Cbz protecting groups were removed by hydrogenolysis with Pd/C catalysis using formic acid as a hydrogen donor [118]. However, this hydrogenolysis process cleaved of a small portion of ester bonds thereby reducing the degree of polymerisation. However, the poly(proline ester) 84 complexed and condensed plasmid DNA and also possessed an excellent *in vitro* cytotoxic profile relative to polylysine and PEI.

Glycopolymers have attracted considerable attention as a result of their unique structures and properties. As a result they have been used in various medicinal applications including drugs and drug delivery systems [119]. Recent developments of new synthetic methodologies have enhanced the production of polysaccharides [120], however, the synthesis of well-defined polysaccharides still requires complicated synthetic procedures. For example, problems such as control of stereoselectivity of anomeric centres and control of regioselectivity of the hydroxyl functionalities need careful consideration. Therefore, only a few methods, including the glycosylation method, [121] ring-opening polymerisation, [122] and enzymatic polymerisation [123] have been available for precision synthesis of polysaccharides. The hexachlorotriphosphazene 86/pyridine coupling system has been particularly useful for the regioselective polycondensation of D-glucosamine derivatives [124]. The polymerisation of the D-glucosamine AB monomer 87 proceeded in a regioselective manner via coupling of the carboxylic acid group located at the 2-position and the primary hydroxyl at the 6-position to afford poly(saccharide ester)s 88 (Scheme 26).

The polymer structure was confirmed by the model reaction of benzyloxycarbonyl amine protected D-glucosamine with butyric acid using hexachlorotriphosphazene 86, which resulted in selective esterfication at the 6-position of the D-glucosamine derivative. After conversion of the poly(saccharide ester)s to the acetylated derivatives, the weight-average molecular weights of the material were determined (22 kD and 156 kD) by GPC analyses.

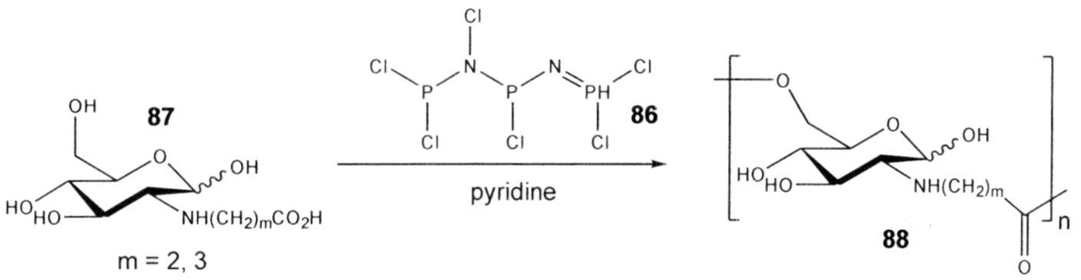

Scheme 26. Synthesis of a poly(amido saccharide) using the coupling reagent hexachlorotriphosphazene [124]

3. Dendrimers

Coupling reagents are used regularly in the stepwise synthesis of dendrimers via both convergent and divergent approaches. However, a detailed summary of coupling reagents in dendrimer synthesis is not the aim of this review –the reader is therefore directed to detailed information on dendrimers and their synthesis in a number of excellent comprehensive reports reported by other authors [125-128]. Nevertheless, in order to demonstrate the wide range of coupling reagents and methods used for the synthesis of dendrimers several key examples have been selected.

Ihre *et al.* have synthesised aliphatic polyester dendrimers via a double exponential growth approach (Scheme 27) [129]. 2,2-*Bis*(hydroxymethyl)propanoic acid was protected as either a cyclic acetonide 89 or a benzyl ester 90. The resulting two complementary species were coupled efficiently using DCC to afford a second generation dendron 91. Selective deprotection of this dendron affords two compatible dendrons that were coupled together again with DCC, rapidly increasing the generation number of the dendron.

The synthesis of aromatic polyester dendrimers using coupling reagents have been reported by a number of research groups [130,131]. Freeman *et al.* have produced polyester dendrimers via an orthogonal convergent approach that demonstrates the high efficiency of

DCC mediated esterfication [131]. Initially, the acid functionalities of 5-(hydroxymethyl) isophthalic acid were protected as benzyl esters in order to form the peripheral surface of the dendrimer. The protected benzyl alcohol was then coupled to 5-(chloromethyl) isophthalic acid using DCC to afford a second generation dendron with a benzyl chloride focal point. Reaction of the second generation dendron with potassium 5-(hydroxymethyl) isophthalate afforded the third generation dendron with a benzyl alcohol focal point. DCC mediated coupling of the third generation dendron and trifunctional core, benzene-1,3,5-tricarboxylic acid afforded the third generation dendrimer in a yield of 85%.

Aromatic polyamide dendrimers were first reported by Miller and Neenan in 1990 [132]. Since this seminal report numerous polyamide dendrimers have been developed, a large portion of which were constructed using coupling reagent chemistries. Kakimoto and co-workers have synthesised aromatic polyamide dendrimers via an orthogonal growth approach employing 3,5-diaminobromobenzene and 3,5-diaminobenzoic acid as the building blocks [133]. Both TPP and DBOP mediated coupling methodologies were utilized in this approach. Aromatic polyamide dendrimers have also been synthesised by Ueda and co-workers [134] from unprotected AB_2 type building blocks using a repetitive DBOP coupling strategy.

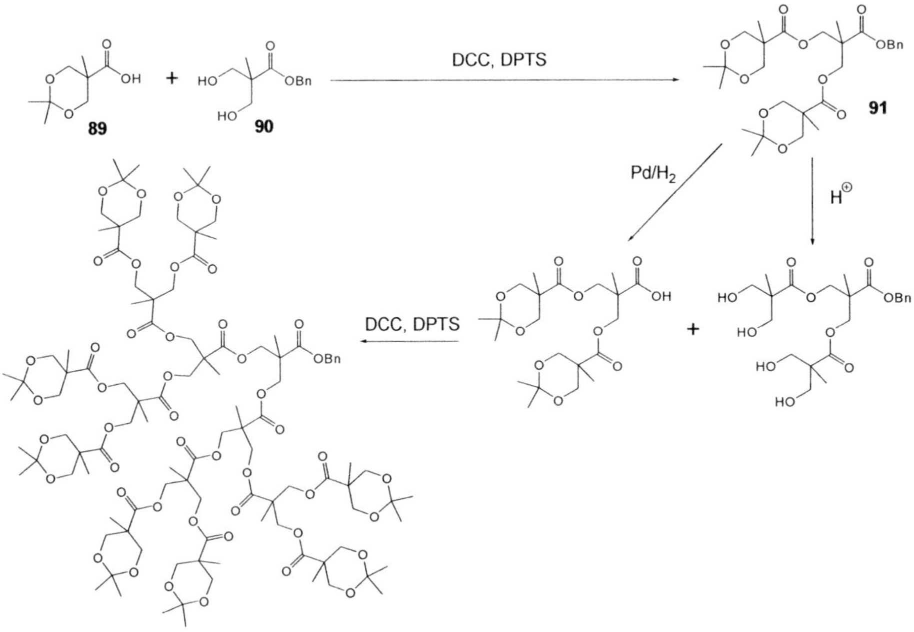

Scheme 27. Synthesises of aliphatic polyester dendrimers via a double exponential growth approach [129]

Romagnoli *et al.* have synthesised achiral and chiral poly(ester amide) dendrimers 92 derived from the AB_2 type monomer 1,3-diamino-2-hydroxypropane, a rigid linker group, 4-carboxybenzaldehyde and the trifunctional core, *tris*(2-ethylamino)-amine (Scheme 28) using a convergent approach [135]. The dendron generation number was increased via a three step process involving amide formation, ester formation and finally aldehyde oxidation. Dendrimers were synthesised up to the third generation utilising a variety of coupling reagents. DCC was found to be appropriate for all esterifications, however, it was beneficial to use different coupling reagents (diphenylphosphorylazide (DPPA) [136] and BOP) for amide formation at different stages of dendron growth.

Aromatic and aliphatic polyamide dendrimers have been synthesized by Aulenta *et al.* [137] for use in fragrance delivery applications. The aromatic polyamide dendrimers were constructed in a convergent manner using 3,5-diaminocinnamic acid as an AB_2 type building block. The dendrons were synthesised via a repetitive HATU mediated coupling strategy and attached to di- and trifunctionalised amine core molecules using BOP (1st generation) or HATU (2nd generation). Aulenta *et al.* have also synthesized aliphatic poly(amide ether) dendrimers [137] derived from *N,N-bis*(2-aminoethyl)amine, diglycolic anhydride and multifunctional amine cores via a convergent growth strategy employing CDI and BOP coupling reagents. Rannard and co-workers have synthesized a variety of amide [138], urethane [139] and carbonate [140] dendrimers using selective CDI coupling methodologies that alleviate the need for protection/deprotection chemistries [24, 25]. The synthesis a poly(amino urethane) dendrimers 93 [139] present in Scheme 29 demonstrates the highly selective nature of CDI and CDI activated intermediates.

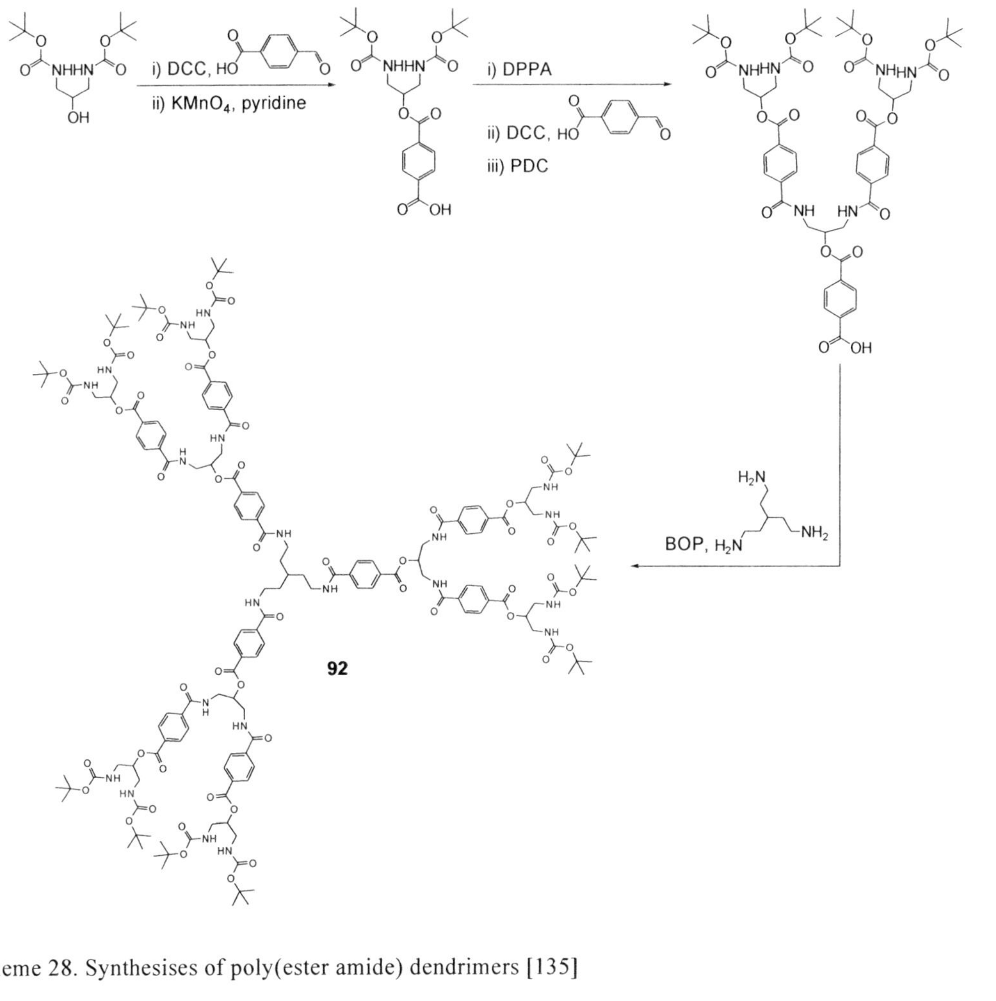

Scheme 28. Synthesises of poly(ester amide) dendrimers [135]

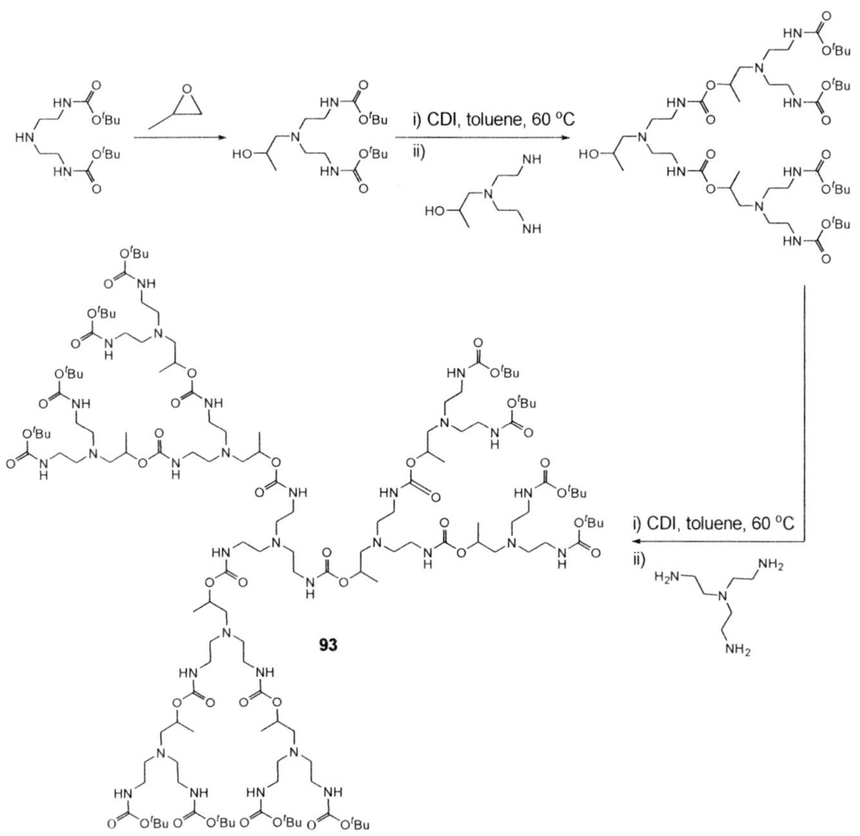

Scheme 29. Synthesis of polyurethane dendrimers using the coupling reagent CDI [139]

4. Hyperbranched Polymers

4.1. Synthesis of Hyperbranched Polyamides using Coupling Reagents

A variety of coupling reagent synthesised hyperbranched polyamides have been reported in the literature [141-147]. These polyamides were constructed from five different AB_2 type monomers (Figure 6) using TPP and DBOP as the coupling reagents.

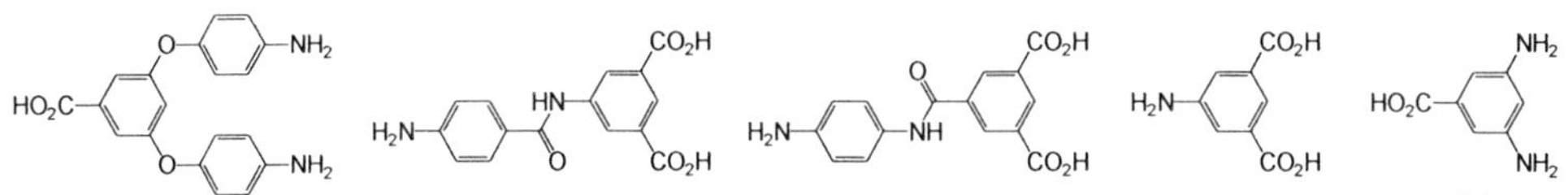

95 94

Figure 6. AB_2 type monomers used to synthesise hyperbranched polyamides [141-147]

Hyperbranched polyamides derived from the AB_2 monomers (Figure 6) were soluble in amide type solvents, possessed inherent viscosities in the range 0.1 to 0.4 dL g^{-1} and featured degree of branching (DB) values of *ca.* 0.5. An exception to this trend was the polyamide derived from 3,5-diaminobenzoic acid 94 [144] which featured a DB value of 0.32. This value suggested that steric hindrance had occurred during the polycondensation. The AB_2 type monomer 94 was polymerised via thermal means and used the coupling reagent TPP in the presence of pyridine. The weight-average molecular weight, polydispersities and DB values of polyamides obtained by thermal polycondensation of 94 were in the range 26-59 kD, 1.9-2.8 and 0.41-0.43, respectively. In comparison, coupling reagent mediated polycondensation of 94 afforded a polyamide with weight-average molecular weight and polydispersity of 155 kD and 3.0, respectively. Polycondensation of 95 (Figure 6) was conducted thermally [148] and using TPP [142] as a coupling reagent. Thermal polycondensation afforded a polyamide with weight-average molecular weight of 74.6 kD, polydispersity of 2.6 and DB value of 0.46. In comparison, the coupling reagent mediated polycondensation afforded lower molecular weight material (36.8 kD) with a narrower polydispersity (1.8) and DB value of 0.48.

Hyperbranched polyamides synthesised from dendron-type AB_4 (96) and AB_8 (97) monomers (Figure 7) using DBOP and TPP have been reported by Ishida *et al.* [144]. The polyamides were synthesised from 96 and 97 using the coupling reagents TPP and DBOP, and also by thermal polycondensation approaches.

TPP (1 equivalent) mediated polycondensation of the AB_4 type monomer 96 afforded a polyamide with weight-average molecular weight and polydispersity of 139 kD and 2.7, respectively. The same polycondensation conducted with DBOP afford a polyamide with higher weight-average molecular weight (193 kD) but possessed a similar polydispersity value (3.3). The polyamides synthesised using both coupling reagents had similar DB values (0.67-0.72) which corresponded well to the minimum DB value (0.67) proposed for polycondensation of AB_4 monomers [145]. Interestingly, thermal polycondensation of 96 at 240-270 °C afforded polyamides with considerably lower weight-average molecular weights (15-24 kD), polydispersities of 1.7-2.0 and DB values in the range 0.71-0.76. Attempts to polymerise the AB_8 type monomer 97 using TPP resulted initially in gelation, however, the DBOP mediated polycondensation was more successful providing a soluble polyamide with a weight-average molecular weight of 109 kD, a polydispersity of 2.6 and DB value of 0.76. In comparison, thermal polycondensation of 97 afforded a polyamide with weight-average molecular weights of only 16.5 kD, a polydispersity of 2.0 and DB value of 0.76.

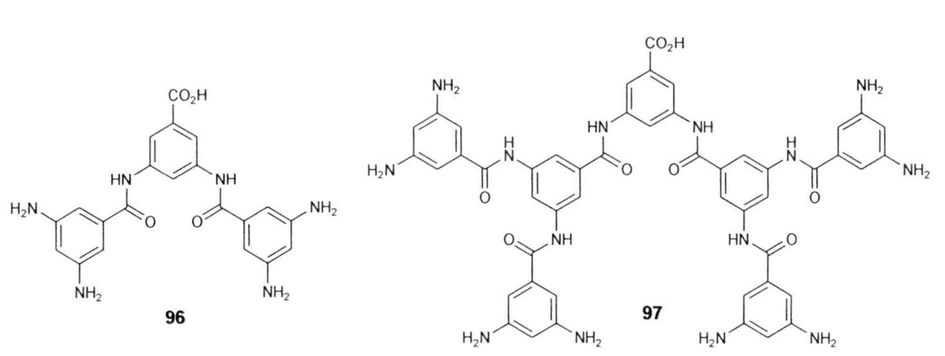

Figure 7. AB_4 and AB_8 monomers used to synthesis hyperbranched polyamides [145]

This example [144] highlights the difficulties of synthesising high molecular weight aromatic polyamides (as a result of low reactivity of aromatic amines) using thermal polycondensation methods and clearly demonstrates the advantages of coupling reagent mediated polycondensations. Co-polymerisation of 96 and 97 using TPP [144] afforded polyamides with weight-average molecular weights and polydispersities between 78-101 kD and 1.9-2.3, respectively, depending of the ratio of the two monomers used. The weight-average molecular weights and polydispersities were found to increase when the amount of 97 was increased and the DB value could be controlled from 0.32 to 0.72 by changing the ratio of the monomers.

Kakimoto and co-workers have reported hyperbranched polyamides [149,150] based upon the direct and stepwise copolymerisation of the AB_2 and AB type monomers using TPP as the coupling reagent. The molar feed ratios of the AB_2 type monomer, 3,5-*bis*(4-aminophenoxy)benzoic acid 95 (Figure 6) and AB type monomer, 3-(4-aminophenoxy)benzoic acid were varied to afford copolymers with weight-average molecular weights ranging from 100 to 330 kD and polydispersities of 2.3-4.1. The polydispersities of the co-polymers was independent of the feed ratio of the monomer, although inherent viscosities generally increased with increasing percentage of AB monomer. The solubility of the polyamide copolymers was dependent on the molar feed ratio of the two monomers and was a considerable improvement when compared to the solubility of the linear polyamide prepared from the AB monomer. Hyperbranched polyamide co-polymers have also been constructed from the AB_2 monomers, 3,5-*bis*(4-aminophenoxy)benzoic acid or 3,5-diaminobenzoic acid 94 and AB monomers 4-(4-aminophenoxy)benzoic acid, 3-aminobenzoic acid or 4-aminobenzoic acid using TPP as the coupling reagent [151].

Co-polymerisation of A_2 and B_3 monomers can lead to not only highly branched structures but also to cross-linking and cyclisation reactions resulting in the polymer becoming a gel or highly cross-linked material [152]. However, if the first condensation step between the A_2 and B_3 monomers is fast when compared to the following propagation step then an intermediate AB_2 type monomer accumulates in solution, thereby preventing gelation. Kakimoto *et al.* have produced aromatic hyperbranched polyamides 98 [153] through intermediate AB_2 molecules 99 formed from A_2 and B_3 monomers using TPP as the coupling reagent. The diamines, *p*-phenylene diamine (PD) and ODA 24 were polymerised in a 1:1 ratio with trimesic acid 100 (Scheme 30) using an excess of coupling reagent in NMP to afford polyamides soluble in aprotic polar solvents.

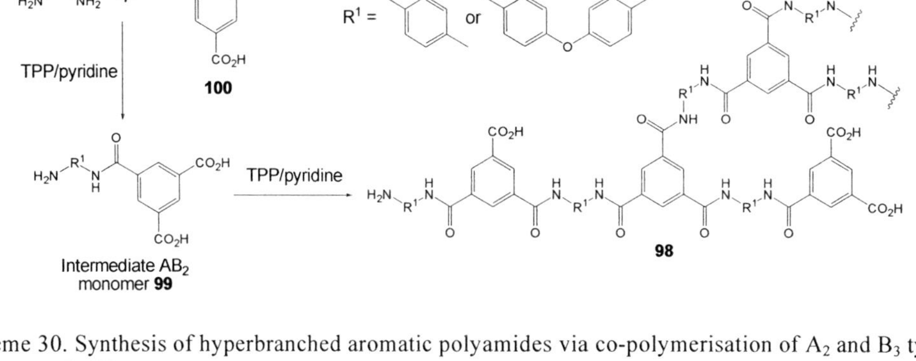

Scheme 30. Synthesis of hyperbranched aromatic polyamides via co-polymerisation of A_2 and B_3 type monomers [153]

However, precise calculation of the resultant polyamides 98 weight-average molecular weight was hindered because of adsorption of the material on to polystyrene based GPC columns used. Therefore, in order to evaluate the weight-average molecular weight of the resultant hyperbranched species, the polyamides 98 were capped with *p*-anisidine in the presence of TPP. GPC analysis of the resulting end capped polymers revealed weight-average molecular weights of 285 and 86.7 kD and polydispersities of 5.26 and 4.93 for the PD and ODA 24 containing polyamides, respectively.

Chang and Shu [154] have reported the synthesis of hyperbranched poly(amide imide)s from A_2 and BB'_2 (101) type monomers using TPP as the coupling reagent (Scheme 31).

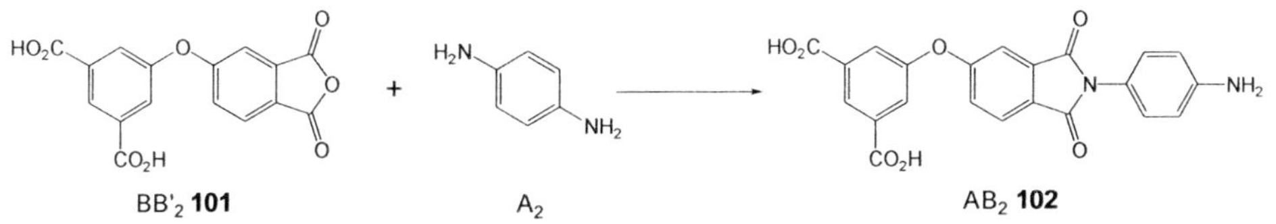

Scheme 31. Synthesis of AB_2 monomer from A_2 and BB'_2 monomers [154]

The difference in reactivities of the B and B' functionalities led to the predominant formation of an AB_2 type monomer 102 prior to polycondensation, hence preventing gelation. As a consequence the resulting poly(amide imide) possessed a structure very similar to that prepared directly from the AB_2 type monomer 102 using TPP. As a result of adsorption of the poly(amide imide) onto the polystyrene based GPC columns used the terminal carboxylic acids were reacted with aniline to afford an amide terminated polymer. The resulting poly(amide imide) possessing a weight-average molecular weight, polydispersity and DB value of 44 kD, 2.7 and 0.60, respectively.

4.2. Synthesis of Hyperbranched Polyimides using Coupling Reagents

Aromatic polyimides are well known as important high performance polymers [155]. Chang *et al.* have synthesised hyperbranched aromatic poly(ether imide)s from the AB_2 type monomer 4-{4-[di(4-aminophenyl)methyl]-phenoxy}phthalic acid 103 (Figure 8) in the presence of TPP [156]. The hyperbranched poly(ether imide)s were soluble in common organic solvents and had DB values of *ca.* 0.56. The weight-average molecular weight and polydispersity of the poly(ether imides) were *ca.* 75 kD and 4.6, respectively.

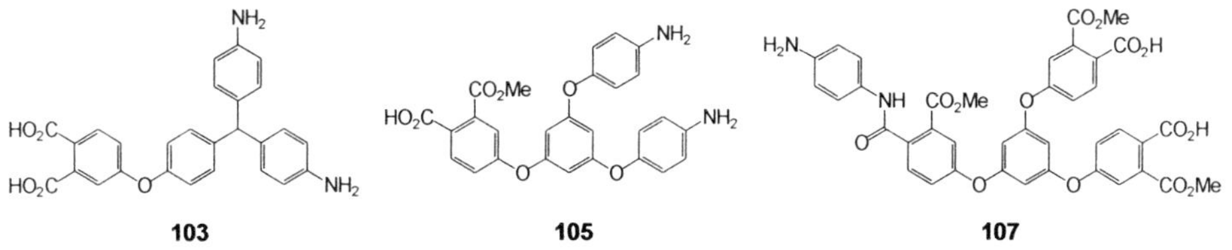

Figure 8. AB_2 type monomers used for the synthesis of hyperbranched poly(ether imides)s [157-158]

Kakimoto and co-workers have produced hyperbranched poly(ether imide)s from both AB_2 monomers [157-160] and combinations of A_2 and B_3 type monomers [160,161] using DBOP as the coupling reagent at room temperature. Poly(ether imide)s 104 comprised of the AB_2 monomer 3,5-*bis*(4-aminophenoxy)diphenyl ether-3',4'-dicarboxylic acid monomethyl ester 105 (Figure 8) were constructed via a two-step procedure involving self-polycondensation using DBOP to afford a poly(amic acid methyl ester) (PAAME) 106 followed by cyclodehydration in the presence of acetic anhydride and pyridine (Scheme 32).

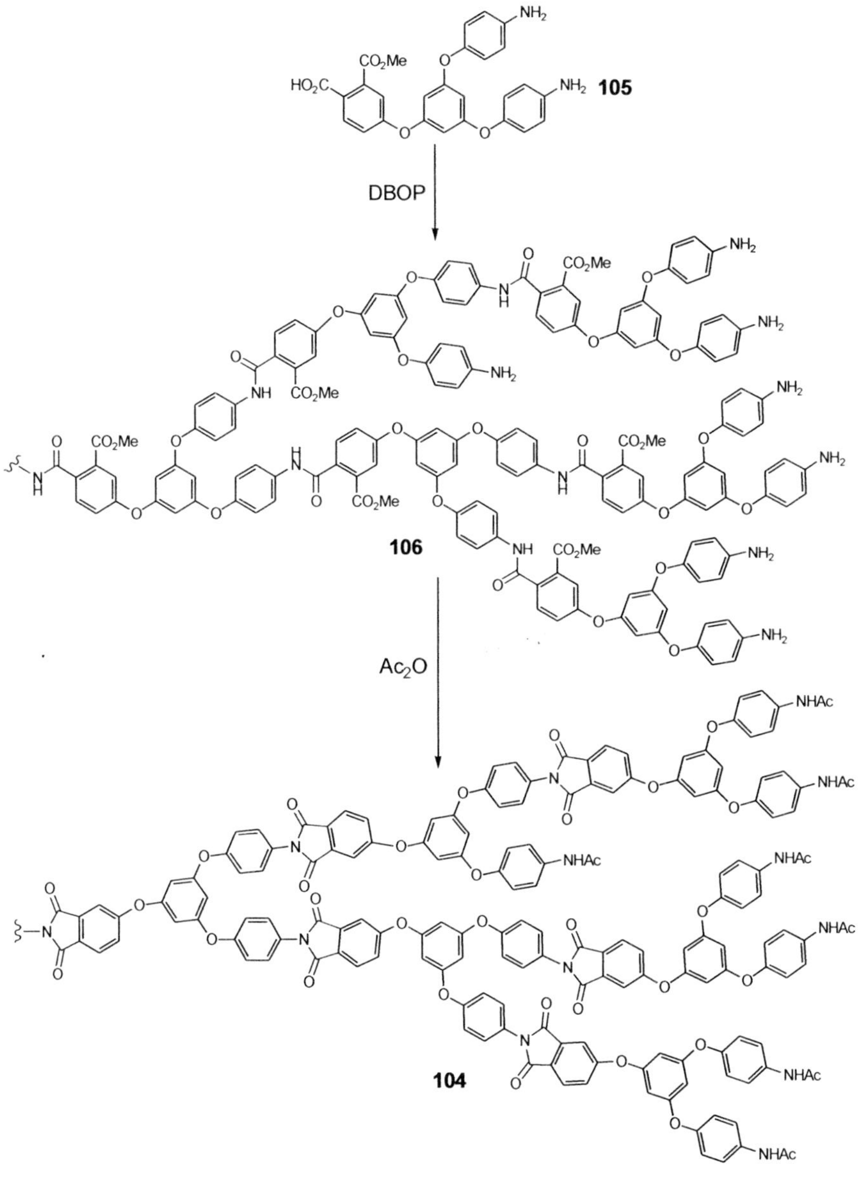

Scheme 32. Synthesis of acetyl amide terminated hyperbranched poly(ether imides)s [158]

The resulting acetyl amide terminated poly(ether imide)s 104 possessed DB values of 0.48 and weight-average molecular weights of *ca.* 188 kD. Alternatively, the PAAME terminal amine groups could be modified [158] prior to cyclodehydration imidazation. This method provided poly(ether imide)s with various amide and imide terminal groups and weight-average molecular weights ranging from 38 to 50 kD and polydispersities of *ca.* 1.6. Poly(ether imide)s with terminal *n*-heptanoamide and 4-methylphthalimide functionalities exhibited excellent solubility in common organic solvents including tetrahydrofuran (THF). End-group modification of the poly(ether imide)s 105 [157] with the branching agent 3,5-*bis*(4-nitrophenoxy)diphenyl ether-3',4'-dicarboxylic anhydride afforded poly(ether imide)s that were devoid of linear units and possessed a DB value and weight-average molecular weight of 1 and 98 kD, respectively [159].

Hyperbranched poly(ether imide)s have also been synthesised from the A_2 type monomer, 1,4-diaminobenzene and the tri(phthalic acid methyl ester) based B_3 type monomer 1,3,5-tri[(3-methyloxycarbonyl-4-carboxyl)phenyloxy]benzene in a 1:1 ratio using the coupling reagent DBOP [161]. The poly(ether imide)s were synthesised [157] by employing a two-step procedure in which a PAAME is constructed by polycondensation of the A_2 and B_3 type monomers using DBOP followed by imidazation. If the end-groups were not modified prior to imidazation a poly(ether imide) 108 (Figure 9) that possessed phthalic anhydride terminal groups was obtained. The resultant anhydride terminated poly(ether imide)s 108 (Figure 9) possessed weight-average molecular weights and DB values ranging from 33.6-125 kD and 0.52-0.56, respectively. Hyperbranched poly(ether imide)s with the same repeating unit as those obtained from the polymerisation of A_2 and B_3 type monomers were also prepared from the AB_2 type monomer 107 (Figure 9) using DBOP.

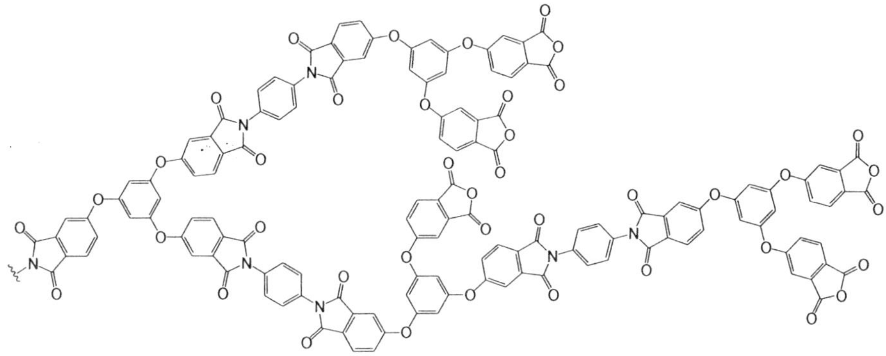

108

Figure 9. Anhydride terminated poly(ether imide)s 108

The poly(ether imide)s prepared from the AB_2 monomer 107 features weight-average molecular weights ranging from 11-173 kD and DB values of *ca.* 0.5. Comparisons between the plots of intrinsic viscosity *versus* weight-average molecular weight for the poly(ether imide)s 108 elucidated that molecular shape was very different depending on the monomers used. The polymers prepared from the AB_2 monomer had a compact highly branching structure, whilst those prepared from the A_2 and B_3 type monomers had a low branching density topography and to a certain extent chain entanglement.

4.3. Synthesis of Hyperbranched Polyesters using Coupling Reagents

Hedrick *et al.* have produced hyperbranched aliphatic polyesters 109 based upon AB_2 type macromonomers 110 synthesised via living ring opening polymerisation (ROP) of ε-caprolactone using aluminium benzyloxide as the initiator [162]. The aluminium benzyloxide initiated polymers were functionalised with benzylidene protected 2,2'-*bis*(hydroxymethyl) propionic acid and deprotected subsequently to form the α-carboxylic-ω-dihydroxy functional AB_2 type macromonomers 110. The macromonomers were condensed into hyperbranched polyesters through reaction with DCC and DPTS. This synthetic approach allowed the polyesters to be constructed with excellent control over the number of monomer units, the size of the macromonomer and the molecular weight. For example, it was possible to create AB_2 macromonomers with very low polydispersities and degrees of polymerisation targeted by variation of the benzylalcohol/initiator to monomer ratio. Through the coupling reagent mediated polycondensation of the AB_2 macromonomers 110 Hedrick *et al.* were able to produce different molecular weight hyperbranched polyesters 109 (ranging from 25 kD to 201 kD) consisting of acid focal points and randomly distributed hydroxyl end groups (Scheme 33).

The weight-average molecular weight was found to increase in a regular fashion depending upon the DP of the macromonomer used, although the polydispersities varied between 1.89 and 2.72. The degree of branching was determined to be 0.37 and was independent of the weight-average molecular weight of both the macromonomer 110 and the final polyester.

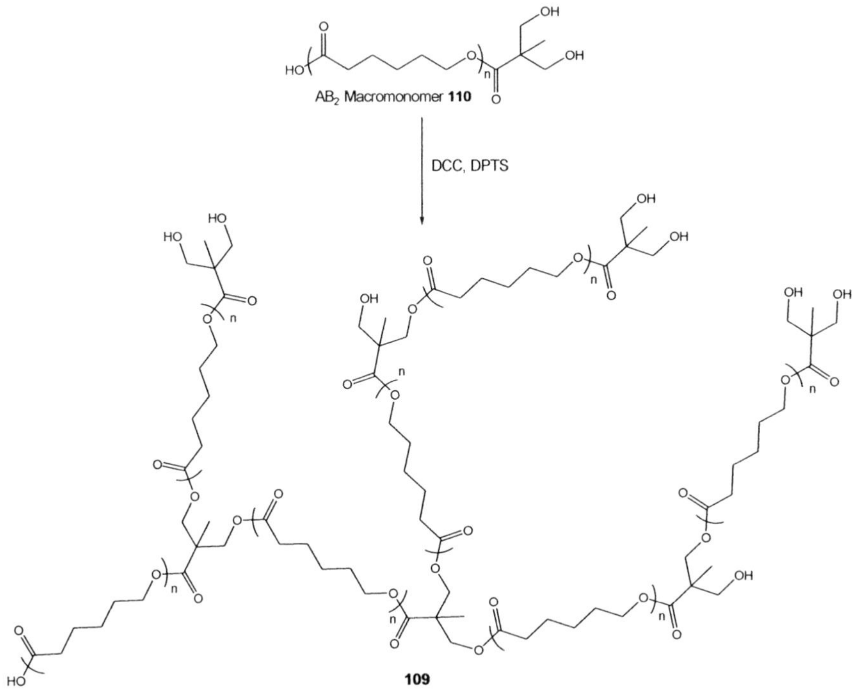

Scheme 33. Polycondensation of AB_2 macromonomers 110 using DCC [162]

Hedrick and co-workers have also utilized macromonomers for the coupling reagent mediated synthesis of hyperbranched polyesters [163]. In this example, the benzyl protected AB_2 and AB_4 type macromonomers were synthesized through living ROP of 1,4,9-trioxaspiro[4,6]-9-undecanone (TOSUO) and/or ε-caprolactone initiated from the *bis*(hydroxymethyl)groups of the benzyl ester of 2,2'-*bis*(hydroxymethyl)propionic acid in the presence of $Sn(Oct)_2$ [164]. The benzyl ester groups were removed readily by catalytic hydrogenolysis to give the requisite acid functionality and AB_2 and AB_4 macromonomers capable of self-polymerisation. The synthesis of the macromonomers was living in nature and allowed for accurate control of the molecular weights and polydispersities of *ca.* 4-15 kD and 1.15-1.20, respectively. Polymerisation of the macromonomers was accomplished using DCC and DPTS to afford hyperbranched polyesters with average molecular numbers from 19 kD to 93 kD and polydispersities consistent with a condensation polymerisation (*ca.* 2). Deacetalisation and reduction of the TOSUO containing polymers [165] produced a hyperbranched polyester with significant hydroxy functionality at the periphery of the macromolecules as well as dispersed along the chain backbone.

Blencowe *et al.* have synthesised a novel hyperbranched polyester 111 [166] that incorporates the rigid *bis*-acetylenic monomer, 3,5-*bis*(3-hydroxylprop-1-ynyl)benzoic acid 112 [167] using the coupling reagents DIC and DCC (Scheme 34).

Attempts to thermally homopolymerise 112 using organometallic esterfication catalysts [168] or activation of hydroxyl [169] and/or acid [169,170] functionalities resulted in decomposition of the monomer and the formation of insoluble residues. The coupling reagents CDI and EDCI were also found to be inappropriate for the polymerisation of 112, resulting in the formation of insoluble low molecular weight species. ^{1}H NMR spectroscopic analysis of the material obtained from polymerisation of monomer 112 using CDI indicated the presence of a variety of anhydride, carbamate and ester linkages. Polymerisation of the monomer 112 using the coupling reagent EDCI resulted in the formation of poorly soluble dark brown solids.

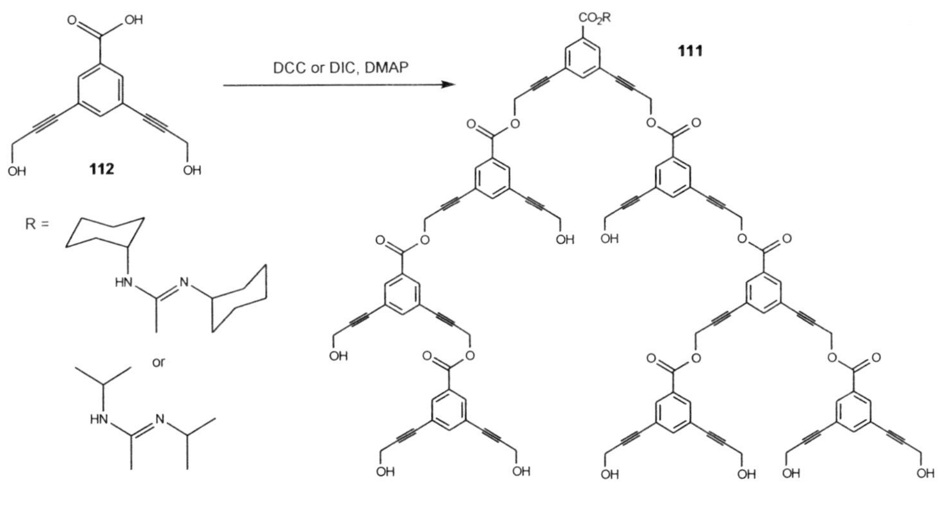

Scheme 34. Polycondensation of the diacetylene AB_2 type monomer 112 using DCC and DIC [167]

MALDI-TOF mass spectroscopic analyses of the solids revealed signals separated by regularly repeating intervals that corresponded to progressive monomer units that emanated from a residue possessing a molecular weight of 173 *m/z*. This peak was found to correspond to a protonated EDCI unit that was bound to the focal point of each hyperbranched macromolecule. A direct consequence of this focal point functionality was the materials poor solubility and precipitation from the reaction solution, preventing the continuation of polymerisation and resulting in the formation of low molecular weight polyesters.

In contrast to the use of CDI and EDCI, the polyesterifications mediated by DCC and DIC afforded soluble (THF) hyperbranched polyesters 111 with molecular weights and degrees of polydispersity ranging between 2.5-11 kD and 1.5-2.5, respectively. The optimum concentration of DCC or DIC coupling reagent was found to be approximately 1.5 equivalents as the molecular weight of the polyesters decreased significantly when the concentration of coupling reagent to monomer was equal, doubled or tripled. The concentration of the monomer 112 in solution and reaction time was found to have little effect on the molecular weight or DB values of 111. In fact the maximum molecular weight material was usually obtained with in the first hour of the reaction, with it then only increasing slightly over a further 1 to 15 hours. The DB values of the polyesters 111 ranged from 0.22-0.33, which is significantly lower than the statistical value of 0.5 calculated by Frey, [171] for the direct homopolymerisation of AB_2 monomers. It was hypothesized that the lower DB values obtained for this type of polyester occurred as a consequence of the rigid conformation of the monomer 112 and the use of bulky coupling reagents, which in turn creates steric hindrance during the coupling steps. Consequently, the hyperbranched polyesters produced consisted of a large proportion of linear units. Interestingly, the polydispersities of 111 were found to vary considerably when DCC was used as the coupling reagent, although that was found to be the opposite for the polyesters 111 synthesised using DIC, which generally had polydispersities of *ca.* 1.7.

3-Carboxyl-5-hydroxybenzoic acid and 3,5-dihydroxybenzoic acid have also been polymerised using CDI as the coupling reagent [172]. Reaction of CDI with the acid functional groups of the monomers results in the selective formation of an imidazolide accompanied by the evolution of carbon dioxide. Subsequent heating results in polymerisation of the imidazolide intermediate to afford hyperbranched polyesters possessing weight-average molecular weights of up to 3.5 kD and featuring either phenol of acid terminal groups.

4.4. Synthesis of Hyperbranched Polyurethanes and Ureas using Coupling Reagents

Rannard *et al.* have reported the selective activation of an unprotected AB_2 type monomer using CDI and its subsequent polymerisation to form a novel class of water-soluble hyperbranched polyurethanes 113 [173,174]. CDI was added at room temperature to a THF solution of *N*-(3-aminopropyl)-diethanolamine 114 (APDE) with the corresponding selective formation of an active carboxamide 115 [24] and no detectable reaction with the hydroxyl functional groups (Scheme 35). After removal of the solvent the residue was melt polymerised at 90 °C for several hours to afford hyperbranched polyurethanes 113 with weight-average molecular weights up to *ca.* 6.5 kD and branching factors of *ca.* 0.6.

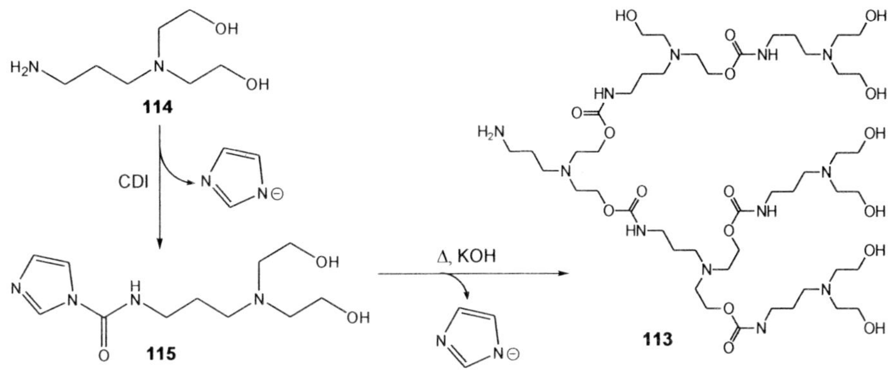

Scheme 35. Polymerisation of APDE using CDI [173,174]

When the polymerisation was conducted at 60 °C the desired reaction was not detected. In contrast, polymerisations conducted at 140 °C resulted in high conversions of monomer but resulted in a decrease in the number-average molecular weight compared to those polyurethanes obtained at 90 °C. MALDI-TOF mass spectrometric analysis of the polyurethanes 113 revealed several distributions corresponding to protonated, sodium and potassium adducts of the hyperbranched polyurethane with an amine focal point. However, there also existed distributions of peaks corresponding to hyperbranched polyurethanes featuring carboxamide and urea focal points. The urea focal point was formed as a result of the reaction of diethanolamine (an impurity present in APDE) with an active carboxamide.

The reaction of CDI and its derivatives with amines was found to be dependent on the nature of the amine. For example, primary amines react with CDI very quickly at room temperature, whereas secondary amines react very slowly at room temperature. This difference in reactivity has proven to be very useful for the selective activation of amine containing AB_2 type monomers to form hyperbranched polyureas. Rannard *et al.* have polymerised successfully the AB_2 type monomers, *bis*(3-aminopropyl)amine 116, *bis*(6-aminohexyl)amine 117 and *tris*(2-aminoethyl)amine 118 (Figure 10) using CDI [175].

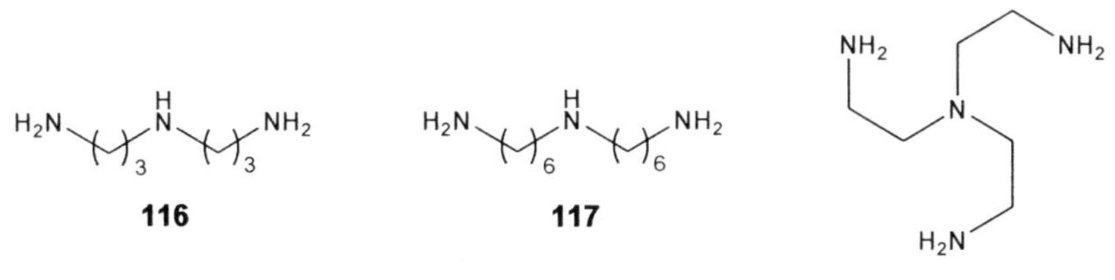

Figure 10. Amine monomers used for CDI mediated synthesis of polyureas [175]

Initially, CDI reacted with the primary amine present in the monomers to form a carboxamide. Subsequent heating to 50-60 °C leads to the reaction of the primary amine carboxamide with the secondary amine groups to afford hyperbranched polyureas.

5. Further Uses of Coupling Reagents in Polymer Chemistry

Although coupling reagents have been applied to the synthesis of a large number of polymeric architectures, their use in polymer post-modification and surface chemistries is even more widespread. However, it is outside the scope of this article to review all research conducted using coupling reagents in this manner. Therefore, a few choice examples have been selected that demonstrate the applicability of coupling reagents in polymer related applications.

5.1. Post-Modification of Polymers Using Coupling Reagents

Coupling reagents have proven to be extremely useful for the post-modification of polymers [176] to form self assembled monolayers or incorporate additional functionalisation. Numerous research groups have utilised coupling reagents for end group functionalisation of hyperbranched macromolecules in order to change the bulk properties of the polymeric material. For example, Sunder *et al.* have introduced mesogenic units onto the end hydroxyl groups of hyperbranched polyglycerol 119 [177] (Scheme 36) resulting in the formation of polymers 120 that exhibit liquid crystalline phases [178].

Initially the hyperbranched polyglycerols 119 were constructed by via ring-opening multibranching polymerisation [179] and then the mesogenic acids were coupled to the free hydroxyl chain ends using DIC. However, even reaction times in excess of one week did not afford complete functionalisation of the polyether as a result of the bulkiness of the mesogenic groups and the difference in reactivity of the primary and secondary hydroxyl groups of the polyglycerol 119.

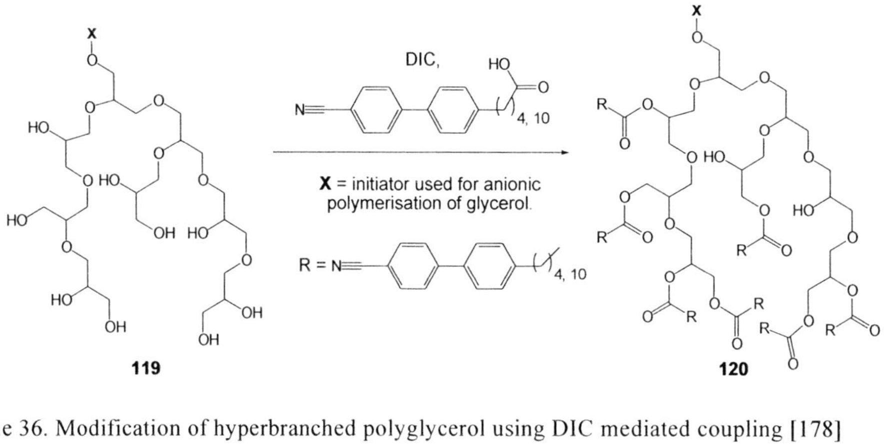

Scheme 36. Modification of hyperbranched polyglycerol using DIC mediated coupling [178]

Hall and co-workers have synthesised polymer supported calix[4]arenes [180] for metal extraction applications [181], using Tentagel S NH_2 and aminomethyl polystyrene. In contrast to the usual method used for the synthesis of calix[4]arene functionalised polymers involving acid chloride formation, Hall and co-workers utilised carbodiimide coupling reagents. When DIC was employed the loading of calix[4]arene onto Tentagel S NH_2 and aminomethyl polystyrene was 10 and 27 % by mass, respectively, which was similar to loadings obtained via the acid chloride method. Interestingly, DCC and EDCI mediated coupling of calix[4]arene to Tentagel S NH_2 and aminomethyl polystyrene only afforded loadings of 8 and < 2%, respectively.

Coupling reagents are particularly useful for the incorporation of sensitive biologically active species into polymeric architectures for applications such as drug delivery, affinity separation, diagnostics and bioprocesses. The biologically active molecules that have thus far been utilised for polymer functionalisation include antibodies, carbohydrates, peptides, protein cell receptor ligands and biotin. Coupling reagents have been used to attach biologically active species to both the end-groups present in linear polymers and to the polymer backbone functionalities affording polymers with reactive pendent groups [182]. In some cases biodegradable or hydrolysable linkages have been incorporated [183] to allow *in vitro* cleavage of the biologically active molecules.

Numerous research groups have utilised the mild conditions offered by coupling reagent methodologies for the attachment of enzymes and proteins to polymers [184]. Hoffman and co-workers, have, for example, employed DCC in the presence of NHS to attach the enzyme trypsin to the carboxyl end-group of poly(*N-iso*propylacrylamide) [185] with number-average molecular weights of 5-19 kD. When a large excess of the enzyme was used about eleven out of the fifteen amine groups in trypsin reacted with the activated poly(*N-iso*propylacrylamide). Henneuse *et al.* have used EDCI and CDI in the modification of poly(aryl ether ether ketone)s (PEEK) [186] derivatives with lysine [187]. The ketone moieties present in the PEEK was converted to an amide [188] consisting of a linker group with a free carboxylic acid at one end. The attachment of lysine was then assessed using the coupling reagents EDCI and CDI, and via an acid chloride intermediate. EDCI only produced low loadings of lysine (*ca.* 30 pmol cm^{-2}) compared to CDI (*ca.* 80 pmol cm^{-2}) and acid chloride formation (*ca.* 170 pmol cm^{-2}), however, CDI coupling was found to be the most appropriate method as it limited detrimental side-reactions. Lee and co-workers have adopted the reverse procedure where a sensitive anticancer drug, camptothecin was attached to the vinylic monomer, 3,6-endomethylene-1,2,3,6-tetrahydrophthalic anhydride via a linker using EDCI [189]. The modified monomer was co-polymerised with acrylic acid to give an acid functionalised polymer with pendent drug groups. Many other drugs including doxorubicin, taxol, neocarzinostatin and podophyllotoxin have been attached to polymers [190], although generally after the polymer has been synthesised in a separate step. Kim and co-workers have used CDI to construct an amphiphilic polymeric conjugate based on cholic acid [191] for the preparation of nanoparticles with uses as drug delivery devices [192]. Cholic acid was coupled to PEG with an average-weight molecular weight of 2 kD using 1.5 equivalents of CDI to afford a polymeric conjugate that was capable of forming clonazepam-loaded polymeric nanoparticles.

Metallopolymers are promising materials for the construction of photochemical conversion devices and light energy storage devices [193]. Modification of styrene *p*-(aminomethyl)styrene copolymers 121 with acid functionalised metal complexes 122 has been achieved using coupling reagents to afford soluble metallopolymers 123 [194]. Attachment of various ruthenium or osmium based chromophores via the formation of an amide link was achieved by coupling the amino polymer 121 to the carboxylic acid derivative of a polypyridyl complex by using BOP (Scheme 37).

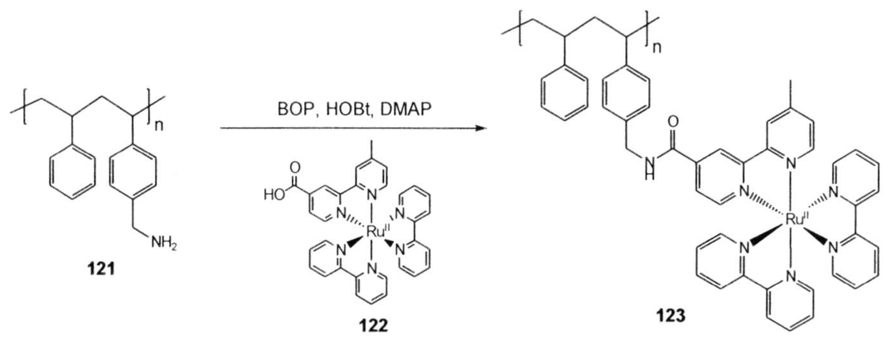

Scheme 37. Synthesis of metallopolymers [195]

The resulting metallopolymers 123 were found to be efficient for long range electron and energy transfer [195] as well as featuring the ability to store multiple redox equivalents on a single polymer chain [196].

Generally, fullerene derivatives are very hydrophobic and as a consequence are insoluble in most polar solvents. Incorporation of fullerene derivatives into polymeric structures [197] is a particularly rewarding approach for modification of the fullerenes solubility. Sun and co-workers have attached methano[60]fullerene carboxylic acid (MFCA) and methano[60]fullerene phenylpropyl carboxylic acid (MFPCA) to linear poly(propionyl-ethylenimine-*co*-ethylenimine) (PPEI) [198] via amide linkages using EDCI. The resulting pendent type polymers (MFCA-PPEI and MFPCA-PPEI) were found to be highly soluble in water and other polar solvents.

5.2. Polymer-Polymer Grafting Using Coupling Reagents

Tobiesen *et al.* have used efficiently EDCI to graft poly(acrylic acid) (PAA) onto nylon-6,6 films via the naturally occurring end groups of the nylon [199]. PAA with weight-average molecular weights of 5 kD and 250 kD were dissolved in water and nylon-6,6 with a number-average molecular weight of 12 kD was added as a thin film. EDCI and NHS were then added and the reaction stirred vigorously for a period of 0.5 to 18 hours at 0, 21 or 60 °C. Removal of the nylon film, followed by extensive washing afforded the PAA modified surface. Although the bulk properties of the nylon were unaffected by the grafting process the procedure resulted in a highly polar surface with increased wetting characteristics as a result of the high number of free carboxylic acids groups present at the surface. To determine whether the carboxylic acid groups of the PAA were still available for further reaction after grafting the PAA onto nylon, hexylamine was grafted onto the free carboxylic acid groups using EDCI. Water no longer wetted the surface of the film thereby providing evidence of the conversion of the free carboxylic groups in to the desired hydrophobic amides.

Hennink and co-workers have used a CDI mediated coupling approach to attach dextrin to polylactide copolymers [200] that are sensitive to hydrolysis, allowing for their use as biodegradable hydrogels for delivery systems [201]. The polylactide co-polymers are biocompatible since the degradation products are lactate (an endogenous compound) and dextrin, which is used as a plasma expander [202].

5.3. Dendritically Modified Surfaces Using Coupling Reagents

Dendritically modified silicas for enhanced chromatography applications [203-206] or catalysis [207] have been topics of great interest in recent times. In the process of developing stationary phases for HPLC applications, Smith and co-workers [208] have assessed and used coupling reagents to modify the surface of aminopropylsilica (AMPS) with dendrons [209] comprised of enantiomerically pure L-lysine. As the dendrons are enantiomerically pure the modified silicas have the potential to act as chiral stationary phases. The coupling reaction between the carboxylic acid at the focal point of the dendrons and AMPS was optimized to obtain the best possible loading. In order to achieve this, three different coupling reagents were assessed, DCC/HOBt, TBTU and HATU. After 3 days, the DCC mediated coupling of the 1st generation L-lysine dendron and AMPS was found to give the highest loadings as determined by elemental analysis. Surprisingly, DCC provided higher substrate loadings than the more reactive uronium salts. As the dendron generation was increased the loading of dendron onto the silica was found to decrease. The decrease in loading was related to the increase in steric bulk of the dendrons as steric hindrance decreases the effectiveness of the coupling reaction. In a similar fashion, Mathews *et al.* have reported [206] the synthesis of stationary phases based on L-glutamic acid containing dendritic molecules grafted onto silica.

Guo and Yu [205] have grafted dendritic polyethers [210,211] onto nanometre silica surfaces functionalised with amine groups. The 1st, 2nd and 3rd generation polyether dendrons with carboxylic acid focal points [211] were grafted onto the functionalised silica (3-aminopropyltriethoxysilane modified silica) using DCC mediated coupling in the presence of DMAP. Zhai *et al.* have synthesised self-assembled films by grafting conjugated dendritic poly(phenylene vinylene) on to the surface of quartz using DCC mediated coupling [212].

CONCLUSION

Over a period of fifty years coupling reagents have been applied to a diverse range of applications with great success. The success of these reagents has arisen as a result of the ability to control reactions efficiently through selective condensation processes and increase the rate at which the desired condensation reaction occurs. Only in relative recent times have coupling reagents have been adopted in polymer synthesis, although the impact of these reagents in this field of chemistry has proven to be substantial. Not only do these reagents offer alternatives to classical methods of polymer synthesis and subsequent modification, but have enabled the construction of a diverse range of polymeric materials that would otherwise have been very difficult or impossible to synthesise. It is, therefore, unsurprising that these highly convenient and beneficial reagents are becoming ever increasingly popular. Indeed, coupling reagents are hugely important and powerful tools for the polymer chemist to have at their disposal.

References

[1] Sheehan, J. C.; Hess, G. P. *J. Am. Chem. Soc.*, 1955, 77, 1067.
[2] Jones, J. H., 1979, 'The Formation of Peptide Bonds: A General Survey', in Gross, E. and Meienhofer, J. (eds.), *The Peptides: Analysis, Synthesis, Biology*, Academic Press, Inc., New York, pp. 65-104.
[3] Sheehan, J. C.; Cruickshank, P. A.; Boshart, G. L. *J. Org. Chem.*, 1961, 26, 2525.
[4] a) Bhalay, G.; Dunstan, A.; Glen, A., *Synlett*, 2000, 12, 1846. b) Shen, C.-M.; Weinshenker, N. M., *Tetrahedron Lett.*, 1972, 3281. c) Shen, C.-M.; Weinshenker, N. M., *Tetrahedron Lett.*, 1972, 3285. d) Desai, M. C.; Stramiello, S., *Tetrahedron Lett.*, 1993, 7685.
[5] a) Déprez, B. P.; Pop, I. E.; Tartar, A. L., *J. Org. Chem.*, 1997, 62, 2594. b) Dendrinos, K.; Huang, E.; Jeong, J.; Kalivretenos, G., *Chem. Commun.*, 1998, 499.
[6] König, W.; Geiger, R. *Chem. Ber.*, 1970, 103, 2024.
[7] Carpino, L. A. *J. Am. Chem. Soc.*, 1993, 115, 4397.
[8] Woodward, R. B.; Olofson, R. A.; Mayer, H. *J. Am. Chem. Soc.*, 1961, 83, 1010.
[9] Wünsch, E.; Drees, F. *Chem. Ber.*, 1966, 99, 110.
[10] Knorr, R.; Trzeciak, A.; Bannwarth, W.; Gillessen, D. *Tetrahedron Lett.*, 1989, 30, 1927.
[11] a) Chen, S.; Xu, J. *Tetrahedron Lett.*, 1992, 33, 647. b) Li, P.; Xu, J. *Chem. Lett.*, 1999, 1163. c) Li, P.; Xu, J. *Tetrahedron Lett.*, 1999, 40, 3605.
[12] Dourtoglou, V.; Ziegler, J.-C.; Gross, B. *Tetrahedron Lett.*, 1978, 15, 1269.
[13] Ehrlich, A.; Rothemund, S.; Brudel, M.; Beyermann, M.; Carpino, L. A.; Bienert, M. *Tetrahedron Lett.*, 1993, 34, 4781.
[14] Castro, B.; Dormoy, J. R.; Evin, G.; Selve, C. *Tetrahedron Lett.*, 1975, 14, 1219.
[15] Carpino, L. A.; El-Faham, A.; Albericio, F. *Tetrahedron Lett.*, 1994, 35, 2279.
[16] Coste, J.; Le-Nguyen, D.; Castro, B.; Dormoy, J. R.; Evin, G.; Selve, C. *Tetrahedron Lett.*, 1990, 31, 205.
[17] Li, P.; Xu, J. *Tetrahedron Lett.*, 1999, 40, 8301.
[18] Diago-Meseguer, J.; Palomo-Coll, A. L.; Fernandez-Lizarbe, J. R.; Zugaza-Bilbao, A. *Synthesis*, 1980, 547.
[19] Coste, J.; Dufour, M.-N.; Pantaloni, A.; Castro, B.; *Tetrahedron Lett.*, 1990, 31, 669.
[20] Ramage, R.; Hopton, D.; Parrott, M.J.; Richardson, R. S.; Kenner, G. W.; Moore, G. A. *J. Chem. Soc., Perkin Trans. I*, 1985, 461.
[21] Akaji, K.; Kuriyama, N.; Kiso, Y. *Tetrahedron Lett.*, 1994, 35, 3315.
[22] a) Coste, J.; Frerot, E.; Jouin, P.; Castro, B. *Tetrahedron Lett.*, 1991, 32, 1967. b) Coste, J.; Frerot, E.; Jouin, P. *J. Org. Chem.*, 1994, 59, 2437.
[23] Staab, H. A., *Justus Liebigs Ann. Chem.*, 1957, 609, 75.
[24] Rannard, S. P.; Davis, N. J., *Org. Lett.*, 2000, 2, 2117.
[25] Rannard, S. P.; Davis, N. J., *Org. Lett.*, 1999, 1, 933.
[26] Cais, R. E.; Sloane, N. J. A., *Polymer*, 1983, 24, 179.
[27] Ueda, M.; Kameyama, A.; Hashimoto, K., *Macromolecules*, 1988, 21, 19.
[28] a) Higashi, F.; Taguchi, Y.; Kokubo, N., *J. Polym. Sci., Part A: Polym. Chem.*, 1981, 19, 2745. b) Higashi, F.; Taguchi, Y., *J. Polym. Sci., Part A: Polym. Chem.*, 1981, 19, 3345.

[29] Dias, M. L.; Silva, A. P. F., *Polym. Eng. Sci.*, 2000, 40, 1777.

[30] a) Mitin, Y. V.; Glinskaya, O. V.; *Tetrahedron Lett.*, 1969, 60, 5267. b) Hodges, R. V.; McDonnell, T. J.; Beauchamp, J. L., *J. Am. Chem. Soc.*, 1980, 102, 1327.

[31] Liu, R.; Orgel, L. E., *Origins Life Evol. Biosphere*, 1998, 28, 47.

[32] Ehler, K. W.; Orgel, L. E., *Biochem. Biophys. Acta*, 1976, 434, 233.

[33] Hill Jr., A. R.; Orgel, L. E., *Origins Life Evol. Biosphere*, 1996, 26, 539.

[34] a) Greenstein, J. P.; Winitz, M., 1961, *Chemistry of Amino Acids, Volume 2*, John Wiley and Sons, Inc., New York, London, pp 782. b) Brack, A., *Origins Life Evol. Biosphere*, 1987, 17, 367.

[35] a) Geiger, T.; Clarke, S., *J. Biol. Chem.*, 1987, 262, 785. b) Brennan, T. V.; Clarke, S., 1995, 'Deamidation and Isoasparatate Formation in Model Synthetic Peptides: The Effects of Sequence and Solution Environment', in Aswad, D. W. (ed), *Deamidation and Isoasparatate Formation in Peptides and Proteins*, CRC Press, Boca Raton, pp. 65-90.

[36] Kricheldorf, H. R., 1987, *[Alpha]-Amino acid-N-Carboxy-Anhydrides and Related Heterocycles: Synthesis, Properties, Peptide Synthesis, Polymerization*, Springer-Verlag, Berlin.

[37] Kricheldorf, H. R.; Mulhaupt, R., *Makromol. Chem.*, 1979, 180, 1419.

[38] Ferris, J. P.; Hill Jr., A. R.; Liu, R.; Orgel, L. E., *Nature*, 1996, 381, 59.

[39] Kurzer, F.; Douraghi-Zadeh, K., *Chem. Rev.*, 1967, 67, 107.

[40] Liu, R.; Orgel, L. E., *J. Am. Chem. Soc.*, 1997, 119, 4791.

[41] Pascual, S.; Haddleton, D. M.; Heywood, D. M.; Khoshdel, E., *Eur. Polym. J.*, 2003, 39, 1559.

[42] a) Staros, J. V., *Biochemistry-US*, 1982, 21, 3950. b) Sehgal, D.; Vijay, I. K., *Anal. Biochem.*, 1994, 218, 87. c) Anderson, G. W.; Zimmerman, J. E.; Callahan, F. M., *J. Am. Chem. Soc.*, 1964, 86, 1839.

[43] Liu, R.; Orgel, L. E., *Origins Life Evol. Biosphere*, 1998, 28, 245.

[44] Hill Jr., A. R.; Bohler, C.; Orgel, L. E., *Origins Life Evol. Biosphere*, 1998, 28, 235.

[45] Orgel, L. E., *Origins Life Evol. Biosphere*, 1998, 28, 227.

[46] a) Urry, D. W., *Res. and Dev.*, 1989, 30, 57. b) Urry, D. W.; Parker, T. M.; Reid, M. C.; Gowda, D. C., *J. Bioactive Compatible Polym.*, 1991, 6, 263. c) Urry, D. W.; Gowda, D. C.; Harris, C.; Cox, B. A., *Am. Chem. Soc. Poly. Chem. Div.*, 1992, 33, 84. d) Urry, D. W., *Angew. Chem. Int., Ed. Engl.*, 1992, 32, 819.

[47] a) Sandberg, L.; Leslie, J.; Leach, C.; Tres, V.; Smith, A.; Smith, D., *Pathol. Biol.*, 1985, 33, 266. b) Indik, Z.; Yeh, H.; Ornstein-Goldstein, N.; Sheppard, P.; Anderson, N.; Rosenbloom, J.; Peltonen, L.; Rosenbloom, J. C., *J. Proc. Natl. Acad. Sci. USA*, 1987, 84, 5680. c) Smith, D. W.; Sandberg, L. B.; Leslie, B. H.; Wolt, T. E.; Minton, S. T.; Myers, B.; Rucker, R. B., *Biochem. Biophys. Res. Commun.*, 1981, 103, 880.

[48] Sandberg, L. B.; Soskel, N, T.; Leslie, J. B., *N. Engl. J. Med.*, 1981, 304, 566.

[49] a) Urry, D. W.; Nicol, A.; Gowda, D. C.; Hoban, L. D.; McKee, A.; Williams, T.; Olsen, D. B.; Cox, B. A., 1993, in Gebelein, C. G., (ed.), *'Medical Applications of Bioelastic Materials'*, Technomic Publishing, Atlanta, Georgia, pp. 82-103. b) Urry, D. W.; Nicol, A.; McPherson, D. T.; Xu, J.; Shewry, P. R.; Harris, C. M.; Parker, T. M.; Gowda, D. C., 1995, in *'Handbook of Biomaterials and Applications'*, Marcel Dekker Inc., New York, pp. 2645-2699.

[50] a) Urry, D. W.; Prasad, K. U., 1985, in Williams, D. W., (ed.), *'Biocompatibility of Tissue Analogues'*, CRC Press Inc., Boca Raton, Florida, pp. 89-116. b) Prasad, K. U.; Iqbal, M. A.; Urry, D. W., *Int. J. Pept. And Protein Res.*, 1985, 25, 408.
[51] Baba, A. R.; Gowda, D. C., *Lett. Peptide Sci.*, 2002, 8, 309.
[52] a) Morgan, P. W., *Chemtech*, 1979, 9, 316. b) Cassidy, P. E., (ed), 1980, *Thermally Stable Polymers*, Marcel Dekker, New York.
[53] a) Kricheldorf,. H. R.; Schmidt, B., *Macromolecules*, 1992, 25, 5465. b) Kricheldorf,. H. R.; Schmidt, B., *Macromolecules*, 1992, 25, 5471. c) Kricheldorf,. H. R.; Burger, R.; *J. Polym. Sci., Part A, Polym. Chem.*, 1994, 32, 355. d) Yang, H. H., (ed), 1989, *Aromatic High-Strength Fibers*, Wiley, New York.
[54] a) Bruma, M.; Mercer, N. N.; Katimoto, M., *J. Polym. Sci., Part A, Polym. Chem.*, 1995, 56, 527. b) Imai, Y.; Malder, N. N.; Katimoto, M., *J. Polym. Sci., Part A, Polym. Chem.*, 1985, 26, 235.
[55] Liou, G. S.; Katimoto, M.; Imai, Y., *J. Polym. Sci., Part A, Polym. Chem.*, 1993, 31, 3265.
[56] Yang, C. P.; Chen, Y.-P.; Woo, E. M., *J. Polym. Sci., Part A, Polym. Chem.*, 2004, 42, 3116.
[57] a) Yang, C. P.; Lin, J. H.; *Polym. Int.*, 1995, 38, 335. b) Yang, C. P.; Hsiao, S. H.; Chou, W. L., *J. Polym. Res.*, 1995, 2, 179.
[58] Ueda, M.; Mochizuki, A., *Macromolecules*, 1985, 18, 2353.
[59] a) Ueda, M.; Morosumi, T.; Kakuta, M.; Sato, R., *Polym. J.*, 1991, 22, 733. b) Ueda, M.; Kakuta, M.; Morosumi, T.; Sato, R., *Polym. J.*, 1991, 23, 167. c) Ueda, M.; Morosumi, T.; Kakuta, M., *Polym. J.*, 1991, 23, 1151.
[60] Ueda, M.; Morosumi, T.; Kakuta, M.; Sugiyama, J.-I., *Macromolecules*, 1992, 25, 6580.
[61] Ueda, M.; Sugiyama, H., *Macromolecules*, 1994, 27, 240.
[62] Ueda, M.; Takabayashi, A.; Seino, H., *Macromolecules*, 1997, 30, 363.
[63] Yu, S.; Seino, H.; Ueda, M., *Macromolecules*, 1999, 32, 1027.
[64] Ueda, M.; Seki, K.; Imai, Y., *Macromolecules*, 1982, 15, 17.
[65] Ogata, N.; Sanui, K.; Tan, S., *Polym. J.*, 1984, 16, 569.
[66] Culbertson, B. M.; Dietz, S., *J. Polym. Sci., Polym. Lett. Ed.*, 1968, 5, 247.
[67] Kroto, H. W.; Allaf, A. W.; Balm, S. P., *Chem. Rev.*, 1991, 91, 1213.
[68] a) Loy, D. A.; Assink, R. A., *J. Am. Chem. Soc.*, 1992, 114, 3977. b) Taki, M.; Takigami, S.; Watanbe, Y.; Nakamura, Y.; Nishimura, J., *Polym. J.*, 1997, 29, 1020. c) Ozawa, M.; Li, J.; Nakahara, K.; Xiao, L.; Sugawara, H.; Kitazawa, K.; Kinbara, K.; Saigo, K., *J. Polym. Sci., Polym. Chem. Ed.*, 1998, 17, 3139.
[69] Xiao, L.; Shimotani, H.; Ozawa, M.; Li, J.; Dragoe, N.; Saigo, K.; Kitazawa, K., *J. Polym. Sci., Part A: Polym. Chem.*, 1999, 37, 3632.
[70] Huang, D.; Zhang, C.; Dalton, L. R.; Weber, W. P., *Designed Monomers and Polymers*, 2000, 3, 95.
[71] a) Green, G. D.; Hall, H. K.; Mulvaney, J. E.; Noonan, J.; Williams, D. J., *Macromolecules*, 1987, 20, 716. b) Green, G. D.; Weinschenk, J. I.; Mulvaney, J. E.; Hall, H. K., *Macromolecules*, 1987, 20, 722. c) Fusi, F.; Padias, A. B.; Hall, H. K., *Macromolecules*, 1991, 24, 1710. d) Wright, M. E.; Mullick, S.; Lackritz, H. S.; Liu, L.-Y., *Macromolecules*, 1994, 27, 3009. e) Liu, L.-Y.; Lackritz, H. S.; Wright, M. E.; Mullick, S., *Macromolecules*, 1995, 28, 1912. f) Kohler, W.; Robello, D. R.; Willand,

C. S.; Williams, D. J., *Macromolecules*, 1991, 24, 4589. g) Heldmann, C.; Warner, M., *Macromolecules*, 1998, 31, 3519.

[72] a) Robello, D. R.; Dao, P. T.; Phelan, J.; Revelli, J.; Schildkraut, J. S.; Scozzafava, M.; Ulman, A.; Willand, C. S., *Chem. Mater.*, 1992, 4, 425. b) Nemoto, N.; Miyata, F.; Nagase, Y.; Abe, J.; Hasegawa, H.; Shirai, Y., *Macromolecules*, 1996, 29, 2365. c) Nemoto, N.; Nagase, Y.; Abe, J.; Hasegawa, H.; Shirai, Y., *Chem. Mater.*, 1997, 9, 304.

[73] Katz, H. E.; Schilling, M. L., *J. Am. Chem. Soc.*, 1989, 111, 7554.

[74] Liaw, D.-J.; Liaw, B.-Y., *J. Polym. Sci., Part A: Polym. Chem.*, 1998, 36, 2301.

[75] Liaw, D.-J.; Liaw, B.-Y.; Tseng, J.-M., *J. Polym. Sci., Part A: Polym. Chem.*, 1999, 37, 2629.

[76] Liaw, D.-J.; Liaw, B.-Y.; Kang, E.-T., *Macromol. Chem. Phys.*, 1999, 200, 2402.

[77] Liaw, D.-J.; Liaw, B.-Y., *Polymer*, 2001, 42, 839.

[78] Swift, G.; Creamer, M.; Wei, X.; Yocom, K., *Macromolecules*, 1998, 130, 379.

[79] Mochizuki, M.; Hirami, M., *Polym. Adv. Tech.*, 1997, 8, 203.

[80] Vert, M.; Li, M. S.; Spenlehauer, G.; Guerin, P., *J. Mater. Sci.–J. Mater. Med.*, 1992, 3, 731.

[81] Saotome, Y.; Miyazawa, T.; Endo, T., *Chem. Lett.*, 1991, 21.

[82] Saotome, Y.; Miyazawa, T.; Endo, T., *Chem. Lett.*, 1991, 153.

[83] Arabuli, N.; Tsitlanadze, G.; Edilashvili, L.; Kharadze, D.; Goguadze, T.; Beridze, V.; Gomurashvili, Z.; Katsarava, R., *Macromol. Chem. Phys.*, 1994, 195, 2279.

[84] Paredes, N.; Rodriguez-Galan, A.; Puiggali, J., *J. Polym. Sci., Part A: Polym. Chem.*, 1998, 36, 1271.

[85] Fan, Y.; Kobayashi, M.; Kise, H., *Polym. J.*, 2000, 32, 817.

[86] Fey, T.; Keul, H.; Höcker, H., *e-Polymers (www.e-polymers.org)*, 2003, 1.

[87] Higashi, F.; Yamada, Y.; Hoshio, A., *J. Polym. Sci., Polym. Chem. Ed.*, 1984, 22, 2181.

[88] a) Higashi, F.; Akiyama, N.; Takahashi, I., *J. Polym. Sci., Polym. Chem. Ed.*, 1984, 22, 1653. b) Higashi, F.; Takahashi, I.; Akiyama, N., *J. Polym. Sci., Polym. Chem. Ed.*, 1984, 22, 3607. c) Higashi, F.; Mashimo, T.; Takahashi, I.; Akiyama, N., *J. Polym. Sci., Polym. Chem. Ed.*, 1985, 23, 3095.

[89] Panayotov, I. M.; Belcheva, N.; Tsvetanov, C., *Makromol. Chem.*, 1987, 188, 2821.

[90] a) Hassner, A.; Axexanian, V., *Tetrahedron Lett.*, 1978, 19, 4475. b) Scriven, E. F. V., *Chem. Soc. Rev.*, 1983, 12, 129.

[91] a) Holmberg, K.; Hansen, B., *Acta Chem. Scand.*, 1979, B33, 410. b) Boden, E. P.; Keck, G, E., *J. Org. Chem.*, 1985, 50, 2394.

[92] Moore, J. S.; Stupp, S. I., *Macromolecules*, 1990, 23, 65.

[93] Moore, J. S.; Stupp, S. I., *J. Am. Chem. Soc.*, 1992, 114, 3429.

[94] Moore, J. S.; Stupp, S. I., *J. Org. Chem.*, 1990, 55, 3374.

[95] Wooley, K. L.; Fréchet, J. M. J.; Hawker, C. J., *Polymer*, 1994, 35, 4489.

[96] Wooley, K. L.; Hawker, C. J.; Fréchet, J. M. J., *Polymer J.*, 1994, 26, 187.

[97] a) Miller, T. M.; Kwock, E. W.; Neenan, T. X., *Macromolecules*, 1992, 25, 3143. b) Miller, T. M.; Neenan, T. X., *Chem. Mater.*, 1990, 2, 346. c) Miller, T. M.; Neenan, T. X.; Zayas, R.; Bair, H. E., *J. Am. Chem. Soc.*, 1992, 114, 1018. d) Kim, Y. H.; Webster, O. W., *Macromolecules*, 1992, 25, 5561.

[98] Higashi, F.; Watabiki, M.; Kim, J.-H., *J. Polym. Sci., Part A, Polym. Chem.*, 2001, 39, 1726.

[99] Higashi, F.; Hoshio, A.; Kiyoshige, J., *J. Polym. Sci., Part A, Polym. Chem.*, 1983, 21, 3241.
[100] Jin, J.-I.; Kang C.-S.; Lee, I.-H.; Yun, Y.-K., *Macromolecules*, 1994, 27, 2664.
[101] a) Nakamura, T.; Hitomi, S.; Watanabe, S.; Shimizu, Y.; Jamshidi, K.; Hyon, S. H.; Ikada, Y., *J. Biomed. Mater. Res.*, 1989, 23, 115. b) Agrawal, C. M.; Haas, K. F.; Leopold, D. A.; Clark, H. G., *Biomaterials*, 1992, 13, 176.
[102] a) Dorough, G. L., *Chem. Abstr.*, 1935, 29, 3422. b) Gruber, P. R.; Hall, E. S.; Kolstad, J. J.; Iwen, M. L.; Benson, R. D.; Borchardt, R. L., *Chem. Abstr.*, 1994, 120, 108011*q*.
[103] Akutsu, F.; Inoki, M.; Uei, H.; Sueyoshi, M.; Kasashima, Y.; Naruchi, K.; Yamaguchi, Y.; Sunahara, M., *Polym. J.*, 1998, 30, 421.
[104] Stadler, P. A., *Helv. Chim. Acta*, 1978, 62, 1675.
[105] Fujisawa, T.; Mori, T.; Fukumoto, K.; Sato, T., *Chem. Lett.*, 1982, 1891.
[106] Chen, W.; Luo, W.; Wang, S.; Bei, J., *Polym. Adv. Technol.*, 2003, 14, 245.
[107] a) Vincenzi, V.; Ferruti, P.; Ford, J.; Duncan, R., *Macromol. Biosci.*, 2001, 1 164. b) Lesniak, M. S.; Brem, H., *Nature*, 2004, 3, 499. c) Tomlinson, R.; Klee, M.; Garrett, S.; Heller, J.; Duncan, R.; Brocchini, S., *Macromolecules*, 2002, 35, 473. d) Aulenta, F.; Hayes, W.; Rannard, S., *Eur. Polym. J.*, 2003, 39, 1741.
[108] Huh, K. M.; Bae, Y. H.; *Polymer*, 1999, 40, 6147.
[109] Lee, S. Y.; Chin, I. J.; Jung, J. S., *Eur. Polym. J.*, 1999, 35, 2147.
[110] Wu, G. Y.; Wu, C. H., *J. Biol. Chem.*, 1987, 262, 4429.
[111] Boussif, O.; Lezoualc'h, F.; Zanta, M. A.; Mergny, M. D.; Scherman, D.; Demeneix, B.; Behr, J.-P., *Proc. Natl. Acad. Sci. U.S.A.*, 1995, 92, 7297.
[112] Zanta, M. A.; Boussif, O.; Adib, A.; Behr, J.-P., *Bioconjugate Chem.*, 1997, 8, 839.
[113] a) Choksakulnimitr, S.; Masuda, S.; Tokuda, H.; Takakura, Y.; Hashida, M., *J. Controlled Release*, 1995, 34, 233. b) Brazeau, G. A.; Attia, S.; Poxon, S.; Hughes, J. A., *Pharm. Res.*, 1998, 15, 680. c) Kotecha, B.; Richardson, G. P., *Hear Res.*, 1994, 73, 173.
[114] Putnam, D.; Langer, R., *Macromolecules*, 1999, 32, 3658.
[115] Kohn, J.; Langer, R., *J. Am. Chem. Soc.*, 1987, 109, 817.
[116] Kwon, H. Y.; Langer, R., *Macromolecules*, 1989, 22, 3250.
[117] a) Rawal, V. H.; Cava, M. P., *Tetrahedron Lett.*, 1985, 26, 6141. b) Wasserman, H. H.; Berger, G. D., *Tetrahedron*, 1983, 39, 2459. c) Wasserman, H. H.; Berger, G. D.; Cho, K. R., *Tetrahedron Lett.*, 1982, 23, 465.
[118] Zhou, Q.-X.; Kohn, J., *Macromolecules*, 1990, 23, 3399.
[119] a) Davis, B. G.; Robinson, M. A., *Current Opinion in Drug Discovery and Development*, 2002, 5, 279. b) Burger, A., 2003, 'Carbohydrate-Based Therapeutics', in Abraham, D. A., (ed.), *'Medicinal Chemistry and Drug Discovery'*, John Wiley and Sons Inc., pp. 203-248.
[120] Kadokawa, J.; Tagaya, H.; Chiba, K., *Synlett*, 1999, 12, 1845.
[121] a) Schurch, F., *Adv. in Carbohydr. Chem.*, 1981, 39, 157. b) Kochetkov, N. K.; Bochkov, A. F.; Yazlovetsky, I. G., *Carbohydr. Res.*, 1967, 5, 243. c) Kochetkov, N. K., *Tetrahedron*, 1987, 39, 2389.
[122] a) Sumitomo, H.; Okada, M., 1987, in Ottenbrite, R. M.; Utracki, L. A.; Inoue, S., (eds.), *'Current Topics in Polymer Science'*, Vol. 1, Hanser, Munich, pp. 15. b) Uryu, T., 1990, in Penczek, S., (ed.), *'Models of Biopolymers by Ring-Opening Polymerization'*, CRC Press, Boca Raton, pp. 133.

[123] a) Kobayashi, S.; Shoda, S.; Uyama, H., *Adv. Polym. Sci.*, 1995, 121, 1. b) Kobayashi, S.; Shoda, S.; Uyama, H., 1997, Kobayashi, S., (ed.), *'Catalysis in Precision Polymerization'*, Wiley, Chichester, Chapter 8.
[124] Kadokawa, J.; Kata, M.; Kobayashi, N.; Karasu, M.; Tagaya, H.; Chiba, K., *Macromol. Rapid Commun.*, 1994, 15, 971.
[125] Grayson, S. M.; Fréchet, J. M. J., *Chem. Rev.*, 2001, 101, 3819.
[126] Fréchet, J. M. J.; Hawker, C. J., 1996, 'Synthesis and Properties of Dendrimers and Hyperbranched Polymers', in Aggarval, S. L., Russo, S. (eds), *Comprehensive Polymer Science*, 2nd Suppl., Pergamon, Oxford, pp. 71-131.
[127] Vogtle, F.; Gestermann, S.; Hesse, R.; Schwierz, H.; Windisch, B., *Prog. Polym. Sci.*, 2000, 987.
[128] Jikei, M.; Kakimoto, M.-A., *J. Polym. Sci., Part A: Polym. Chem.*, 2004, 42, 1293.
[129] Ihre, H.; Hult, A.; Fréchet, J. M. J.; Gitsov, I., *Macromolecules*, 1998, 31, 4061.
[130] a) Bhyrappa, P.; Young, J. K.; Moore, J. S.; Suslick, K. S., *J. Am. Chem. Soc.*, 1996, 118, 5708. b) Bhyrappa, P.; Young, J. K.; Moore, J. S.; Suslick, K. S., *J. Mol. Catal. A: Chem.*, 1996, 113, 109. c) Wooley, K. L.; Hawker, C. J.; Fréchet, J. M. J., *J. Am. Chem. Soc.*,1993, 115, 11496. d) Hawker, C. J.; Wooley, K. L.; Fréchet, J. M. J., *Macromol. Symp.*, 1994, 77, 11. e) Hawker, C. J.; Fréchet, J. M. J., *J. Am. Chem. Soc.*,1992, 114, 8405.
[131] Freeman, A. W.; Fréchet, J. M. J., *Org. Lett.*, 1999, 1, 685.
[132] Miller, T. M.; Neenan, T. X., *Chem. Mater.*, 1990, 2, 346.
[133] Ishida, Y.; Jikei, M.; Kakimoto, M., *Macromolecules*, 2000, 33, 3202.
[134] a) Yamakawa, Y.; Ueda, M.; Takeuchi, K.; Asai, M., *Macromolecules*, 1999, 32, 8363. b) Hayakawa, T.; Yamakawa, Y.; Nomura, M.; Okazaki, M.; Takeuchi, K.; Asai, M.; Ueda, M., *Polym. J.*, 2000, 32, 784. c) Okazaki, M.; Washio, I..; Shibasaki, Y.; Ueda, M., *J. Am. Chem. Soc.*, 2003, 125, 8120.
[135] Romagnoli, B.; Ashton, P. R.; Harwood, L. M.; Philp, D.; Price, D. W.; Smith, M. H.; Hayes, W., *Tetrahedron*, 2003, 59, 3975.
[136] Pan, Y.; Ford, W. T., *J. Polym. Sci. Part A: Polym. Chem.*, 2000, 38, 1533.
[137] Aulenta, F., 2003, *PhD Thesis*, University of Reading.
[138] Rannard, S.; Davis, N.; McFarland, H., *Polym. Int.*, 2000, 49, 1002.
[139] a) Feast, W. J.; Rannard, S. P.; Stoddart, A., *Macromolecules*, 2003, 36, 9704. b) Feast, W. J.; Rannard, S. P.; Stoddart, A., *Polymeric Mater. Sci. Eng.*, 2001, 84, 701.
[140] Rannard, S.; Davis, N., *J. Am. Chem. Soc.*, 2000, 122, 11729.
[141] Yang, G.; Jikei, M.; Kakimoto, M., *Proc. Jpn. Acad. B*, 1998, 74, 188.
[142] Yang, G.; Jikei, M.; Kakimoto, M., *Macromolecules*, 1999, 32, 2215.
[143] Russo, S.; Boulares, A.; Mariani, A., *Macromol. Symp.*, 1998, 128, 13.
[144] Ishida, Y.; Sun, A. C. F.; Jikei, M.; Kakimoto, M., *Macromolecules*, 2000, 33, 2832.
[145] Holter, D.; Frey, H., *Acta Polym.*, 1997, 48, 298.
[146] Haba, O.; Tajima, H.; Ueda, M.; Nagahata, R., *Chem. Lett.*, 1998, 333.
[147] Monticelli, O.; Mendichi, R.; Bisbano, S.; Mariani, A.; Russo, S., *Macromol. Chem. Phys.*, 2000, 2021, 2123.
[148] Yang, G.; Jikei, M.; Kakimoto, M., *Macromolecules*, 1998, 31, 5964.
[149] Jikei, M.; Fujii, K.; Yang, G.; Kakimoto, M., *Macromolecules*, 2000, 33, 6228.
[150] Jikei, M.; Fujii, K.; Kakimoto, M., *J. Polym. Sci., Part A: Polym. Chem.*, 2001, 39, 3304.

[151] Jikei, M.; Fujii, K.; Kakimoto, M., *Macromol. Symp.*, 2003, 199, 223.
[152] a) Flory, P. J., 1953, *Principles of Polymer Chemistry*, Cornell University Press, New York, Chapter 9. b) Flory, P. J., *J. Am. Chem. Soc.*, 1952, 74, 2718. c) Odian, G., 1991, *Principles of Polymerisation*, John Wiley and Sons, New York, pp. 108.
[153] Jikei, M.; Chon, S.-H.; Kakimoto, M.; Kawauchi, S.; Imase, T.; Watanebe, J., *Macromolecules*, 1999, 32, 2061.
[154] Chang, Y.-T.; Shu, C.-F., *Macromolecules*, 2003, 36, 661.
[155] a) Wilson, D.; Stenzerberzer, H. D.; Hergenrother, P. M., 1990, *Polyimides*, Wilson, D.; Stenzerberzer, H. D.; Hergenrother, P. M., (eds.), Blackie, New York. b) Ghosh, M. K.; Mittel, K. L., 1996, *Polyimides: Fundmentals and Applications*, Ghosh, M. K.; Mittel, K. L., (eds.), Marcel Dekker, New York. C) Sroog, C. E., 1997, *Macromolecular Design of Polymeric Materials*, Hatada, K.; Kitayama, T.; Vogl, O., (eds.), Marcel Dekker, New York, pp. 713.
[156] Chang, Y.-T.; Shu, C.-F.; Leu, C.-M.; Wei, K.-H., *J. Polym. Sci., Part A: Polym. Chem.*, 2003, 41, 3726.
[157] Yamanaka, K.; Jikei, M.; Kakimoto, M., *Macromolecules.*, 2000, 33, 1111.
[158] Yamanaka, K.; Jikei, M.; Kakimoto, M., *Macromolecules.*, 2000, 33, 6937.
[159] Yamanaka, K.; Jikei, M.; Kakimoto, M., *Macromolecules.*, 2000, 33, 3910.
[160] Hao, J.; Jikei, M.; Kakimoto, M., *Macromolecules.*, 2003, 36, 3519.
[161] Hao, J.; Jikei, M.; Kakimoto, M., *Macromolecules.*, 2002, 35, 5372.
[162] Trollsås, M.; Atthoff, B.; Claesson, H.; Hedrick, J. L., *Macromolecules*, 1998, 31, 3439.
[163] Trollsås, M.; Hedrick, J. L.; Mecerreyes, D.; Jéröme, R.; Dubois, P., *J. Polym. Sci., Part A: Polym. Chem.*, 1998, 36, 3187.
[164] a) Trollsås, M.; Hedrick, J. L.; Mecerreyes, D.; Dubois, P.; Jéröme, R.; Ihre, H.; Hult, A., *Macromolecules*, 1997, 30, 8508. b) Trollsås, M.; Hedrick, J. L.; Mecerreyes, D.; Dubois, P.; Jéröme, R.; Ihre, H.; Hult, A., *Macromolecules*, 1998, 31, 2756.
[165] a) Tian, D.; Dubois, P.; Grandfils, C.; Jéröme, R., *Macromolecules*, 1997, 30, 406. b) Tian, D.; Halleux, O.; Dubois, P.; Grandfils, C.; Jéröme, R.; Sobry, R.; Van den Bossche, G., *Macromolecules*, 1998, 31, 924.
[166] Blencowe, A.; Davidson, L.; Hayes, W., *Eur. Polym. J.*, 2003, 39, 1955.
[167] a) Davidson, L.; Freebairn, K. W.; Russell, A. T.; Trivedi, H. S.; Hayes, W., *Synlett*, 2002, 2, 251. b) Davidson, L.; Blencowe, A.; Drew, M. G. B.; Freebairn, K. W.; Hayes, W., *J. Mater. Chem.*, 2003, 13, 758.
[168] a) Kambouris, P.; Hawker, C. J., *J. Chem. Soc., Perkin Trans. 1*, 1993, 2717. b) Feast, W. J.; Stainton, N. M., *J. Mater. Chem.*, 1995, 5, 405.
[169] a) Chu, F.; Hawker, C. J.; Pomery, P. J.; Hill, D. J. T., *J. Polym. Sci., Part A: Polym. Chem.*, 1997, 35, 1627. b) Turner, S. R.; Voit, B. L.; Mourey, T. H., *Macromolecules*, 1993, 26, 4617.
[170] a) Hawker, C. J.; Lee, R.; Fréchet, J. M. J., *J. Am. Chem. Soc.*, 1991, 113, 4583. b) Hahn, S.-W.; Yun, Y.-K.; Jin J.-I., *Macromolecules*, 1998, 31, 6417. c) Kricheldorf, H. R.; Bolender, O.; Wollheim, T., *Macromolecules*, 1999, 32, 3878.
[171] Hölter, D.; Burgath, A.; Frey, H., *Acta. Polym.*, 1997, 48, 30.
[172] Rannard, S.; Davis, N., *Polym. Mater. Sci. Eng.*, 2001, 84, 2.
[173] Davis, N. J.; Rannard, S. P., *Abstr. Pap. Am. Chem. S.*, 1997, 158.
[174] Rannard, S. P.; Davis, N. J.; Herbert, I., *Macromolecules*, in press.
[175] Rannard, S.; Davis, N., *Polym. Mater. Sci. Eng.*, 2001, 84, 2.

[176] a) Marchand-Brynaert, J.; Deldime, M.; Dupont, I.; Dewez, J.-L.; Schneider, Y.-J., *J. Colloid Interface Sci.*, 1995, 173, 236. b) Ito, Y.; Liu, S.-Q.; Nakabayashi, M.; Imanishi, Y., *Biomaterials*, 1992, 13, 789. c) Deldime, M.; Dewez, J.-L.; Schneider, Y.-J.; Marchand-Brynaert, J., *Appl. Surface Sci.*, 1995, 90, 1. d) Boxus, T.; Deldime-Rubbens, M.; Mougenot, P.; Schneider, Y.-J.; Marchand-Brynaert, J., *Polym. Adv. Technol.*, 1996, 7, 589. e) Alexander, M. N.; Wright, P. V.; Ratner, B. D., *Surface Interface Anal.*, 1996, 24, 217.
[177] Sunder, A.; Quincy, M.-F.; Mülhaupt, R.; Frey, H., *Angew. Chem., Int. Ed. Engl.*, 1999, 38, 2928.
[178] Bunning, T. J.; Kreuzer, F.-H., *Trends Polym. Sci.*, 1995, 3, 318.
[179] Sunder, A.; Hanselmann, R.; Frey, H.; Mülhaupt, R., *Macromolecules*, 1999, 32, 4240.
[180] Hall, C. W.; Cockayne, J. S.; Kan, M. J.; Nicholson, G. P., *Green Chem.*, 2001, 3, 114.
[181] a) Roundhill, D. M., *Prog. Inorg. Chem.*, 1996, 43, 533. b) Yilmaz, M., *Handbook. Eng. Polym. Mater.*, 1997, 339.
[182] a) Maeda, H.; Takeshita, J.; Kanamaur, R., *Int. J. Pept, Protein Res.*, 1979, 14, 81. b) Maeda, H.; Matsumoto, T.; Konno, T.; Iwai, T.; Ueda, M., *J. Protein Chem.*, 1984, 3, 181.
[183] Krinick, N. L.; Sun, Y.; Joyner, D.; Spikes, J. D.; Straight, R. C.; Kopecek, J., *J. Biomater. Sci. Polym. Ed.*, 1994, 5, 303.
[184] a) Hoffman, A. S., *J. Controlled Release*, 1987, 6, 297. b) Okamura, K.; Ikura, K.; Yoshiwaka, M.; Sakaki, R.; Chiba, H., *Agric. Biol. Chem.*, 1984, 48, 2435. c) Taniguchi, M.; Kobayashi, M.; Fujii, M., *Biotechnol. Bioeng.*, 1989, 34, 1092. d) Taniguchi, M.; Hoshino, A. S.; Watanabe K, Sugai, K., Fujii, M., *Biotechnol. Bioeng.*, 1992, 39, 287. e) Chen, G. H.; Hoffman, A. S., *Polym. Prepr.*, 1992, 33, 468. f) Chen, G. H.; Hoffman, A. S., *Bioconjugate Chem.*, 1993, 4, 509. g) Takei, Y.; Aoki, T.; Sanui, K.; Ogata, N.; Okano, T.; Sakurai, Y., *Bioconjugate Chem.*, 1993, 4, 42. h) Takei, Y.; Aoki, T.; Sanui, K.; Ogata, N.; Okano, T.; Sakurai, Y., *Bioconjugate Chem.*, 1993, 4, 341. i) Takei, Y.; Matsukata, M.; Aoki, T.; Sanui, K.; Ogata, N.; Kikuchi, A.; Sakurai, Y.; Okano, T., *Bioconjugate Chem.*, 1994, 5, 577. j) Matsukata, M.; Takei, Y.; Aoki, T.; Sanui, K.; Ogata, N.; Kikuchi, A.; Sakurai, Y.; Okano, T., *J. Biochem.*, 1994, 116, 682.
[185] Ding, Z.; Chen, G.; Hoffman, A. S., *Bioconjugate Chem.*, 1996, 7, 121.
[186] a) Noiset, O.; Henneuse, C.; Schneider, Y.-J.; Marchand-Brynaert, J., *Macromolecules*, 1997, 30, 540. b) Marchand-Brynaert, J.; Pantano, G.; Noiset, O., *Polymer*, 1997, 38, 1387.
[187] Henneuse, C.; Goret, B.; Marchand-Brynaert, J., *Polymer*, 1998, 39, 835.
[188] Henneuse, C.; Boxus, T.; Tesolin, L.; Pantano, G.; Marchand-Brynaert, J., *Synthesis*, 1996, 495.
[189] Lee, N.-J.; Ju, S.-S.; Cho, W.-J.; Kim, S.-H.; Kang, K.-T.; Brady, T.; Theodorakis, E. A., *Eur. Polym. J.*, 2003, 39, 367.
[190] a) Vasey, P. A.; Kaye, S. B.; Morrison, R.; Twelve, C.; Wilson, P.; Duncan, R., *Clin. Cancer Res.*, 1999, 5, 83. b) Pendri, A.; Conover, C. D.; Greenwald, R. B., *Anti-Cancer Drug Des.*, 1998, 13, 387. c) Terwogt, J. M. M.; Huinink, W. W. B.; Schellens, J. H. M.; Schot, M.; Mandjes, I. A. M.; Zurlo, M. G., *Anti-Cancer Drugs*, 2001, 12, 315. d) Maeda, H., *Med. Res. Rev.*, 1991, 6, 181. e) Greenwald, R. B.; Conover, C. D.; Pendri, A.; Choe, Y. H.; Martinez, A.; Wu, D., *J. Control Release*, 1999, 61, 281.
[191] Kim, I.-S.; Kim, S.-H., *Int. J. Pharm.*, 2001, 226, 23.

[192] a) Kreuter, J., *J. Controlled Release*, 1991, 16, 169. b) Kubitschko, S.; Spinke, J.; Bruckner, T.; Pohl, S.; Oranth, N., *Anal. Biochem.*, 1997, 253, 112. c) Leroux, J. C.; Cozens, R.; Roesel, J. L.; Galli, B.; Kubel, F.; Doelker, E.; Gurny, R., *J. Pharm. Sci.*, 1995, 84, 1387. d) Demoy, M.; Andreux, J. P.; Weingarten, C.; Gouritin, B.; Guilloux, V.; Couvreur, P., *Pharm. Res.*, 1999, 16, 37. e) De Jaeghere, F.; Allemann, E.; Feijen, J.; Kissel, T.; Doelker, E.; Gurny, R., *J. Drug Target.*, 2000, 8, 143.

[193] a) Jones, W. E.; Watkins, D. M.; Fox, M. A., *Chem. Eng. News*, 1993, 71, 38. b) Guillet, J., 1985, *'Polymer Photophysics and Photochemistry'*, Cambridge University Pres, Cambridge. c) Webber, S. E., *Chem. Rev.*, 1990, 90, 1469. d) Stoessel, S. J.; Stille, J. K., *Macromolecules*, 1992, 25, 1832. e) Baxter, S. M.; Jones, W. E.; Danielson, E.; Worl, L. A.; Younathan, J.; Strouse, G. F.; Meyer, T. J., *Coord. Chem. Rev.*, 1991, 111, 47.

[194] Dupray, L. M.; Meyer, T. J., *Inorg. Chem.*, 1996, 35, 6299.

[195] a) Younathan, J. N.; Jones, W. E.; Meyer, T. J., *J. Phys. Chem.*, 1991, 95, 488. b) Olmstead, J.; McClanahan, S. F.; Danielson, E.; Younathan, J. N.; Meyer, J., *J. Am. Chem. Soc.*, 1987, 109, 3297. c) Jones, W. J., Jr.; Baxter, S. M.; Strouse, G. F.; Meyer, J., *J. Am. Chem. Soc.*, 1993, 115, 7363. d) Strouse, G. F.; Worl, L. A.; Younathan, J. N.; Meyer, J., *J. Am. Chem. Soc.*, 1989, 111, 9101.

[196] Worl, L. A.; Strouse, G. F.; Younathan, J. N.; Baxter, S. M.; Meyer, T. J., *J. Am. Chem. Soc.*, 1990, 112, 7571.

[197] a) Manolovsa, N.; Rashkov, I.; Beguin, F.; van Damme, H., *J. Chem. Soc., Chem. Commun.*, 1993, 1725. b) Manolovsa, N.; Rashkov, I.; van Damme, H.; Beguin, F., *Polym. Bull.*, 1994, 33, 175. c) Sun, Y.-P.; Liu, B.; Lawson, G. E., *Photochem. Photobiol.*, 1997, 66, 301. d) Sun, Y.-P.; Liu, B.; Moton, D. K., *J. Chem. Soc., Chem. Commun.*, 1996, 2699. e) Nakajima, Y. N.; Yagami, T.; Fukuhara, K.; Sueyoshi, S.; Miyata, N., *J. Chem. Soc., Chem. Commun.*, 1994, 517.

[198] Sun, Y.-P.; Lawson, G. E.; Huang, W.; Wright, A. D.; Moton, D. K., *Macromolecules*, 1999, 32, 8747.

[199] Tobiesen, F. A.; Michielsen, S., *J. Polym. Sci.: Part A: Polym. Chem.*, 2002, 40, 719.

[200] van Dijk-Wolthuis, W. N. E.; Tsang, S. K. Y.; Kettenes-van den Bosch, J. J.; Hennink, W. E., *Polym.*, 1997, 38, 6235.

[201] a) Gombotz, W. R.; Pettit, D. K., *Bioconjugate, Chem.*, 1995, 6, 332. b) Kim, S. W.; Bae, Y. H.; Okano, T., *Pharm. Res.*, 1992, 9, 283. c) Cheng, J.; Jo, S.; Park, K., *Carbohydr. Polym.*, 1995, 28, 69.

[202] *Martindale, The Extra Pharmacopeia*, 30th edn., Reynolds, J. E. F., (ed.), The Pharmaceutical Press, London, 1993, pp. 650.

[203] Newkome, G. R.; Yoo, K. S.; Kabir, A.; Malik, A., *Tetrahedron Lett.*, 2001, 42, 7537.

[204] Sakai, K.; Teng, T. C.; Katada, A.; Harada, T.; Uemura, S.; Asami, Y.; Sakata, M.; Kunitake, M.; Hirayama, C., *Chem. Lett.*, 2001, 510.

[205] Guo, Z. X.; Yu, J., *J. Mater. Chem.*, 2002, 12, 468.

[206] a) Mathews, B. T.; Beezer, A. E.; Snowden, M. J.; Hardy, M. J.; Mitchell, J. C., *New. J. Chem.*, 2001, 25, 807. b) Mathews, B. T.; Beezer, A. E.; Snowden, M. J.; Hardy, M. J.; Mitchell, J. C., *Chromatographia.*, 2001, 53, 147.

[207] a) Spange, S., *Prog. Polym. Sci.*, 2000, 25, 781. b) Guidotti, B. R.; Herzog, E.; Bangerter, F.; Caseri, W. R.; Suter, U. W., *J. Colloid Interface Sci.*, 1997, 191, 209. c) Nitzan, B.; Margel, S., *J. Polym. Sci. Part A: Polym. Chem.*, 1997, 35, 171.

[208] Driffield, M.; Goodall, D. M.; Klute, A. S.; Smith, D. K.; Wilson, K., *Langmuir*, 2002, 18, 8660.

[209] a) Tam, J. P.; *Proc. Natl. Acad. Sci.*, 1988, 85, 5409. b) Roy, R.; Zanini, D.; Meunier, S. J.; Romanowska, A., *J. Chem. Soc., Chem. Commun.*, 1993, 1869. c) Grandjean, C.; Rommens, C.; Gras-Masse, H.; Melnyk, O., *Angew. Chem., Int. Ed. Engl.*, 2000, 39, 1068.

[210] a) Hawker, C.; Fréchet, J. M. J., *J. Chem. Soc., Chem. Commun.*, 1990, 1010. b) Hawker, C.; Fréchet, J. M. J., *J. Am. Chem. Soc.*, 1990, 112, 7638.

[211] a) Kawa, M.; Fréchet, J. M. J., *Chem. Mater.*, 1998, 10, 286. b) Tully, D. C.; Trimble, A. R.; Fréchet, J. M. J., *Chem. Mater.*, 1999, 11, 2892. c) Bo, Z.; Zhang, X.; Yi, X.; Yang, M.; Shen, J.; Rehn, Y.; Xi, S., *Polym. Bull.*, 1997, 38, 257.

[212] a) Zhai, J.; Bai, F. L.; He, Q. G.; Li, Y. S.; Jiang, L.; Lin, T., *Chinese Chemical Lett.*, 2000, 11, 819. b) Zhai, J.; Li, Y. S.; He, Q. G.; Jiang, L.; Bai, F. L., *J. Phys. Chem. B.*, 2001, 105, 4094.

INDEX

A

B

C

D

E

N

O

P

T

U

V

W

X

Y

Z